3D Laser Microfabrication

Edited by
Hiroaki Misawa and
Saulius Juodkazis

Related Titles

Bordo, V. G., Rubahn, H.-G.

Optics and Spectroscopy at Surfaces and Interfaces

281 pages with 144 figures
2005
Softcover
ISBN 3-527-40560-7

Bäumer, S. (ed.)

Handbook of Plastic Optics

199 pages with 135 figures and 37 tables
2005
Hardcover
ISBN 3-527-40424-4

Franssila, S.

Introduction to Microfabrication

408 pages
2004
Hardcover
ISBN 0-470-85105-8

Sinzinger, S., Jahns, J.

Microoptics

447 pages with 209 figures and 12 tables
2003
Hardcover
ISBN 3-527-40355-8

Schleich, W. P.

Quantum Optics in Phase Space

716 pages with 220 figures
2001
Hardcover
ISBN 3-527-29435-X

3D Laser Microfabrication

Principles and Applications

Edited by
Hiroaki Misawa and Saulius Juodkazis

WILEY-VCH Verlag GmbH & Co. KGaA

Editors

Hiroaki Misawa
Research Institute for Electronic Science
Hokkaido University, Japan
misawa@es.hokudai.ac.jp

Saulius Juodkazis
Research Institute for Electronic Science
Hokkaido University, Japan
saulius@es.hokudai.ac.jp

Cover

Periodic structure of a body-centered cubic lattice
numerically simulated by interference of four plane
waves.

All books published by Wiley-VCH are
carefully produced. Nevertheless, authors,
editors, and publisher do not warrant the
information contained in these books,
including this book, to be free of errors.
Readers are advised to keep in mind that
statements, data, illustrations, procedural
details or other items may inadvertently
be inaccurate.

Library of Congress Card No.: applied for

British Library Cataloguing-in-Publication Data
A catalogue record for this book is available
from the British Library.

**Bibliographic information published by
Die Deutsche Bibliothek**
Die Deutsche Bibliothek lists this publication
in the Deutsche Nationalbibliografie; detailed
bibliographic data is available in the Internet at
<http://dnb.ddb.de>.

Typesetting   Kühn & Weyh, Satz und Medien,
Freiburg
Printing   Strauss GmbH, Mörlenbach
Bookbinding   Litges & Dopf Buchbinderei GmbH,
Heppenheim

Printed in the Federal Republic of Germany.
Printed on acid-free paper.

ISBN-13:   978-3-527-31055-5
ISBN-10:   3-527-31055-X

Contents

3D Laser Microfabrication. Principles and Applications.
Edited by H. Misawa and S. Juodkazis
Copyright © 2006 WILEY-VCH Verlag GmbH & Co. KGaA, Weinheim
ISBN: 3-527-31055-X

List of Contributors

Selcuk Akturk
Georgia Institute of Technology
Georgia Center for Ultrafast Optics
School of Physics
837 State St.
Atlanta, GA 30332
USA

Niko Baersch
Laser Zentrum Hannover eV
Hollerithallee 8
30419 Hannover
Germany

Thorsten Bauer
Laser Zentrum Hannover eV
Hollerithallee 8
30419 Hannover
Germany

John Buck
Georgia Institute of Technology
School of Electrical and Computer
Engineering
Van Leer Electrical Engineering
Building
777 Atlantic Drive NW
Atlanta, GA 30332-0250
USA

Qiang Cao
Georgia Institute of Technology
Georgia Center for Ultrafast Optics
School of Physics
837 State St.
Atlanta, GA 30332
USA

Boris N. Chichkov
Laser Zentrum Hannover eV
Hollerithallee 8
30419 Hannover
Germany

Hiroshi Fukumura
Department of Chemistry
Graduate School of Science
Tohoku University
Sendai 980-8578
Japan

Eugenijus Gaižauskas
Laser Research Center
Department of Quantum Electronics
University of Vilnius
Sauletekio al.9
10222 Vilnius
Lithuania

3D Laser Microfabrication. Principles and Applications.
Edited by H. Misawa and S. Juodkazis
Copyright © 2006 WILEY-VCH Verlag GmbH & Co. KGaA, Weinheim
ISBN: 3-527-31055-X

Eugene G. Gamaly
Laser Physics Centre
Oliphant Building No. 60
Research School of Physical Sciences
and Engineering
The Australian National University
Canberra ACT 0200
Australia

Min Gu
Swinburne University of Technology
School of Biophysical Sciences and
Electrical Engineering
Centre for Micro-Photonics
P.O. Box 218
Hawthorn
Victoria 3122
Australia

Xun Gu
Georgia Institute of Technology
Georgia Center for Ultrafast Optics
School of Physics
837 State St.
Atlanta, GA 30332
USA

Richard F. Haglund, Jr.
Vanderbilt University
Department of Physics and Astronomy
and W M Keck Free Electron Laser
Center
Box 1807, Station B 6301 Stevenson
Nashville, TN 37235
USA

Koji Hatanaka
Department of Chemistry
Graduate School of Science
Tohoku University
Sendai 980-8578
Japan

Henry Helvajian
The Aerospace Corporation
Space Materials Laboratory, M2/241
2350 El Segundo Blvd.
El Segundo, CA 90245
USA

Saulius Juodkazis
Research Institute for Electronic
Science
Hokkaido University
North 21 – West 10, CRIS Bldg.
Sapporo 001-0021
Japan

Guenther Kamlage
Laser Zentrum Hannover eV
Hollerithallee 8
30419 Hannover
Germany

Peter G. Kazansky
Optoelectronics Research Centre
University of Southampton
Southampton
SO17 1BJ
United Kingdom

Ulrich Klug
Laser Zentrum Hannover eV
Hollerithallee 8
30419 Hannover
Germany

Juergen Koch
Laser Zentrum Hannover eV
Hollerithallee 8
30419 Hannover
Germany

Frank Korte
Laser Zentrum Hannover eV
Hollerithallee 8
30419 Hannover
Germany

Frank E. Livingston
The Aerospace Corporation
Space Materials Laboratory, M2/241
2350 El Segundo Blvd.
El Segundo, CA 90245
USA

Barry Luther-Davies
Laser Physics Centre
Oliphant Building No. 60
Research School of Physical Sciences
and Engineering
The Australian National University
Canberra ACT 0200
Australia

Shigeki Matsuo
Graduate School of Engineering
Department of Ecosystems Engineering
The University of Tokushima
2-1 Minamijosanjima
770-8606 Tokushima
Japan

Hiroaki Misawa
Research Institute for Electronic
Science
Hokkaido University
North 21 – West 10, CRIS Bldg.
Sapporo 001-0021
Japan

Vygantas Mizeikis
Research Institute for Electronic
Science
Hokkaido University
North 21 – West 10, CRIS Bldg.
Sapporo 001-0021
Japan

Andreas Ostendorf
Laser Zentrum Hannover eV
Hollerithallee 8
30419 Hannover
Germany

Andrei Rode
Laser Physics Centre
Oliphant Building No. 60
Research School of Physical Sciences
and Engineering
The Australian National University
Canberra ACT 0200
Australia

Arpana Shreenath
Georgia Institute of Technology
Georgia Center for Ultrafast Optics
School of Physics
837 State St.
Atlanta, GA 30332
USA

Jesper Serbin
Laser Zentrum Hannover eV
Hollerithallee 8
30419 Hannover
Germany

Rick Trebino
Georgia Institute of Technology
Georgia Center for Ultrafast Optics
School of Physics
837 State St.
Atlanta, GA 30332
USA

Guangyong Zhou
Swinburne University of Technology
School of Biophysical Sciences and
Electrical Engineering
Centre for Micro-Photonics
P.O. Box 218
Hawthorn
Victoria 3122
Australia

1
Introduction

Hiroaki Misawa and Saulius Juodkazis

Three-dimensional (3D) laser micro-fabrication has become a fast growing field of science and technology. The very first investigations of the laser modifications and structuring of materials immediately followed the invention of the laser in 1960. Starting from the observed photomodifications of laser rod materials and ripple formation on the irradiated surfaces as unwanted consequences of a high laser fluence, the potential of material structuring was tapped. The possibility arose of having a highly directional light beam, easily focused into close to diffraction-limited spot size, and with the arrival of pulsed lasers with progressively shorter pulse durations, the field of laser material processing emerged. Understanding the physical and chemical mechanisms of light–matter interactions and ablation (from *lat. ablation,* removal) at high irradiance began the ever widening number of scientific and industrial applications.

The arrival of ultrashort (subpicosecond) lasers had expanded the field of material processing into the real 3D realm. Even though ablation can be used to fabricate 3D microstructures the real 3D structuring of materials (from the inside) requires highly nonlinear light–matter interaction in terms of light intensity for which the ultrashort pulses are indispensable. The focused ultrashort pulses inside solid-state materials introduces a novel, not fully exploited, tool for nano-micro-structuring. The irradiance at the focus can reach $\sim$ 100 TW cm^{-2} (1 TW = 10^{12} W) at which any material, including dielectrics, is ionized within several optical cycles, i.e., optically-induced dielectric breakdown ensues. It is noteworthy, that such irradiance is reached using low pulse energies < 0.5 µJ at typical pulse durations < 200 fs (1 fs = 10^{-15} s). Combining this with an inherently micro-technological approach, it makes this high-irradiance technology attractive in terms of its "green", environmentally friendly aspect due to high precision and effectiveness. In theory, the inherently 3D structuring avoids a lengthy and wasteful multi-step approach based on lithography with subsequent solid-state materials growth and processing. A continuing trend of reduction of the "photon cost" of femtosecond lasers and an increase of the average pulse power at higher repetition rates places the 3D laser micro(nano)technology firmly on the list of future movers. We believe that 3D laser micro-fabrication will implement the visionary top-down

3D Laser Microfabrication. Principles and Applications.
Edited by H. Misawa and S. Juodkazis
Copyright © 2006 WILEY-VCH Verlag GmbH & Co. KGaA, Weinheim
ISBN: 3-527-31055-X

approach of nano-technology and will show solutions for merger with the bottom-up self-assembly route currently advancing fast in its own right.

In this book, we have made a first attempt to review the state-of-the-art of the interdisciplinary field of 3D laser microfabrication for material micro-nano-structuring. The distinction of this field in terms of physical mechanisms is a 3D enclosure of the processed microvolume inside the material. In the case of dielectric breakdown by ultrashort pulses, the matter can be transferred from a solid to liquid, gaseous, and a multiply-ionized plasma state, which still possesses the solid state density. This creates unique conditions as long as the surrounding medium can hold the high temperature and high pressure microvolume of ionized material, without crack formation. Obviously, this is most feasible when the ionized volume is minute (of sub-micrometer cross-section). The chemical modifications are also radically different for in-bulk laser processing, indeed, the compositional stoichiometry of the focal volume is conserved over the radical phase changes which the matter endures at the focus. When the pressure of the ionized material at the focus becomes higher than the "cold" pressure, the Young modulus, the shock and rarefaction waves emerge. The shock-modified compressed region has unique altered physical and chemical properties and can be confined within sub-micrometer cross-sections. Material can be also chemically and structurally altered by properly chosen exposure (not necessarily by fs-pulses) and post-exposure treatment for designed properties and funcionality, e.g., 3D structuring of ceramic glasses.

Research in the 3D laser microfabrication field is prompted by an increasing number of prospective applications; however, it is usually approached as an engineering "optimization problem". The required processing conditions can be more easily found by a fast trial-and-error method using powerful computers, experiment automation tools, and software based on intelligent self-learning algorithms. Hence, understanding the underlying physical principles and review of the results achieved could help further progress in this field. This is a shared view of the group of authors who teamed up for this project. The scope of this book ranges from the principles of 3D laser fabrication to its application. Direct 3D laser writing is the main topic of the book. The mechanisms of light–matter interaction are discussed, applications are reviewed, and future prospects are outlined.

The idea of this project stemmed from the importance of nonlinear light–matter interaction; thus, first of all, the intensity (irradiance) should be known and controlled. We address the issues of light delivery to the photo-modification site, describe the mechanism of light–matter interaction, and show the versatility of the phenomena together with the broad field of application. The correct estimate of the pulse energy, duration, and the focal volume are crucial, which in the case of ultrashort pulses, is not trivial. The book starts with the theoretical Chapter 2 (E. E. Gamaly et al.) on the light–matter interaction at high irradiance inside the bulk of the dielectric. It is based on multi-photon and avalanche ionization theories developed more than 50 years ago. However, their predictions are now applied to 3D laser fabrication. Chapter 3 (M. Gu and G. Zhou) addresses light-focusing issues relevant to energy delivery and spatial distribution. Chapters 4 and 5 (X. Gu et al. and J. Buck and R. Trebino) describe the basic principles of pulse duration measurements and nonlinear optics, respectively. Chapter 6 (E. Gaižauskas) discusses a mechanism of filament formation in dielectric media. Chapter 7 (R. Haglund) surveys photo-physical and photochemical aspects of light-meter interactions. Micro- and nano-in-bulk structuring of glass is described in Chapter 8 (P. Kazansky). X-ray generation by ultrashort pulses and their potential for time-resolved structural characterization is discussed in Chapter 9 (K. Hatanaka and H. Fukumura). The applications of 3D laser microfabrication are further explored in Chapters 10–12. Fabrication of photonic crystals and their templates are described in Chapter 10 (S. Juodkazis et al.). Flexibility and versatility of 3D micro-structuring of glass ceramic is highlighted in Chapter 11 (F. Livingston and H. Helvajian). Setups, fabrication principles and different examples of 3D micro-structuring are described in Chapter 12 (A. Ostendorf et al.) by one of the leading group in the field from Hannover Laser Zentrum. All chapters are self-inclusive and can be read in any order.

Since the field of 3D laser microfabrication is growing so quickly, we tried to focus on the topics which are more general and have high potential for future advance; at the same time striking a balance between the theory and applications. Unfortunately, there was no possibility of highlighting many other very important developments, namely, non-thermal melting, imaging of photoinduced movement of atoms by ultrafast X-ray pulses, Monte-Carlo simulations of atomic movement at high excitation, laser intracell and DNA surgery, or waveguide recording.

Seminal contributions have been cited throughout the book. However, some references just show a relevant example rather than stress the first demonstration or priority (this purpose is better served in the original papers). There are a number of excellent books on the related subjects of ablation [1], laser processing of materials [2], laser damage [3], and on the basics of light–matter interactions [4, 5] which covers closely related topics. Here, however, we would like to stress that 3D laser processing stands out, with its own unique physical and chemical mechanisms of photo-structuring and an obviously increasing field of application.

We are grateful indeed for the support and help from our colleagues, reviewers, and publishing team. We are also grateful to all the contributors with whom we have had a number of discussions over recent years.

H. Misawa
S. Juodkazis   Sapporo, December 1, 2005

References

1 J.C. Miller ed., "Laser Ablation", Springer, Berlin, 1994.

2 S. M. Metev and V. P. Veiko, eds., "Laser-Assisted Microtechnology", Springer, Berlin, 1994.

3 R. M. Wood, "Laser Damage in Optical Materials", Adam Hilger, Bristol, 1986.

4 D. Bäuerle, "Laser processing and Chemistry", Springer, Berlin, edition. 3rd, 2000.

5 Ya. B. Zel'dovich and Yu. P. Raizer, "Physics of Shock Waves and High-Temperature Hydrodynamic Phenomena", eds. W. D. Hayes and R. F. Probstein, Dover, Mineola, New York, 2002.

2
Laser–Matter Interaction Confined Inside the Bulk of a Transparent Solid

Eugene Gamaly, Barry Luther-Davies and Andrei Rode

Abstract

In this chapter we discuss the laser–matter interaction physics that occurs when an intense laser beam is tightly focused inside a transparent dielectric such that the interaction zone where high energy density is deposited is confined inside a cold and dense solid. Material modifications produced in this manner can form detectable nanoscale structures as the basis for a memory bit.

We describe the single-pulse-laser-solid interaction in two limiting cases. In the low-intensity case the deposited energy density is well below the damage threshold but it is sufficient to trigger a particular phase transition that in some conditions may become irreversible. At high energy density, the material is ionized early in the pulse, all bonds are broken, the material is converted into hot and dense plasma, and the pressure in the interaction zone may be much greater than the strength of the surrounding solid. The restricted material expansion after the end of the pulse, shock wave propagation, the compression of the cold solid, and the formation of a void inside the target, are all described. The theoretical approach is extended to the case where multiple pulses irradiate the same volume in the solid. We discuss the properties of the laser-affected material and the possibility of detecting it using a probe beam. We compare the results with experiments and draw conclusions.

2.1
Introduction

There is a fundamental difference in the laser–matter interaction when a laser beam is tightly focused inside a transparent material from when it is focused onto the surface, because the interaction zone containing high energy density is confined inside a cold and dense solid. For this reason the hydrodynamic expansion is insignificant when the energy density is lower than the structural damage threshold and above this threshold it is highly restricted. The deposition of high energy density in a small confined volume results in a change (reversible or irre-

3D Laser Microfabrication. Principles and Applications.
Edited by H. Misawa and S. Juodkazis
Copyright © 2006 WILEY-VCH Verlag GmbH & Co. KGaA, Weinheim
ISBN: 3-527-31055-X

versible) in the optical and structural properties in the affected region, thereby creating a zone that can be detected afterwards by an optical probe. If the structure is very small ($\ll$ μm^3 in volume) it can be used as a memory bit for high-density 3D optical storage.

The laser–matter interaction physics when the focus is confined within transparent dielectrics is drastically different at low laser intensity (below the ionization threshold) and for relatively long wavelength ($\lambda \geq 500$ nm) from that at high intensity. The interaction of a laser with dielectrics at an intensity above the ionization threshold proceeds in the laser-plasma interaction mode in a similar way for all the materials [1].

With a further increase in the energy density, a strong shock wave is generated in the interaction region and this propagates into the surrounding cold material. Shock-wave propagation is accompanied by compression of the solid material at the wave front and decompression behind it, leading to the formation of a void inside the material. In some solids chemical decomposition may occur at a relatively low temperature or beam intensity. The decomposed matter can be released in the gas phase and then expands, producing a bubble inside the material, but it is qualitatively different from the high-intensity plasma phenomenon described above.

Transparent dielectrics have several distinctive features. Firstly, they have a wide optical band-gap (it ranges from 2.2–2.4 eV for chalcogenide glasses, and up to 8.8 eV for sapphire) that ensures they are transparent in the visible or near-infrared at low intensity. Therefore, in order to induce material modification with moderate energy pulses, the laser intensity should be increased to induce a strongly nonlinear response from the material, such as plasma formation. At intensities in excess of 10^{14} W cm^{-2} most dielectrics can be ionized early in the laser pulse and afterwards, therefore, the interaction proceeds in the laser-plasma mode which is similar for all materials.

A second feature of dielectrics is their relatively low thermal conductivity characterized by the thermal diffusion coefficient, D, which is typically $\sim 10^{-3}$ cm^2 s^{-1} compared with $\sim$cm^2 s^{-1} for metals. Therefore micron-sized regions will cool in a time of $\sim$10 microsecond ($t \sim l^2/D \sim 10^{-5}$ s). Hence, the laser effect of multiple laser pulses focused into the same point in a dielectric, will accumulate if the period between the pulses is shorter than the cooling time. Thus, if the single pulse energy is too low to produce any modification of the material, a change can be induced using a high pulse repetition rate, because of this accumulation phenomenon. The local temperature rise resulting from energy accumulation eventually saturates as the energy inflow from the laser is balanced by heat conduction, this typically taking a few thousand pulses at a repetition rate in the 10–100 MHz range. This effect has been experimentally demonstrated from measurements of the size of a void produced inside a dielectric by a high-repetition-rate laser [2]. The size of a damage zone increased with the number of pulses hitting the same point in the material. The accumulation effect has also been demonstrated during ablation of chalcogenide glass by a high repetition rate, 76 MHz, laser [3]. In the latter case, a single laser pulse heats the target surface only by several tens of Kelvin, which is insufficient to produce any phase change. However, the energy density rises above the ablation threshold due to energy accumulation when a few hundred pulses hit the same spot. This is also accompanied by a marked change in the interaction physics from a laser–solid to a laser–plasma interaction. Thus repetition rate becomes another means of controlling the size of the structure produced by the laser.

In the following we describe the laser–solid interaction when a laser beam is tightly focused inside a transparent dielectric in two limit cases: the case of low intensity well below the ablation threshold (nondestructive interaction); and the high-energy-density case when a material is ionized, and all bonds are broken (destructive interaction). Then we make a comparison with experiment, discuss and draw conclusions.

2.2
Laser–matter Interactions: Basic Processes and Governing Equations

Two particular properties of transparent dielectrics, namely their large absorption length and low thermal conductivity, define the mode of the laser–matter interaction for these materials at relatively low intensity ($< 10^{12}$ W cm^{-2}) and for a laser wavelength of $\lambda > 300$ nm. One can easily see that, in order to produce some detectable structure inside the material, it must be reasonably transparent. On the other hand, transparency means that the absorption length is large. In practical terms, to create a high intensity using low pulse energy, requires that energy to be focused to the smallest possible volume, with dimensions of the order of the laser wavelength $\sim \lambda$.

The full description of the laser–matter interaction process and laser-induced material modification from first principles, embraces the self-consistent set of equations that includes the Maxwell equations for the laser field coupling with matter complemented with the equations describing the evolution of the energy distribution functions for electrons and phonons (ions) and ionization equations. This is a formidable task even for modern supercomputers. Therefore, this complicated problem is usually split into the sequence of simpler interconnected problems: the absorption of laser light, ionization, energy transfer from electrons to ions, heat conduction, and hydrodynamic expansion which we are describing below.

Let us consider first the intensity distribution in a focal volume with tight focusing using high numerical aperture optics.

2.2.1
Laser Intensity Distribution in a Focal Domain

The intensity distribution in the focus of an axially symmetric ideal Gaussian beam, $I(r,z)$, is well known [4]. The complex amplitude of the electric field then reads:

$$E(r, z, t) = E_0(t) \frac{1}{(1 + z^2/z_0^2)^{1/2}} \exp\left\{ -\frac{r^2}{r_0^2(1 + z^2/z_0^2)} \right\}$$

$$\times \exp\left\{ -ikz - ik\frac{r^2}{2z(1 + z_0^2/z^2)} + i\tan^{-1}(z_0^2/z^2) \right\} \tag{1}$$

Here r_0 is a minimum waist radius at $z = 0$ and z_0 is the Rayleigh length. Both lateral and axial space scales are connected by the familiar relation, $z_0 = \pi r_0^2 n_0/\lambda$, here n_0 is the real part of the refractive index in the material. The minimum waist radius in the case of diffraction-limited focusing by numerical aperture lens (NA) is given by $r_0 = 1.22\lambda/NA$. The cylindrical focal volume (where the electric field decreases e-fold along the radius and by a factor of two along the z-axis) is then expressed as:

$$V_{foc,E} = 2\pi r_0^2 z_0 = \lambda z_0^2 \tag{2}$$

In the following we relate the absorbed energy to the incident laser intensity, averaged over many laser light periods: $I_0 = \frac{c}{8\pi}|E_0|^2$, where $E_0(r = 0; z = 0)$ is properly related to the incident laser electric field. Correspondingly, the volume where the intensity decreases approximately twice ($z_{1/2} = (\sqrt{2}-1)z_0$; $r_{1/e} = \sqrt{2}\, r_0$) reads:

$$V_{foc,I} = 2\pi r_{i/e}^2 z_{1/2} = (\sqrt{2} - 1)\pi r_0^2 z_0 = (\sqrt{2} - 1)^2 \lambda z_0^2 \tag{3}$$

In real experimental cases, corrections for the short pulse length, the use of a thick lens, the nonlinearity of the material, aberrations in the focusing optics, etc., should be made. Even with tight focusing, the focal volume is usually of the order of the cube of the wavelength, $V_{foc,1} \approx \lambda^3$. The electric field in the tight focus represents a complex interference pattern where the notion of polarization is hard to introduce. Therefore, in the following we consider only the effects of the total intensity on the matter in a focal volume.

2.2.2
Absorbed Energy Density Rate

The absorbed laser energy per unit time and per unit volume, Q_{abs}, is related to the gradient of the energy flux in a medium (the Pointing vector) [5] as follows:

$$Q_{abs} = -\nabla S = -\nabla\left(\frac{c}{4\pi}E \times H\right) \tag{4}$$

Time averaging (4) over many laser periods and replacing the space derivatives by the time derivative from the Maxwell equations [5] results in the form:

$$Q_{abs} = \frac{\omega}{8\pi} = \varepsilon''|E_a|^2 \tag{5}$$

Here $\varepsilon = \varepsilon' + i\varepsilon''$ is a complex dielectric function. E_a denotes the electric field in the medium averaged over the short timescale (ω^{-1}) but it maintains, of course, the temporal dependence of the field (intensity) of the incident laser pulse at $t \gg \omega^{-1}$. The spatial dependence of the field and the intensity inside the solid is determined by the focusing conditions. The absorbed energy should be related to the incident laser flux intensity averaged over the many laser light periods: $I_0 = \frac{c}{8\pi}|E_0|^2$, where E_0 is the incident laser electric field. The value of the electric field at the sample–vacuum interface, $E_a(0)$, is related to the amplitude of the incident laser field, E_0, by the boundary conditions:

$$|E_a(0)|^2 = \frac{4}{|1 + \varepsilon^{1/2}|^2}|E_{in}|^2 \tag{6}$$

Finally, the expression of the absorbed energy density by the incident laser flux reads:

$$Q_{abs} = \frac{\omega}{c}\frac{4\varepsilon}{|1 + \varepsilon^{1/2}|^2}I(r, z, t) \equiv \frac{2A}{l_{abs}}I(r, z, t) \tag{7}$$

l_{abs} is the electric field absorption depth :

$$l_s = \frac{c}{\omega\kappa} \tag{8}$$

A is the absorption coefficient defined by the Fresnel formula [5] as the following

$$A = 1 - R = \frac{4n}{(n + 1)^2 + k^2} \tag{9}$$

n and k are, respectively, the real and imaginary parts of the complex refractive index for the medium:

$$N \equiv \varepsilon^{1/2} = n + i\kappa \tag{10}$$

The intensity in (7) is defined by (1) as $I(r, z, t) = I_0(t) \cdot I(r, z)$, where I_0 relates to the intensity at the beam waist.

It was implicitly assumed in this derivation that the optical parameters of the medium are space- and time-independent and that they are not affected by laser–material interaction. One should also note that the relations between the Pointing vector, the intensity and absorption presented above, are rigorously valid only for the plane wave. However, comparison with experiment has shown that they are also valid with sufficient accuracy for tightly focused beams.

2.2.3
Electron–phonon (ions) Energy Exchange, Heat Conduction and Hydrodynamics: Two-temperature Approximation

First we recall that the fundamental interaction of light with matter involves the following physical processes. The incident laser radiation first penetrates the target and induces oscillations of the optical electrons. These electrons gain energy from the oscillating field by the disruption of the oscillating phase due to random collisions with atoms. The electron oscillation energy thereby converts to electron excitations. Following this, the electrons transfer energy to the lattice (ions) by means of electron–phonon (electron–ion) collisions over a period characterized by the temperature equilibration time, t_{e-L}, and by the means of electron heat conduction with a characteristic time, t_{th}.

The processes of the redistribution and transfer of the absorbed energy take place either during the energy deposition time or after the end of the laser pulse, depending on the relations between the laser pulse duration and the characteristic times for the energy exchange between the electrons and the lattice and energy transport into the bulk of a target. The magnitude of the energy loss in the collisions of the identical particles (electron–electron and ion–ion collisions) can be of the same order as the energy itself. Therefore, such collisions lead to fast energy equilibration inside each species of identical particles, i.e., to the establishment of equilibrium distribution over velocities characterized by the temperature of a particular species. Afterwards, the distribution functions adiabatically follow the changes in energy of each subsystem with time.

This is the physical reason for the description of an electron–phonon (ion) system as a mixture of two liquids with two different temperatures, T_e for electrons, and T_L for lattice (ions). A conventional two-fluid approximation for a plasma containing the electrons and ions of one kind can be rigorously derived by reducing the set of electron–ion kinetic equations to the coupled equations for the successive velocity moments [6]. The infinite set of moment equations is conventionally truncated at the second moment by assuming the existence of the equation of state – the relation between pressure, density and temperature for each type of particle. The resulting set of coupled equations comprises mass, momentum and energy conservation equations for electrons and ions (see Appendix). Hydrodynamic motion commences after the electrons transfer the absorbed energy to ions in excess of that necessary for breaking inter-atomic bonds. A similar approach can also be applied to the electron–phonon interaction in a solid when electrons are excited by a laser [7] at relatively low intensity, insufficient for destruction of the material.

There are no mass or momentum changes in the course of nondestructive laser–matter interaction at low intensity. A material modification can be described by the coupled energy equations for electrons and phonons in the two-temperature approximation introduced in [7]. The energy conservation law is expressed by the following set of coupled equations for the electron and lattice temperature [7, 8] which are similar to those for a plasma [6]:

$$
\begin{aligned}
C_e n_e \frac{\partial T_e}{\partial t} &= Q_{abs} - C_e \, n_e \, v_{en} \, (T_e - T_L) + \nabla q_{el} \\
C_L n_a \frac{\partial T_L}{\partial t} &= C_e n_e \, v_{en} \, (T_e - T_L) + \nabla q_{phonon}
\end{aligned}
\tag{11}
$$

Here C_e, C_L, n_e and n_a represent the electron and lattice specific heat and number density of free carriers and atoms respectively; Q_{abs}, q_{el}, q_{phonon} are the absorbed energy density rate, the electron heat conduction flux and the phonon heat conduction flux, and v_{en} is the electron-to-phonon energy exchange rate. Usually the electron heat conduction dominates the energy transfer in solid as well as in a plasma. In the following we neglect the phonon (ions) heat conduction.

2.2.4
Temperature in the Absorption Region

The ultra-short pulse interaction takes place when the laser pulse duration, t_p, is shorter than both the electron–phonon energy exchange and the heat conduction time, $t_p < \{t_{e-L}, t_{th}\}$ and generally requires the pulse duration to be less than a pico-second. Generally in dielectrics, $t_{e-L} \ll t_{th}$. Therefore for a long period the temperature in a laser-affected solid remains the same as that after the electron-lattice equilibration time. It is instructive, for further study, to express the temperature in the interaction region during the pulse as a function of the laser and material parameters. Let us consider a pulse where $t_{e-L} < t_p < t_{th}$, and the density of the solid remains essentially unchanged during the laser pulse. Hence, we assume that electron–phonon temperature equilibration occurs during the laser pulse (thus the electron and lattice temperatures are the same, $T_e = T_L = T$), and the heat losses are negligible. The energy conservation law (11) takes a simple form:

$$Cn_a \frac{\partial T}{\partial t} = Q_{abs} \tag{12}$$

Here C, n_a are the lattice specific heat and the atomic density, respectively. The temperature in a focal volume after integration of (12) with the help of (7) by time can be obtained as follows:

$$T(r, z, t) = \frac{2A\, F(t)}{l_{abs} C\, n_a} I(r, z)$$

$$F(t) = \int_0^t I_0(t)dt \tag{13}$$

Here F is the laser fluence – the laser energy delivered per unit area at the waist of the beam. At times longer than the pulse duration, t_p, the temperature depends on the total fluence per pulse $F(t = t_p) = F_p$. The temperature is a maximum at the beam waist ($z = 0$; $r = 0$) at $t = t_p$:

$$T_m = \frac{2A\, F_p}{l_{abs} C\, n_a} \tag{14}$$

The maximum temperature in the focal volume is expressed by the laser energy per pulse, $E_{pulse} = F_p \cdot \pi r_0^2$, and target parameters as the following:

$$T_m = \frac{2A\,E_{pulse}}{C\,n_a}\frac{1}{\pi r_0^2 l_{abs}} \tag{15}$$

The absorption volume from the above formula, $V_{abs} = \pi r_0^2 l_{abs}/2$, defines the deposited energy density and temperature in the laser-affected volume. As we show later, due to a change in the interaction regime (e.g., from laser–solid to laser–plasma) this volume can be made much smaller than the focal volume $V_{foc,I} = \left(\sqrt{2}-1\right)^2 \pi r_0^2 z_0$.

From (15) one can see two ways of increasing the temperature to a level where a change in the optical or structural properties can be achieved. The first way involves a moderate increase in the laser energy per pulse and (or) focusing conditions. This applies in the low-intensity limit for nondestructing phase transitions (crystal-to-crystal, crystal-amorphous, etc.). The interaction volume in this case is comparable with the focal volume. The second way involves a marked increase in absorption. In the high-intensity limit, the deposited energy density increases above the ionization and damage threshold and the interaction changes to the laser–plasma mode. As will be seen later, this can rapidly increase the absorption and result in a decrease in the interaction volume compared with the focal volume because of the reduction in l_{abs}.

2.2.5
Absorption Mechanisms

The light absorption mechanisms in solids can be classified into several types [9]:

1. Intraband transitions, mainly comprising the contribution of free charge carriers (electrons) in metals and semi-conductors.
2. Inter-band transitions (for example, single and multi-photon absorption).
3. Absorption by excitons.
4. Absorption on impurities and defects.

The first mechanism plays a major role in metals and in plasma. We discuss it later in connection with laser–solid interactions at high intensity.

The second mechanism is of major importance in electromagnetic field interaction with dielectrics at low intensity when the laser light does not change the material parameters during the interaction. Note that the majority of transparent media are dielectrics. A simple estimate of the light absorption coefficient due to direct inter-band transitions can be presented in a form following [9]:

$$\alpha_d(\omega) = \frac{1}{I}\frac{dI}{dx} \approx \frac{e^2}{\hbar c/r_B}\frac{\hbar\omega\cdot\left(\hbar\omega - E_g\right)^{1/2}}{(2\,\mathrm{Ry})^{3/2}} \tag{16}$$

Here ω is the frequency of the incident light, E_g is the band gap in the dielectric, r_B is the Bohr radius, e is the electron charge. The energy is measured in

Rydbergs, $Ry = \frac{e^4 m}{2\hbar^2} = 13.6$ eV. The absorption coefficient for indirect transition is a factor of $\left(\frac{m}{N}\right)^{1/2} \sim 10^{-2}$ less than that for the direct transition.

2.2.6
Threshold for the Change in Optical and Material Properties ("Optical Damage")

Any change in the optical and material properties depends on the energy density (temperature) in the focal region and, therefore, on the intensity in this area. The maximum temperature at the end of the pulse is a function of fluence (combination of an average intensity during the pulse, focal spot size, and pulse duration). In terms of the temperature (intensity) one can establish different stages of material modification or damage. For example, the change in the optical properties of a dielectric can be achieved without mechanical damage by an increase in the temperature. The change in optical properties can also be associated with a phase transition. For example, a transition from the crystalline to the amorphous state in silicon or in chalcogenide glasses, results in a change in the refractive index. The phase equilibrium relations describe the relation between the temperature-density in a material and the phase state for known phase transitions.

One can introduce a logical sequence of material modification following an increase in the intensity (temperature) in the focal region. First, changes in optical properties occur due to temperature changes, this is followed by phase transitions (crystal–crystal, crystal–amorphous, solid–liquid, etc.); then material decomposition breaking interatomic bonds, and ionization. The direct relation between the phase transition temperature and the threshold laser fluence that must be achieved is given by equation (15). Any further increase in the intensity above the ionization threshold results in heating the plasma which has been created.

2.3
Nondestructive Interaction: Laser-induced Phase Transitions

We consider in this section the laser–matter interaction and laser-induced phase transformations in conditions when the energy density in the interaction volume is below the ionization and structural damage thresholds. Reversible phase transitions are attractive for applications in 3D optical memories because they allow, in principle, the design of read-write-erase devices.

In this section we aim to establish a relation between the laser and material parameters necessary for a structural phase transition to be completed. Electrons initially absorb the laser energy. They then transfer it to the lattice during the electron-to-lattice energy exchange time. The structural transition occurs when atoms are moved into the positions corresponding to a new phase during the phase transition time. Let us first estimate the relevant timescales of these processes.

2.3.1
Electron–Phonon Energy Exchange Rate

The electron-to-phonon energy exchange rate, v_{en}, is expressed by the electron–phonon momentum exchange rate, (often referred to as the optical or transport rate) as follows [9]:

$$v_{en} \approx \frac{3}{4} v_{opt} \left(\frac{m^*}{M} \right)^{1/2} \tag{17}$$

m^*, M are the effective (re-normalized) electron mass, and the atomic mass, respectively. The momentum exchange rate in the adiabatic approximation $\left((\hbar \omega_D / J_i)^{1/2} << 1 \right)$ can be estimated by the atomic frequency:

$$v_{opt} \approx \left(\frac{m^*}{M} \right)^{1/2} \frac{J_i}{\hbar}, \quad T_L \sim T_D \tag{18}$$

Here J_i, T_D, ω_D are the ionization potential, the Debye temperature, and the Debye frequency. Hence, the electron-to-lattice energy exchange rate can be estimated in the form:

$$v_{en} \approx \frac{3}{4} \frac{m^*}{M} \frac{J_i}{\hbar} \tag{19}$$

For example, the effective electron–phonon optical frequency and the electron–phonon energy exchange rate for chalcogenide glass, As_2S_3,(Sulphur: $J_i = 10.36$ eV; $M = 32$ au.; Arsenic: $J_i = 9.81$ eV; $M = 74.92$ au) are $v_{opt} = 3.5 \times 10^{13}$ s^{-1} and $v_{en} = 1.6 \times 10^{11}$ s^{-1} respectively. Thus, the electron and lattice temperatures equilibrate within laser-irradiated chalcogenide glass after ~ 6 picoseconds.

2.3.2
Phase Transition Criteria and Time

Let us associate the transition from the crystalline to the amorphous state with a loss of long-range order due to short-wavelength fluctuations in the atomic displacements. At the phase transition temperature, the symmetry changes from that of a crystalline space group to the rotationally invariant state of a disordered solid. Let us apply to this transition the criterion similar to the Lindemann criterion for melting: the transition occurs when the atomic mean square displacement due to thermal vibrations is a significant fraction of the interatomic distance (i.e., the lattice constant). The total displacement of a single atom is the sum of contributions from all independent phonon modes [10]:

$$\langle \Delta r^2 \rangle = \frac{3 \hbar^2}{M k_B T_D} \frac{T}{T_D} \tag{20}$$

Taking the phase transition temperature of the same order of magnitude as the Debye temperature, one obtains that, during a crystal-to-amorphous phase transition in a chalcogenide glass, the single atom displacement comprises less than

10% of the interatomic distance. Therefore the time for this displacement to occur, $t_{displ} \approx \langle \Delta r^2 \rangle^{1/2} / v_{sound}$, is of the order of 10 femtoseconds. Thus, the time for the energy transfer from the electrons to the lattice dominates in a laser-induced phase transition. Now we consider laser-induced phase transitions in different materials which can result in the formation of diffractive structures suitable for optical memories.

2.3.3
Formation of Diffractive Structures in Different Materials

2.3.3.1 Modifications Induced by Light in Noncrystalline Chalcogenide Glass

Amorphous-to-crystalline phase transition
Noncrystalline chalcogenide glasses are intrinsically metastable solids without long-range order. The chalcogen atoms (S, Se, Te) possess lone pair electrons, which are normally nonbonding. The states associated with nonbonding electrons lie at the top of the valence band and hence they are preferentially excited by illumination. A two-coordinated initial state changes to a single- or three-coordinated state due to excitation and this results in structural and optical changes. The structure and bond configuration can be changed either by photon absorption or by heating [11].

For example, illumination of amorphous As_2S_3 by photons with energy near or above the bandgap (~2.4 eV) changes the refractive index from 2.447 to 2.569 (almost 5 %) due to an *amorphous-crystalline phase transition* (a change from short-range to long-range order). Electronic excitation by photon absorption is believed to be responsible for bond switching which is also considered as a reason for other light-induced changes such as photo-darkening [12]. The mechanisms of these transitions are not yet fully understood.

The glass transition temperature for As_2S_3 is $T_g \sim 430$ K; whilst the melting temperature is $T_m \sim 580$ K. Therefore, the laser parameters should be carefully chosen to induce desirable changes. Let us obtain the single pulse laser parameters necessary to achieve the crystal-to-amorphous transition temperature in As_2S_3 ($n_a = 3.9$–4.26×10^{22} cm^{-3}; $C = 2.7$ k$_B$) by the action of the 515 nm laser beam (imaginary part of refractive index $k = 0.04$; absorption coefficient $A = 0.826$; absorption depth $l = c/\omega k = 2.08 \times 10^{-4}$ cm). The fluence necessary to induce a temperature of $T_g \sim 430$ K is 8×10^{-2} J/cm^{-2} per pulse in accordance with (15). If the laser beam is tightly focused down to a waist of 5×10^{-9} cm^2 ($r_0 = 0.4$ micron) inside the bulk, then the energy per laser pulse is $E_{pulse} \sim 0.4$ nano Joules (intensity of 0.8×10^{12} W cm^{-2} for a 100 femtosecond pulse). We note that the optical properties of thin films of chalcogenide glasses prepared by different methods (thermal evaporation, laser deposition, etc.) are significantly different. For example, the absorption coefficient for As_2S_3 at 514.5 nm reported in [13] gives an absorption length five times higher than that presented above.

One should remember that, if the femtosecond pulses are used for excitation, the phase transition occurs several picoseconds after the end of the pulse once the energy transfer from electrons to atoms is completed.

Photo-darkening and photo-bleaching effects

The irradiation of As_2S_3 by near-band-gap photons (~2.4 eV) results in the band-gap shift (decrease in the bandwidth) towards lower energy, up to the saturation limit [12]. The effect is a maximum at $T = 14$ K; and the band-gap decreases by up to $\Delta E = -0.2$ eV. This shift in the band-gap decreases as the temperature increases, disappearing when the temperature approaches the glass transition point, T_g. Photo-darkening in As_2S_3 is accompanied by a volume increase of up to ~ 0.4%, and the change in refractive index can reach $\Delta n = 0.1$. It was observed [12] that the temperature needed to induce such transitions in thin films was lower than that in the bulk.

The photo-darkening effect has been employed for writing waveguides in laser-deposited As_2S_3 chalcogenide, PMMA-coated, films. The waveguides were created using either a focused beam from a 514.5 nm CW Ar laser or 532 nm from a CW frequency-doubled Nd laser [3] at intensities up to 2.5×10^3 W cm^{-2}. The measured change in refractive index was $\Delta n = + 0.01 - 0.04$.

2.3.3.2 Two-photon Excitation of Fluorescence

Detectable structures have been obtained by transforming a photochromic material (~1% weight) embedded into the host polymer matrix by two-photon excitation [14]. Two diffraction-limited laser beams (1064 nm and 532 nm) which overlapped in time and space, were focused inside the polymer. Two-photon excitation (which is equivalent in this case to the absorption of a 355 nm photon with an energy of 3.5 eV) of the photochromic molecule transforms the original molecule into a different form (merocyanine in [14]), which can emit red-shifted fluorescence when excited by green light. Thus, the laser-affected zone (memory bit) can afterwards be detected (read) by the irradiation of light in the red-green region of the visible spectrum. Irradiation for 5 s at 355 nm at a total fluence of 4 mJ cm^{-2} and for 60 s with two beams of 532 nm and 1064 nm at 20 mJ cm^{-2} was needed in order to produce a detectable structure. The essential feature of two-photon excitation – a quadratic dependence of the absorption on the laser intensity – ensures that the excited area is confined inside the focal volume.

Two-photon excitation of fluorescence of a dye molecule was also used in [15]. The laser (pulse duration 100 fs, repetition rate 80 MHz average incident laser power $p = 50$ mW), was focused with numerical aperture lens of NA = 1.4 to a diffraction-limited waist of less than a micron in diameter. With these parameters, the absorption of two low-energy photons led to saturation in the fluorescence output.

In references [14] and [15], the use of two-photon excitation for three-dimensional optical storage memory with writing, reading and erasing of the information, was proposed.

2.3.3.3 Photopolymerization

The refractive index in a sub-micrometer volume inside a photopolymer can be increased by two-photon excitation of a photo-initiator at the waist of a tightly focused laser beam [16]. Thus, the information is written as a sub-micrometer memory bit. Information is then read with axial resolution by differential interference contrast microscopy.

The simultaneous absorption of two photons (two-photon polymerization) leads to a density increase in the polymer and a concomitant increase in its index of refraction. On polymerization the refractive index increased from 1.541 to 1.554 (0.8 % change). For example, liquid acrylate ester blend can be solidified by two-photon excitation. Many photoresists are also known to undergo density changes when they are excited by laser light. For example, a laser pulse (100 fs, 620 nm, 100 MHz) was focused by NA = 1.4 to the minimum waist radius of 540 nm. The illumination dwell time per exposed pixel was ~10 ms, at 2–3 mW of average power [16].

Fabrication of three-dimensional microstructures by a tightly focused laser beam in polymerizable resin (refractive index $n = 1.5$) due to two-photon absorption has been reported in [17]. The laser beam (pulse duration 120–150 fs; wavelength 398 nm; energy per pulse 0.6 μJ) was tightly focused with NA = 1.35, corresponding to a minimum radius of the beam waist of 360 nm. The polymerization threshold was found to be less than half the boiling threshold. At the threshold of photomodification (polymerization–solidification) the size of the affected area was found to be less than 200 nm. The reason why the affected area is less than the Gausian beam waist radius relates to the quadratic dependence of two-photon absorption on laser intensity. The threshold of photopolymerization was achieved at a laser energy per pulse of 3 nJ, and a fluence of 2.9 J cm^{-2} (average intensity ~ 2.5 × 10^{13} W cm^{-2}). The author's analysis suggests that the absorption was very low, even at such a high intensity.

2.3.3.4 Photorefractive Effect

The refractive index can be changed locally as a result of photo-refraction [18]. The laser beam is focused by a microscope objective lens into the diffraction-limited spot inside a photorefractive crystal. The electrons in the beam focal spot are excited from the donor level to the conduction band. The number of excited electrons depends on the local intensity distribution. The excited electrons then diffuse and drift until they recombine with the vacant donor sites. The spatial distribution of the recombined electrons generates a local electric field. The electric field produces a nonuniform refractive index spatial distribution due to an electro-optic (Pockels) effect. Thus, the area with a different refractive index in a focal volume (a single bit of data) is created in the photorefractive medium. Usually the refractive index change due to the photorefractive effect is less than 10^{-2}.

An argon laser beam (wavelength 476.5 nm, the intensity on the spot 2 kW cm^{-2}) was focused by a microscope objective lens (NA = 1.0, oil immersion) into the diffraction-limited spot inside Fe-doped LiNbO$_3$ crystal [18]. Each bit was recorded at an exposure time of 25 ms (fluence of 50 J cm^{-2}). The recorded bits have their refractive index lower than that in the unexposed regions. The recorded data points were read with a phase-contrast microscope objective lens. Note, that the index change in a photorefractive crystal is reversible. Therefore, this effect can be used to design read-write-erase memory devices [18].

2.4
Laser–Solid Interaction at High Intensity

The major mechanisms of absorption in the low intensity laser–solid interaction are inter-band transitions. The absorption of a photon beam with near band-gap energy at low intensity is small, which corresponds to a large real and small imaginary part of the dielectric function. The optical parameters (refractive index) in these conditions are only slightly changed during the interaction in comparison with those of the cold material. The absorption can be increased if the photon energy increases above the band-gap value with loss of transparency or/and if the incident light intensity increases to a level where the energy of the electron oscillations in the laser field becomes comparable to the band-gap energy. In such conditions the properties of the material and the laser–material interaction change rapidly during the pulse. In fact, as the intensity increases above the ionization threshold, the material becomes transformed into a plasma which is accompanied by almost complete absorption of the incident light, a decrease in the absorption depth and the creation of a high energy density in the absorption region. High-intensity laser–solid interactions thereby allow the formation of quite different three-dimensional structures inside a transparent solid in a controllable and predictable way.

2.4.1
Limitations Imposed by the Laser Beam Self-focusing

The power in a laser beam which is aimed to deliver the energy to a desirable spot inside a bulk transparent solid should be kept lower than the self-focusing threshold for the medium. The critical value for the laser beam power depends on the nonlinear part of the refractive index, n_2, ($n = n_0 + n_2 I$), as follows [19]:

$$P_{cr} = \frac{\lambda^2}{2\pi \, n_0 \, n_2} \qquad (21)$$

The self-focusing of the beam begins when the power in a laser beam, P_0, exceeds the critical value, $P_0 > P_{cr}$. The Gaussian beam under the above conditions self-focuses after propagating along the distance, L_{s-f} [19]:

$$L_{s-f} = \frac{2\pi \, n_0 \, r_0^2}{\lambda} \left(\frac{P_0}{P_{cr}} - 1 \right)^{-1/2} \qquad (22)$$

Here r_0 is the minimum waist radius of the Gaussian beam. For example, in a fused silica ($n_0 = 1.45$; $n_2 = 3.54 \times 10^{-16}$ cm^2 W^{-1}) for $\lambda = 1000$ nm, the critical power comprises 3 MW, while the self-focusing distance (assuming $P_0 = 2P_{cr}$ and $r_0 \sim \lambda$), equals $\sim 9\lambda$.

2.4.2
Optical Breakdown: Ionization Mechanisms and Thresholds

The optical breakdown of dielectrics and optical damage produced by the action of an intense laser beam has been extensively studied over several decades [20–32]. It is well established [20–22] that two major mechanisms are responsible for the conversion of a material into a plasma: ionization by electron impact (avalanche ionization), and ionization produced by simultaneous absorption of multiple photons [9]. The relative contribution of both mechanisms is different at different laser wavelengths, pulse durations, and intensity, and for different materials. We present the breakdown threshold dependence of laser and material parameters for intense and short (sub-picosecond) pulses where the physics of processes is most transparent.

2.4.2.1 Ionization by Electron Impact (Avalanche Ionization)

Let us first consider the laser–solid interaction in conditions when direct photon absorption by electrons in a valence band is negligibly small. A few (seed) electrons in the conduction band oscillate in the electromagnetic field of the laser and can gain net energy by collisions. Thus, they can be gradually accelerated to an energy in excess of the band-gap. The process of electron acceleration can be represented, in a simplified way, as for Joule heating [22]:

$$\frac{d\varepsilon}{dt} = \frac{e^2 \, v_{e-ph}}{2m^*\left(v_{e-ph}^2 + \omega^2\right)} E^2 = 2\varepsilon_{osc} \frac{\omega^2 \, v_{e-ph}}{\left(v_{e-ph}^2 + \omega^2\right)} \tag{23}$$

Here e, m^*, v_{eff} are respectively the electron charge, the effective mass, and the effective collision rate; ω, E, and $\varepsilon_{osc} = \frac{e^2 E^2}{4m^* \omega^2}$ are the laser frequency, the electric field and the electron oscillation energy in the field. Electrons accelerated to an energy in excess of the band-gap $\varepsilon > \Delta_{gap}$ can collide with electrons in the valence band and transfers sufficient energy to them for excitation into the conduction band. Thus, an avalanche of ionization events can be created. The probability of such event can be estimated with the help of (23) as follows:

$$w_{imp} \approx 2 \frac{\varepsilon_{osc}}{\Delta_{gap}} \frac{\omega^2 \, v_{e-ph}}{\left(v_{e-ph}^2 + \omega^2\right)} \tag{24}$$

In this classical approach (the electron is continuously accelerated) the probability of ionization is proportional to the laser intensity (the oscillation energy). The electron (hole)–phonon momentum exchange rate changes along with the increase in laser intensity (temperature of a solid) in a different way for different temperature ranges. At low intensity when the electron temperature just exceeds the Debye temperature the electron–phonon rate grows with an increase in temperature. For SiO_2 $v_{eff} \sim 5 \times 10^{14}$ s^{-1} [23]. The light frequency (for visible light, $\omega \geq 10^{15}$ s^{-1}) exceeds the collision rate, $\omega > v_{eff}$. It follows from (24) that ionization rate then growth in proportion to the square of the laser wavelength in correspondence with the Monte Carlo solutions of the Boltzmann equation for electrons [23]. With further increase in temperature the effective electron–lattice collision rate responsible for momentum exchange saturates at the plasma frequency ($\sim 10^{16}$ s^{-1}) [1]. At this stage the wavelength dependence of the ionization rate almost disappears due to $\omega < v_{eff}$ as it follows from (24). This conclusion is in agreement with the rigorous calculations of [23].

It is worth noting that "classical" (as opposed to quantum) treatment is valid for very high intensity and an optical wavelength of a few hundred nanometers. It was established [22] that the value of the dimensionless parameter $\gamma = \frac{\Delta\varepsilon}{\hbar\omega} \frac{\varepsilon}{\hbar\omega} \sim 1$ separates the parameter space in two regions where the classical, $\gamma > 1$, or quantum, $\gamma < 1$, approach is valid. Thus, if the energy gain of an electron in one collision, $\Delta\varepsilon$, (which is proportional to the oscillation energy) and the electron energy are both higher than the photon energy, then $\gamma > 1$, and the classical equation (24) is valid. The classical approach applies at the very high intensities which have been recently used in short-pulse-solid interaction experiments. For example, at $I = 10^{14}$ $W\,cm^{-2}$ and $\hbar\omega = (2–3)$ eV, one can see that $\Delta\varepsilon/\hbar\omega \sim 3–4$; $\varepsilon > \Delta\varepsilon$, therefore $\gamma \gg 1$ and the classical approximation applies.

In practical calculations, many researchers [23, 25] use the approximate formula for impact ionization suggested by Keldysh where the threshold nature of the process ($\varepsilon > \Delta_{gap}$) is explicitly presented:

$$w_{imp} = p(s^{-1}) \left[\frac{\varepsilon}{\Delta_{gap}} - 1 \right]^2 \qquad (25)$$

The pre-factor for fused silica is $p = (1.3–1.5) \times 10^{15}$ s^{-1} [23, 25]. It should be noted that impact ionization requires the effective electron collision frequency and initial electron density, n_0, to be relatively high.

The number density of electrons generated by such an avalanche process reads:

$$n_e = n_0\, 2^{w_{imp}t} = n_0\, e^{\ln 2 \cdot w_{imp} \cdot t} \qquad (26)$$

It is generally accepted that breakdown occurs when the number density of electrons reaches the critical density corresponding to the frequency of the incident light $n_c = m_e\omega^2/4\pi e^2$. Thus, laser parameters, (intensity, wavelength, pulse duration) and material parameters (band-gap width and electron–phonon effective rate) at the breakdown threshold are combined by condition, $n_e = n_c$.

2.4.2.2 Multiphoton Ionization

The second ionization mechanism relates to simultaneous absorption of several photons [9]. This process has no threshold and hence the contribution of multi-photon ionization can be important even at relatively low intensity. Multiphoton ionization creates the initial (seed) electron density, n_0, which then grows by the avalanche process. Multiphoton ionization can proceed in two limits separated by the value of the Keldysh parameter $\Gamma^2 = \varepsilon_{osc}/\Delta_{gap} \sim 1$. Tunneling ionization occurs in conditions when $\omega \ll eE/(\Delta_g m)^{1/2}$ or $\Delta_{gap} \ll \varepsilon_{osc}$. The ionization probability in this case does not depend on the frequency of the field and parallels the action of a static field [33, 34]:

$$w_{tunnel} \approx \frac{\Delta_g}{\hbar\omega} \exp\left[-\frac{4}{3}\frac{\Delta_g^{3/2}}{\hbar}\frac{(2m)^{1/2}}{eE}\right] = \frac{\Delta_g}{\hbar\omega} \exp\left[-\frac{4}{3}\frac{\Delta_g}{\hbar\omega}\left(\frac{\Delta_g}{\varepsilon_{osc}}\right)^{1/2}\right] \qquad (27)$$

The multi-quantum photo-effect takes place in the opposite limit $\Delta_{gap} > \varepsilon_{osc}$. Intensities around $I \sim 10^{14}$ W cm^{-2} and photon energy $\hbar\omega = (2-3)$ eV are typical for sub-picosecond pulse interaction experiments with silica [24–30, 32]. The Keldysh parameter for all recently published experiments is around unity depending on the band-gap value (for some materials, such as silicon, it is higher, for silica it is lower than one). Therefore it is reasonable to take the ionization probability (the probability of ionization per atom per second) in the multi-photon form [22]:

$$w_{mpi} \approx \omega n_{ph}^{3/2} \left(\frac{\varepsilon_{osc}}{2\Delta_{gap}}\right)^{n_{ph}} \qquad (28)$$

Here $n_{ph} = \Delta_{gap}/\hbar\omega$ is the number of photons necessary for the electron to be transferred from the valence to the conductivity band. One can see that, with near band-gap energy , $\Delta_{gap} \sim \hbar\omega$, and $\varepsilon_{osc} \sim \Delta_{gap}$, both formulae give an ionization rate of $\sim 10^{15}$ s^{-1}.

2.4.3
Transient Electron and Energy Density in a Focal Domain

The time dependence of the number density of electrons n_e created by the avalanche and multi-photon processes can be obtained with the help of the simplified rate equation [1, 22, 23, 25, 26]:

$$\frac{dn_e}{dt} = n_e w_{imp} + n_a w_{mpi} \qquad (29)$$

Here n_a is the density of neutral atoms, w_{imp} and w_{mpi} are the probabilities (in s^{-1}) for the ionization by electron impact and the multi-photon ionization, respectively. In the above equation the electron losses due to recombination and diffusion from the focal area, as well as the space dependence, are ignored due to sub-picosecond pulse duration and the small size of a focal spot. Recombination time is around 1 picosecond in fused silica in accordance with [23].

Solutions of the rate equation in different conditions and for different materials [23, 25, 29] allow an estimate the relative role and interplay of the impact and multi-photon ionisation. The solution to Eq. (29) with the initial condition $n_e(t = 0)$ = n_0 and under assumption that w_{imp} and w_{mpi} are the time independent, is the following:

$$n_e(I, \lambda, t) = \left\{ n_0 + \frac{n_a w_{mpi}}{w_{imp}} \left[1 - \exp\left(-w_{imp}t\right) \right] \right\} \exp\left(w_{imp}t\right) \qquad (30)$$

The importance of multi-photon ionization at low intensity is clear from (30) even when the avalanche dominates. Multi-photon effects (second term in the brackets) generate the initial number of electrons, which, although small, can be multiplied by the avalanche process. The multi-photon ionization rate dominates, $w_{mpi} > w_{imp}$, for any relationship between the frequency of the incident light and the effective collision frequency in conditions when $\varepsilon_{osc} > \Delta_g \sim \hbar\omega$. However, even at high intensity the contribution of the avalanche process is crucially important: at $w_{mpi} \sim w_{imp}$ the seed electrons are generated by the multi-photon effect, whilst final growth is due to avalanche ionization. Such an interplay of two mechanisms has been demonstrated with the direct numerical solution of kinetic Fokker–Planck equation [25]. In conditions $w_{mpi} \sim w_{imp} \sim 10^{15}$ s^{-1} ($\varepsilon_{osc} \sim \Delta_g \sim \hbar\omega$) the critical density of electrons is achieved in a few femtoseconds. Therefore, for pulse duration $\sim$100 fs the ionization threshold can be reached early in the pulse and afterwards the interaction proceeds in the laser–plasma interaction mode.

2.4.2.1 Ionization and Damage Thresholds

It is conventionally suggested that the ionization threshold (or breakdown threshold) is achieved when the electron number density reaches the critical density corresponding to the incident laser wavelength. The ionization threshold for the majority of materials lies at intensities in between 10^{13} and 10^{14} W cm^{-2} ($\lambda \sim 1$ μm) with a strong nonlinear dependence on intensity. The conduction-band electrons gain energy in an intense short pulse much faster than they transfer energy to the lattice. Therefore the actual structural damage (breaking inter-atomic bonds) occurs after electron-to-lattice energy transfer, usually after the pulse end. It was determined that, in fused silica, the ionization threshold was reached to the end of 100 fs pulse at 1064 nm at the intensity 1.2×10^{13} W cm^{-2} [23]. Similar breakdown thresholds in a range of $(2.8 \div 1) \times 10^{13}$ W cm^{-2} were measured in the interaction of a 120 fs, 620 nm laser with glass, MgF$_2$, sapphire, and fused silica [26]. This behavior is to be expected, since all transparent dielectrics share the same general properties of slow thermal diffusion, fast electron–phonon scattering and similar ionization rates. The breakdown threshold fluence (J cm^{-2}) is an appropriate parameter for characterization conditions at different pulse duration. It is found that the threshold fluence varies slowly if pulse duration is below 100 femtosecond. For example, for the most studied case of fused silica the following threshold fluences were determined: ~ 2 J cm^{-2} (1053 nm; ~ 300 fs) and ~ 1 J cm^{-2}

(526 nm; ~ 200 fs) [25]; 1.2 J cm^{-2} (620 nm; ~120 fs) [26]; 2.25 J cm^{-2} (780 nm; ~ 220 fs) [29]; 3 J cm^{-2} (800 nm; 10–100 fs) [30].

2.4.3.2 Absorption Coefficient and Absorption Depth in Plasma

If ionization is completed early in the pulse the plasma formed in the focal volume has a free-electron density comparable to the ion density of about 10^{23} cm^{-3}. Hence, the laser interaction proceeds with plasma during the remaining part of the pulse. One can consider the electron number density (and thus the electron plasma frequency) as being constant in order to estimate the optical properties of the laser-affected solid. The dielectric function of plasma can be properly described in the Drude approximation when the ions are considered as a neutralizing background [5]:

$$\varepsilon = 1 - \frac{\omega_p^2}{\omega^2 + v_{eff}^2} + i \frac{\omega_p^2}{\omega^2 + v_{eff}^2} \frac{v_{eff}}{\omega} \equiv \varepsilon' + i\varepsilon'' \tag{31}$$

Here the electron plasma frequency, ω_p, is an explicit function of the number density of the conductivity electrons, n_e, and the electron effective mass, m^*, $\omega_p^2 = 4\pi e^2 n_e / m^*$. One can estimate the optical parameters of this plasma, assuming that the effective collision frequency is approximately equal to the plasma frequency, $v_{eff} \sim \omega_p$, for the nonideal plasma which is created. The real and imaginary parts of the dielectric function and refractive index, $N \equiv \sqrt{\varepsilon} = n + ik$, are then expressed as follows:

$$\varepsilon' \approx \frac{\omega^2}{\omega_{pe}^2}; \ \varepsilon'' \approx \frac{\omega_{pe}}{\omega} \left(1 + \frac{\omega^2}{\omega_{pe}^2}\right)^{-1}; \ n \approx k = \left(\frac{\varepsilon''}{2}\right)^{1/2} \tag{32}$$

The absorption length is $l_s = \frac{c}{\omega k}$. The absorption coefficient is estimated by the Fresnel formula [5] as:

$$A = 1 - R \approx \frac{4n}{(n+1)^2 + n^2} \tag{33}$$

Let us estimate, for example, the optical parameters of plasma obtained after the breakdown of silica glass ($\omega_p = 1.45 \times 10^{16}$ s^{-1}) by an 800 nm laser ($\omega = 4.7 \times 10^{15}$ s^{-1}). One obtains with the help of (32) and (33) $\varepsilon' = 0.095$; $\varepsilon'' = 2.79$, $\varepsilon = 2.8$ and, correspondingly, the real and imaginary parts of the refractive index $n \sim \kappa = \varepsilon/\sqrt{2} = 1.985$ thus giving the absorption length of $l_s = 32$ nm, and absorption coefficient $A = 0.62$. Thus, the optical breakdown converts silica into a metal-like medium reducing the energy deposition volume by two orders of magnitude and, correspondingly, massively increasing the absorbed energy density.

2.4.3.3 Electron Temperature and Pressure in Energy Deposition Volume to the End of the Laser Pulse

The electron-to-ion energy transfer time and the heat conduction time in a hot plasma lies in a range of picoseconds. Therefore, in a sub-picosecond laser–solid interaction, the deposited energy is confined to the electron whilst the ions remain cold. Hence, the electron temperature in the focal volume at the end of laser pulse can be estimated by equations (12) and (13) with all losses and space dependence being neglected. The absorption coefficient and length for plasma are taken from the previous section

$$T_e(t_p) = \frac{2A\,F(t_p)}{l_{abs}\,C_e\,n_e} \tag{34}$$

As an example, consider the interaction of 800 nm laser ($C_e \sim 3/2$; the absorption length $l_s = 32$ nm, the absorption coefficient $A = 0.62$) with fused silica assuming that all atoms are singly ionized ($n_e = 6.6 \times 10^{22}$ cm^{-3}) at fluence well above the breakdown threshold, $F = 10$ J cm^{-2} (intensity 10^{14} W cm^{-2} for a 100 fs pulse). One obtains with the help of (34), the maximum electron temperature $T_e = 244$ eV and the corresponding thermal pressure $P_e = n_e\,T_e = 25.8$ Mbar (2.58×10^{12} Pa). The value of the electron temperature is well above the band-gap energy of ~ 9 eV, the ionization potential and the binding energy. Equation of state for this hot and dense plasma can be approximated by the sum of three terms [36]. One term corresponds to the binding forces ("cold" pressure and energy), the second term relates to the thermal part (coinciding with that for an ideal gas), and the third term corresponds to the Coulomb interaction between plasma particles. As follows from the above estimates, the thermal pressure significantly exceeds all moduli for silica, therefore the "cold" terms can be neglected. The ratio of the energy of Coulomb interactions to the thermal energy is characterized by the parameter, the so-called number of particles in the Debye sphere, $N_D = 1.7 \times 10^9 \left(T_e^3/n_e\right)^{1/2}$ (the temperature in electronvolts) [6]. The Coulomb terms can be neglected if $N_D \gg 1$. One can easily see that, using the above estimates, in the case considered, $N_D \sim 25$. Thus, a solid is transformed into a state of ideal gas at solid-state density.

2.4.4
Electron-to-ion Energy Transfer: Heat Conduction and Shock Wave Formation

The electron-to-ion energy exchange rate, ν_{en}, in plasma is expressed via the electron–ion momentum exchange rate, ν_{ei}, in accordance with Landau [5] as follows:

$$\nu_{en} \approx \frac{m_e}{m_i}\nu_{ei} \tag{35}$$

The electron–ion collision rate for the momentum exchange in ideal plasma is well known [6]:

$$\nu_{ei} \approx 3 \times 10^{-6}\ln\Lambda\,\frac{n_e Z}{T_{eV}^{3/2}} \tag{36}$$

Here Z is the ion charge and $\ln\Lambda$ is the Coulomb logarithm. Λ is the ratio of the maximum and minimum impact parameters. The maximum impact parameter is close to the electron Debye length (screening distance). The minimum impact parameter is the larger of either the classical distance of closest approach in the Coulomb collisions, or the DeBroglie wavelength of the electron [6]. Hence, electrons in fused silica heated by a tightly focused laser beam (T_e = 244 eV, n_e = 6.6 × 10^{22} cm^{-3}, $\ln \Lambda$ = 3.7, Z = 1, $(m_i)_a$ = 20) transfer energy to the ions over a time $t_{en} = (v_{en})^{-1}$ = 200 picoseconds that is longer than the laser pulse duration of 100 fs. Hydrodynamic motion, such as the emergence of a shock wave from the focal volume into a cold surrounding material, only starts after completion of the energy transfer.

2.4.4.1 Electronic Heat Conduction

Unlike motion of the ions, energy transfer by nonlinear electronic heat conduction starts immediately after the energy absorption. Therefore, a heat wave can propagate outside the heated area before the shock wave emerges. We estimate the extent of the heated volume to the moment when the shock wave catches up with the heat wave. We approximate the energy deposition volume by the sphere with radius, r_0. E_a relates to the absorbed laser energy. The cooling of a heated volume of plasma is described by the three-dimensional nonlinear equation [36]:

$$\frac{\partial T}{\partial r} = \frac{\partial}{\partial r} r^2 D T^n \frac{\partial T}{\partial r} \tag{37}$$

The thermal diffusion coefficient is defined conventionally as the following:

$$D = \frac{l_e v_e}{3} = \frac{v_e^2}{3 v_{ei}} \tag{38}$$

Here l_e, v_e and v_{ei} are the electron mean free path, the velocity and the collision rate from (36) respectively. It is convenient to express the diffusion coefficient by the temperature at the end of the laser pulse, T_0 (the initial temperature for cooling):

$$D = D_0 \left(\frac{T}{T_0}\right)^{5/2}; \quad D_0 = \frac{2T_0}{3 m_e v_{ei}(T_0)} \tag{39}$$

Here n = 5/2 as for ideal plasma. The energy conservation law complements the heat conduction equation:

$$\int_0^r C n_a T 4\pi r^2 dr = E_{abs} \tag{40}$$

Here C and n_a are the heat capacity and atomic (electron) number density respectively, E_a is the energy deposited in the absorption volume. It is instructive to present the equations (37) and (40) in a scaling form:

$$\frac{r^2}{t} = \frac{r_0^2}{t_0}\left(\frac{T}{T_0}\right)^n ; \quad t_0 = \frac{r_0^2}{D_0}$$

$$T_0\, r_0^3 = T\, r^3$$

(41)

t_0 is the time for a heat wave to travel a distance r_0. Then, the temperature at the heat wave front and the distance traveled by this front, r_{th}, are presented as a function of the absorbed energy and the material parameters in close correspondence to the exact solutions [36]:

$$r_{th} = (Dt)^{\frac{1}{2+3n}}\left(\frac{E_{abs}}{\frac{4}{3}\pi\, Cn_a}\right)^{\frac{n}{2+3n}} \sim t^{\frac{1}{2+3n}}\, E_a^{\frac{n}{2+3n}}$$

(42)

$$T = (Dt)^{\frac{3}{2+3n}}\left(\frac{E_{abs}}{\frac{4}{3}\pi\, Cn_a}\right)^{\frac{2}{2+3n}} \sim t^{-\frac{3}{2+3n}}\, E_a^{\frac{2}{2+3n}}$$

The temperature and heat penetration distances can be expressed in a compact form as a function of the initial temperature:

$$r_{th} = r_0\left(\frac{t}{t_0}\right)^{\frac{1}{2+3n}}$$

$$T = T_0\left(\frac{t_0}{t}\right)^{\frac{3}{2+3n}}$$

(43)

One can see that, at $n = 0$, all the above formulae reduce to those for the linear heat conduction.

2.4.4.2 Shock Wave Formation

Let us estimate the distance travelled by the heat wave to the moment when electrons have transferred their energy to the ions. At this instant the velocity of heat wave compares with the ion velocity, and the ion motion becomes dominant in the energy transfer. Consider, for example, the case of fused silica heated by a tightly focused beam ($T_0 = 244$ eV, $r_0 \sim 0.5$ micron). One gets, with the help of (39) and (41), $D_0 = 1.5 \times 10^3$ cm^2 s^{-1}, and $t_0 = 1.67 \times 10^{-12}$ s. The shock wave leaves the heat wave behind at the time when electrons transfer their energy to ions. This time comprises $t_{en} = (v_{en})^{-1} = 200$ picoseconds. Taking $n = 5/2$ in (43) for electronic heat conduction one obtains that the shock wave emerges at $r_{shock} = 1.65\, r_0$ while temperature decreases to $T_{shock} = 0.22\, T_0$. Correspondingly, the pressure behind the shock equals $P_e = 0.22\, P_0 = 5.74$ Mbar $= 5.74 \times 10^{11}$ Pa. This pressure considerably exceeds the cold silica modulus which is of the order of $P_0 \sim 10^{10}$ Pa. Therefore, a strong shock wave emerges, which compresses the material up to a density $\rho \sim \rho_0\,(\gamma + 1)/(\gamma - 1) \sim 2\rho_0$ ($\gamma \sim 3$ is the adiabatic constant for cold glass).

The material behind the shock wavefront can be compressed and transformed to another phase state in such high-pressure conditions. After unloading, the shock-affected material then has to be transformed into a final state at normal pressure. The final state may possess properties different from those in the initial state. We consider, in succession, the stages of compression and phase transformation, pressure release and material transformation into a post-shock state.

2.4.5
Shock Wave Expansion and Stopping

The shock wave propagating in a cold material loses its energy due to dissipation, and it gradually transforms into the sound wave. The distance at which the shock effectively stops, defines the shock-affected area. This distance can be estimated from the condition that the pressure behind the shock equals the so-called cold pressure [36] of the unperturbed material, P_0. It is possible to estimate P_0 from the initial mass density, ρ_0, and the speed of sound, c_s, in the cold material as follows $P_0 \sim \rho_0 c_s^2$. This value is comparable with the Young modulus of the material. The distance where the shock stops, is expressed by the radius, where the shock initially emerges via the energy conservation condition:

$$r_{stop} \approx r_{shock} \left(\frac{P_{shock}}{P_0} \right)^{1/3} \tag{44}$$

The sound wave continues to propagate at $r > r_{stop}$, apparently not affecting the properties of material.

One can apply the above formula to estimate the shock-affected area in the experiments of [37] where 300 nJ, 100 fs, 800 nm laser pulses were tightly focused to the sub-micrometer region in fused silica. Taking, conservatively, the radius where the shock emerges as $r_{shock} \sim 10^{-4}$ cm, $P_{shock} \sim 5 \times 10^{12}$ erg cm^{-3}, and $P_0 \sim 10^{11}$ erg cm^{-3} one obtains $r_{stop} \sim 3.7 \times 10^{-4}$ cm for a single pulse.

2.4.6
Shock and Rarefaction Waves: Formation of Void

At high energy density, another interesting phenomenon was observed to occur: namely, the formation of a hollow or low-density region within the focal volume [37]. The formation of this void can be understood from simple reasoning. The strong spherical shock wave starts to propagate outside the center of symmetry at radius r_{shock}. The pressure behind the shock decreases as r^{-3}. Therefore, the shock remains strong and compresses material only over a short distance Δr, $(r > \Delta r)$ to an average compression ratio of $\delta = \rho/\rho_0$. A rarefaction wave propagates to the center of the sphere creating a void with maximum radius following from mass conservation:

$$r_{void} < \delta^{1/3} \left[(r_{shock} + \Delta r)^3 - r_{shock}^3 \right]^{1/3} \approx \delta^{1/3} r_{shock} \left(\frac{3\Delta r}{r_{shock}} \right)^{1/3} \tag{45}$$

Thus, at $\delta < 2$ and $\Delta r \sim 0.1$–0.2 the void radius is comparable to that for shock formation. In fact, many other processes after the end of the pulse, during cooling and phase transformation can also affect the final size of the void.

Another estimate of the void radius is based on the assumption of isentropic expansion [38, 39]. The heated material can be considered as a dense and hot gas (absorbed energy per atom exceeds the binding energy) which starts to expand adiabatically with adiabatic constant, γ, after the pulse end (see Appendix). Therefore, the condition PV^γ = constant, holds. The heated area stops expanding when the pressure inside the expanding volume is comparable with the pressure in the cold material, P_0. The adiabatic equation takes a form:

$$PV_{abs}^\gamma = P_0 V_{void}^\gamma \tag{46}$$

The energy deposition volume is estimated as follows $V_{abs} = \pi r_0^2 l_{abs}$. We assume that, in the course of expansion the void attains a spherical shape. Then the void radius reads:

$$r_{void} = \left(\frac{3}{4} r_0^2 l_{abs}\right)^{1/3} \left(\frac{P}{P_0}\right)^{1/3\gamma} \tag{47}$$

Of course, this is only a qualitative estimate because the equation of state of the laser affected material undergoes dramatic changes as the material cycles from a solid to a melt, to a hot gas and back. Accordingly, the adiabatic constant (or rather the Gruneisen coefficient [36]) changes in a range from $\gamma = 5/3$ to $\gamma = 3$. Nevertheless, an estimate by (47) gives a reasonable estimate that the void radius lies in the sub-micron range. Taking $r_0 \sim 10^{-4}$ cm; $l_{abs} \sim 3 \times 10^{-6}$ cm; $P \sim 2.5 \times 10^{12}$ Pa, $P_0 \sim 10^{10}$ Pa; $\gamma \sim 5/3$–2, one obtains $r_{void} \sim 0.6$–0.9 microns. Note that this is the void size *during the interaction*, the final void forms after the reverse phase transition and cooling.

2.4.7
Properties of Shock-and-heat-affected Solid after Unloading

Phase transformations in quartz, silica and glasses induced by strong shock waves have been studied for decades [see 36, 40 and references therein]. The pressure ranges for different phase transitions to occur under shock wave loading and unloading have been established experimentally and understood theoretically [40]. Quartz and silica converts to dense phase of stishovite (mass density 4.29 g cm^{-3}) in the range between 15 and 46 Gpa. The stishovite phase exists up to a pressure of 77–110 Gpa. Silica and stishovite melts at $P > 110$ GPa which is in excess of the shear modulus for liquid silica $\sim 10^{10}$ Pa.

Dense phases usually transform into low-density phases (2.29–2.14 g cm^{-3}) when the pressure releases back to the ambient level. Numerous observations exist of amorphization upon compression and decompression. An amorphous phase denser than the initial silica sometimes forms when unloading occurs from 15 to 46 GPa. Analysis of experiments shows that pressure release and the reverse phase transition follows an isentropic path. In studies of shock compression and decompression under the action of shock waves induced by explosives, the loading and release timescales are of the order of ~ 10^{-9}–10^{-8} s. The heating rate in the shock wave experiments is 10^{12} K s^{-1}, that is, the temperature rises to 10^3 K during one nanosecond.

Let us now compare the results of shock wave experiments with conventional explosives with the shock-induced changes by high-power lasers. The peak pressure at the front of a shock wave driven by the laser, reaches several hundreds of GPa in excess of the pressure value necessary to induce structural phase changes and melting. Therefore, the region where the melting occurs is located very close to where the energy deposition occurs. The zones where structural changes and amorphization occur are located further away. The heating rate by a powerful short-pulse laser is > 3×10^{15} K s^{-1}: and the temperature of the atomic sub-system rises to 50 electronvolts over 200 picoseconds. The cooling time of micron-sized heated region takes tens of microseconds. Supercooling of dense phases may occur if the quenching time is sufficiently short. Short heating and cooling times, along with the small size of the area where the phase transition takes place, can affect the rate of the direct and reverse phase transitions. In fact, phase transitions in these space and timescales have been studied very little.

The refractive index changes in a range of 0.05 to 0.45 along with protrusions surrounding the central void that were denser than silica were observed as a result of laser-induced microexplosion in the bulk of silica [2]. This is the evidence of formation of denser phase during fast laser compression and quenching, however, little is known of the exact nature of the phase.

2.5
Multiple-pulse Interaction: Energy Accumulation

In order to produce an optical breakdown inside a transparent dielectric by a single, tightly focused, short laser pulse one needs the energy per pulse ~50 nJ (average intensity in excess of 10^{12} W cm^{-2}). However, the low heat conduction coefficient within transparent dielectrics implies that another way exists to control the changes induced by lasers in these materials. One can, additionally, vary the number of pulses arriving at the same spot at high repetition rates using a lower energy per pulse laser.

The low heat conduction in glasses (diffusion coefficient ~ 10^{-3} cm^2 s^{-1}) results in a long cooling time with the micron-sized region cooling over ~10 microseconds. Suppose that the laser energy is delivered to the same spot by a succession of short low-energy pulses with the period between them being shorter than the cooling time. In such conditions a sample accumulates energy from many successive pulses and can be heated to very high temperature. This effect was observed experimentally in the bulk heating of glass [2] and in the ablation of transparent dielectrics [3]. Below, we present two simple models for the cumulative heating which allows the dependence of the size of the void on the number of pulses heating the spot.

2.5.1
The Heat-affected Zone from the Action of Many Consecutive Pulses

Let us first consider consecutive heating assuming that the laser-affected region comprises a zone affected by the sum of heat waves produced by the successive pulses. We assume that there are negligible losses between the pulses. This is particularly true when the period between successive pulses is short compared with the cooling time as is the case when 10–100 MHz repetition-rate lasers are used. In the case of N pulses hitting the same place, the absorption energy can be approximated by $E_N = NE_a$. The time duration of N pulses with a repetition rate of R_{rep} is $t_N = N(R_{rep})^{-1}$. Then the propagation of the heat front and the temperature at the front after the action of N pulses is expressed by the single pulse length and the number of pulses in accordance with (42) as the following:

$$r_{th,N} = r_{th,1} N^{\frac{n+1}{2+3n}}$$ (48)

One can see that the exponent in the dependence of the heat-affected range on the number of pulses in (48) changes from 0.5 for the linear heat conduction ($n = 0$) to 0.368 for nonlinear heat transfer at $n = 5/2$. Compared with the experimental results of [2] that the heat-affected area is significantly larger, as expected, than the experimentally observed dependence of the *damaged* region on the number of pulses.

2.5.2
Cumulative Heating and Adiabatic Expansion

Let us consider void formation by the action of multiple pulses in a similar way to the formation of cavity by a single pulse [38, 39]. It is convenient to express the radius of the void created by a single pulse in (47) as a function of the absorbed energy as follows:

$$r_{1,void} = r_{abs} \left(\frac{E_a}{E_0} \right)^{1/3\gamma}$$ (49)

Here $E_0 = P_0 V_{abs} E_0$. At low heat conduction the temperature of the heated focal volume is practically unchanged before the arrival of the next pulse. The next laser pulse adds another E_{abs} to the total energy. Thus the energy in the focal volume equals NE_{abs} after N laser pulses, all losses in the time period between consecutive pulses have been neglected. Correspondingly the radius of the void created by the action of N laser pulses reads:

$$r_{N,void} \approx r_{1,void} N^{1/3\gamma}$$ (50)

Let us compare this model [39] to the experimental data. Structures in the bulk of a zinc-doped borosilicate glass (*Corning 0211*) were produced by a 25 MHz, 30 fs, 5 nJ, 800 nm laser as reported in [2]. The radius of these voids was measured using interference-contrast optical microscopy as a function of the number of pulses incident on the sample. The number of pulses hitting the same spot changed from 10^2 to 10^5. The focal volume was estimated to be $V_{foc} \sim 0.3$ μm^3 ($r_{foc} = 0.4$ μm). The authors [2] suggested a thermal melting model for calculation of the laser-affected zone and approximately fitted it to the experimental data using 30% for the absorption coefficient. As has been demonstrated above the laser–solid interaction is the interplay between many complicated processes. The temperature after a single pulse is ~ 0.3 eV [40] under the conditions of the experiments in [2]. After 50 pulses the absorbed and accumulated energy is enough for ionization of the material and therefore for the change to the laser–plasma interaction mode. Heat conduction then also changes from a linear process (with respect to the temperature) to a nonlinear process. Thus, the molecular bonds in the material are broken, atoms are ionized and therefore we can apply the approach described in the previous sections. Equation (50) can then be applied to describe the radius of the void as a function of the number of pulses. The equation $r_{N,void} \approx 0.65\ N^{1/4}$ μm (with $\gamma = 4/3$) [39] fits the experimental data [2] well up to 10^3 pulses per sample. For a larger number of pulses, the material in the focal volume converts to plasma, the laser–matter interaction mode changes, nonlinear heat conduction takes place, all of which leads to an increase in the size of the affected volume. Thus, the energy conservation and isentropic expansion allows the semi-quantitative description of experiments.

2.6
Conclusions

In this chapter we have attempted to review the physics of the laser interaction inside a transparent solid on the basis of experimental and theoretical studies that have been reported over the past decade. The main focus has been on interactions at high intensity when the material undergoes optical breakdown and is swiftly converted into the plasma state early in the pulse. We would like to draw some conclusions to clarify three issues summarizing: i) what we know on the subject; (ii) what we don't know, and (iii) how the knowledge already gained can be used in applications to create three-dimensional optical memories and the formation of photonic crystals.

   (i) We know how the transition from laser–cold solid interaction to laser–plasma interaction occurs. Optically induced breakdown in transparent dielectrics is the major mechanism leading to plasma formation, which results in strong laser absorption and high concentration of energy in the material. It is well established that the interplay between electron avalanche and multi-photon ionization is the major factor leading to breakdown. The measured threshold for breakdown inside many transparent dielectrics of $\sim 10^{12}$ W cm^{-2} is in close agreement with the theoretical calculations that take into account the avalanche and multi-photon ionization along with modifications of optical properties during the interaction. Heat and shock wave propagation, material compression and pressure release to the normal level can be estimated and provide qualitative agreement with the experimental data. A simple hydrodynamic model based on the point-like explosion allows the prediction of the size of a void produced by single and multiple laser pulses, with a reasonable accuracy. As a result, control over the size of the memory bits formed by single and multiple pulses can be achieved. It is also well established that laser-created three-dimensional diffractive structures can be detected ("read") by the interaction with a weak probe laser beam and, therefore, they can serve as memory "bits".

(ii) Things we do not know include how the laser–matter interaction proceeds in three-dimensional space. This includes the relation between an axially-symmetric focused beam and the real 3D distribution of the absorbed energy density, electron density, temperature, and pressure, etc. Furthermore, we do not know how the transition to spherical symmetry at high intensity occurs. We do not know the real shape of the cavity, the exact phase state and the distribution in space of the laser-modified material, which will be important for the formation of a 3D photonic crystal

(iii) On the basis of knowledge gained from existing experimental and theoretical studies *we can predict* semi-quantitatively (with an accuracy in the range 40–50%) the result of the laser–matter interaction at high intensity ($\sim 10^{14}$ W cm^{-2}); the size of the cavity (not its shape); and the possible material changes (very approximate range for density and refraction index changes). *This knowledge is sufficient* for the prediction and interpretation of experiments aimed for 3D memory applications.

The studies on a tightly focused laser inside a transparent solid constitute a very exciting field for both applied and fundamental science, with many problems yet to be solved. For example, the fundamental problem of phase transitions in conditions when the temperature (pressure) in a volume less than a cubic micron rises and falls in a few picoseconds, is poorly understood. The rate of rise of temperature in these conditions is 10^{16} Kelvin per second, which is impossible to achieve by any other means. In such conditions a new state of matter is most probably created. From the application viewpoint, the question arises of how much smaller the size of the memory bit can be made. One cannot exclude that fundamental and applied problems are closely interrelated. The formation of a new phase in a zone close to the peak intensity, in principle, allows the formation of a diffractive structure whose size is much smaller than the radius of a focal spot.

Acknowledgment

The support of the Australian Research Council through its Centre of Excellence and Federation Fellowship programs is gratefully acknowledged.

References

1 E.G. Gamaly, A. V. Rode, B. Luther-Davies, and V.T. Tikhonchuk, Phys. of Plasmas **9**, 949–957 (2002).

2 C.B. Schaffer, J.F. Garcia, E. Mazur, *Bulk heating of transparent materials using a high-repetition rate femtosecond laser*, Appl. Phys. A **76**, 351–354 (2003).

3 A. Zakery, Y. Ruan, A.V. Rode, M. Samoc, and B. Luther-Davies, *Low-loss waveguides in ultra-fast deposited As_2S_3 chalcogenide films*, JOSA B, **20**, 1–9 (2003).

4 B.E.A. Saleh and M.C. Teich, *Fundamentals of Photonics*, (John Wiley & Sons, NY, 1991) 81–107.

5 L.D. Landau and E.M. Lifshitz, *Electrodynamics of Continuous Media*, (Pergamon Press, Oxford, 1984).

6 W.L. Kruer, *The Physics of Laser Plasma Interactions*, (Addison-Wesley, New York, 1988).

7 M.I. Kaganov, I.M. Lifshitz and L.V. Tanatarov, *Relaxation between Electrons and the Crystalline Lattice*, Sov. Phys. JETP, **4** (2), 173 (1957).

8 P.B. Allen, *Theory of Thermal Relaxation of Electrons in Metals*, PRL, 59, N0. **13**, 1460 (1987).

9 Yu.A. Il'inski and L.V. Keldysh, *Electromagnetic response of Material Media*, (Plenum Press, 1994).

10 A. Zangwill, *Physics of Surfaces*, (Cambridge University Press, 1988).

11 M.A. Popescu, *Non-Crystalline Chalcogenides*, (Kluwer Academic Publishers, 2000).

12 J.P. de Neufville, S.C. Moss, S.R. Ovshinsky, J. Non-Cryst. Solids, **13**, 191, (1973/1974).

13 Ali Saliminia, T.V. Galstian, and A. Villneuve, *Optical-field induced mass transport in As_2S_3 chalcogenide glasses*, Phys. Rev. Lett. **85**, 4112–4115, (2000).

14 D.A. Parthenopoulos and P.M. Rentzepis, *Three-dimensional optical storage memory*, Science, **245**, 843 (1989).

15 W. Denk, J.H. Strickler, W.W. Webb, *Two-photon laser scanning fluorescence microscopy*, Science, **248**, 73 (1990).

16 J.H. Strickler, and W.W. Webb, *Three-dimensional optical data storage in refractive media by two-photon point excitation*, Optics Letters, **16**, 1780 (1991).

17 M. Miwa, S. Juodkazis, T. Kawakami, S. Matsuo, H. Misawa, *Femtosecond two-photon stereo lithography*, Appl. Phys. A, **73**, 561–566, (2001).

18 Y. Kawata, H. Ueki, Y. Hashimoto, and S. Kawata, *Three-dimensional optical memory with a photo-refractive crystal*, Appl. Opt., **34**, 4105–4110, (1995).

19 S.A. Akhmanov, V.A. Vyspoukh, and A.S. Chirkin, *Optics of Femtosecond Laser Pulses*, (Moscow, Nauka, 1988).

20 E. Yablonovitch and N. Bloembergen, *Avalanche ionization and the limiting diameter of filaments induced by light pulses in transparent media*, Phys. Rev. Lett., **29**, 907–910, (1972).

21 D.W. Fradin, N. Bloembergen, and J.P. Letellier, *Dependence of laser-induced breakdown field strength on pulse duration*, Appl. Phys. Lett., **22**, 635 – 637 (1973).

22 Yu. P. Raizer, *Laser Spark and Propagation of Discharges*, Moscow, Nauka (in Russian, 1974).

23 D. Arnold and E. Cartier, *Theory of laser-induced free-electron heating and impact*

ionization in wide-band-gap solids. PRB, **46 (23)**, 15102–15115, (1992).

24 D. Du, X. Liu, G. Korn, J. Squier, and G. Mourou, *Laser-induced breakdown by impact ionization in SiO_2 with pulse width from 7 ns to 150 fs*, Appl. Phys. Lett., **64 (23)**, 3071–3073, (1994).

25 B.C. Stuart, M.D. Feit, A.M. Rubenchick, B.W. Shore, and M.D. Perry, *Laser-induced damage in dielectrics with Nanosecond to picosecond pulses*, PRL, **74**, 2248–2251, (1995).

26 D. von der Linde and H. Schuler, *Breakdown threshold and plasma formation in femtosecond laser-solid interaction*, JOSA B, **13 (1)**, 216–222, (1996).

27 K. Miura, Jianrong Qiu, H. Inouye, and T. Mitsuyu, *Photowritten optical waveguides in various glasses with ultra-short pulse laser*, Appl. Phys. Lett. **71 (23)**, 3329 (1997).

28 P.P. Pronko, P.A. VanRompay, C. Horvath, F. Loesel, T. Juhasz, X. Liu and G. Mourou, *Avalanche ionization and dielectric breakdown in silicon with ultra-fast laser pulses*, Phy. Rev. B, **58**, 2387–2390, (1998).

29 M. Lenzner, J. Kruger, S. Sartania, Z. Cheng, Ch. Spielmann, G. Mourou, W. Kautek, and F. Krausz, *Femtosecond optical breakdown in dielectrics*, Phys. Rev. Lett., **80**, 4076–4079 (1998).

30 An-Chun Tien, S. Backus, H. Kapteyn, M. Murname, and G. Mourou, *Short-pulse laser Damage in transparent materials as a function of Pulse duration*, Phys. Rev. Lett., **82**, 3883–3886 (1999).

31 L. Sudrie, A. Couairon, M. Franko, B. Lamouroux, B. Prade, S. Tzortzakis and A. Mysyrovicz, *Femtosecond laser-Induced Damage and Filamentary Propa-*

gation in Fused silica, Phys. Rev. Lett., **89**, 186601-1 (2002).

32 C.W. Carr, H.B. Radousky, A.M. Rubenchik, M.D. Feit and S.G. Demos, *Localized dynamics during laser-induced damage in optical materials*, Phys. Rev. Lett., **92**, 087401-1 (2004).

33 J. R. Oppenheimer, *Three notes on the quantum theory of a periodic effect*, Phys. Rev., **31**, 66 (1928).

34 L.V. Keldysh, *Ionization in the field of a strong electro-magnetic wave*, Sov. Phys. JETP, **20**, 1307 (1965).

35 E.M. Lifshitz and L.P. Pitaevski, *Physical Kinetics* (Pergamon Press, Oxford, 1984).

36 Ya.B. Zel'dovich and Yu.P. Raizer, *Physics of Shock Waves and High-Temperature Hydrodynamic Phenomena*, (Dover, New York, 2002).

37 E.N. Glezer, M. Milosavljevic, L. Huang, R.J. Finlay, T.-H. Her, J.P. Callan and E. Mazur, *Three-dimensional optical storage inside transparent materials*, Opt. Lett., **21**, 2023–2025 (1996).

38 S. Juodkazis, A.V. Rode, E.G. Gamaly, S. Matsuo, H. Misawa, *Recording and reading of three-dimensional optical memory in glasses*, Appl. Phys. B **77**, 361–368 (2003).

39 E.G. Gamaly, S. Juodkazis, A.V. Rode, B. Luther-Davies, H. Misawa, *Recording and reading 3-D structures in transparent solids*, Proceedings of the 1st Pacific International Conference on Application of Lasers and Optics 2004, Melbourne, 19-21/04/2004.

40 Sheng-Nian Luo, T.J. Arens and P.D. Asimov, *Polymorphism, superheating and amorphization of silica upon shock wave loading and release*, J. Geophys. Res., **108**, 2421 (2003).

Appendix: Two-temperature Approximation for Plasma

Let us assume that the electron–ion plasma is created early in the laser pulse. Then, the processes in nonrelativistic (v ≪ c) nonmagnetic plasma are described by the coupled set of the kinetic equations for electrons and ions:

$$\frac{\partial f_j}{\partial t} + v \frac{\partial f_j}{\partial x} + \frac{q_j}{m_j} E \frac{\partial f_j}{\partial v} = \sum_k \left(\frac{\partial f_{jk}}{\partial t} \right)_c \tag{A1}$$

Here $f_{jz}(x,v,t)$ is a distribution function either for electrons ($j = e$) or ions ($j = i$), q_j, m_j is an electric charge and mass for each species. E is the electrostatic field in the plasma. The last term on the right-hand side represents the rate of change in the distribution function due to collisions with the kth charge species. Because the collisions between the particles of the same kind lead to the fast establishment of the equilibrium distribution within species, one can reduce the above set of kinetic equations to the coupled equations for the successive velocity moments [6]. The infinite set of moment equations is conventionally truncated at the second moment by assumption of the existence of the equation of state – the relation between pressure, density and temperature for each type of particle. This is a conventional two-fluid approximation for plasma containing the electrons and ions of one kind [6]. The essential correction to the equation of state for a description of the dense plasma created inside a bulk solid in comparison to the two-fluid model of [6] is that the "cold" energy and "cold" pressure responsible for the binding forces in a solid should be taken into account [36]. The electrons and ions are described by the coupled set of equations related to the major conservation laws. The conservation of mass (which relates to the zero velocity moment) reads:

$$\frac{\partial \rho}{\partial t} + \frac{\partial}{\partial x} \rho u_i = 0; \quad \rho = m_i n_i \tag{A2}$$

Here u_i, M, n_i and ρ are respectively the ion velocity, mass, number and mass density for ions. Under the assumption of quasi-neutrality, the electron number density relates to that of ions by the relation $n_e = Zn_i$ where Ze is the ion charge. The conservation of momentum (first velocity moment) is expressed as the following:

$$m_e n_e \left(\frac{\partial}{\partial t} + u_e \frac{\partial}{\partial x} \right) u_e = -en_e E - \frac{\partial P_e}{\partial x} - m_e v_{ei} n_e u_e$$

$$m_i n_i \left(\frac{\partial}{\partial t} + u_i \frac{\partial}{\partial x} \right) u_i = eZn_i E - \frac{\partial P_i}{\partial x} + m_e v_{ei} n_e u_e \tag{A3}$$

Here $P_{e,i}$ are the pressures of electrons and ions respectively, v_{ei} is the rate of collision of electrons with ions, E is the electrostatic field of the charge separation. All characteristic time periods are long in comparison with the electronic oscillations, $t \gg \omega_{pe}^{-1}$. Therefore, one can neglect the electron inertia. Then, the electrostatic field from the first equation of (A3) can be related to the electron pressure as the following:

$$en_e E \approx -\frac{\partial P_e}{\partial x} - m_e v_{ei} n_e u_e \tag{A4}$$

The conservation of momentum for ions then can be presented in the form:

$$\frac{\partial \rho u_1}{\partial t} = -\frac{\partial}{\partial x}\left(P_e + P_i + \rho u_i^2\right) \tag{A5}$$

The energy conservation for electrons (ignoring the fast electron motion and electron kinetic energy) reads:

$$\frac{\partial(n_e \varepsilon_e)}{\partial t} = Q_{abs} + \frac{\partial}{\partial x}\kappa_e \nabla T_e - v_{en}n_e(T_e - T_i) \tag{A6}$$

Here Q_{abs} is the absorbed energy density rate. One can see that (A6) formally coincides with equation (11) for conductivity electrons in a solid. The energy conservation for ions takes the form:

$$\frac{\partial}{\partial t}\left(\frac{1}{2}n_i m_i u_i^2 + n_i \varepsilon_i\right) = v_{en}n_e(T_e - T_i) -$$
$$-\frac{\partial}{\partial x}\left\{n_i u_i\left(\frac{m_i u_i^2}{2} + \varepsilon_i + \frac{P_i + P_e}{n_i}\right) - \kappa_i \nabla T_i\right\} \tag{A7}$$

ε_i and ε_e are the electron and ion energy per particle. Equation (A7) when hydrodynamic motion is ignored, so $u_i = 0$, formally coincides with the second equation in (11) for phonons. The ion energy includes the so-called "cold" energy that depends on density and it is independent of temperature [36]. Therefore, the energy change during the phase transition is accounted for.

It is instructive to rewrite the energy equation for ions (A7) in the form:

$$\frac{(P_i + P_e)}{2}\left(\frac{\partial}{\partial t} + u_i\frac{\partial}{\partial x}\right)\ln\frac{(P_i + P_e)}{n_i^{\gamma}} + \frac{\partial}{\partial x}\kappa_i \nabla T_i = v_{en}n_e(T_e - T_i) \tag{A8}$$

Here γ is the adiabatic exponent that, in the solid density plasma, will be used in the form of the density dependent Gruneisen coefficient [36]. It is clear from (A8) that the expansion of the hot solid density plasma, after laser pulse termination, obeys the conventional adiabatic law when the heat losses are negligible, and $T_e = T_i$.

3
Spherical Aberration and its Compensation for High Numerical Aperture Objectives

Min Gu and Guangyong Zhou

Abstract

Over the past few years, two-photon (2p) absorption has found wide application in three-dimensional (3D) micro-fabrication [1–3] and 3D optical data storage [4–6]. Due to the quadratic dependence of the 2p absorption effects on the pump intensity, a chemical and physical reaction can be strictly limited to a tiny region around the focal point [7]. Consequently, a chemical and physical reaction can be created deep inside a thick material. To achieve micro-structures with smaller elements or high recording density, a light beam of a short wavelength and high numerical aperture (NA) objective should be used. For a high-NA objective, a linearly polarized beam of light can become depolarized in the focus of the lens. In other words, if the incident electric field is along the x direction, the components of the electric field in the y and z directions become nonzero near the focus of a high-NA objective [8, 9]. Vectorial Debye theory has been used to study the depolarization effect [9]. Another important issue regarding a high-NA objective is the effect of focal point aberration. The higher the NA, the stronger the aberration effect. For example, spherical aberration can be generated if the incident beam is convergent or divergent for an infinity-corrected objective. Another example is that spherical aberration may occur when a light beam is focused into a medium that has a different refractive index from that of the immersion material [10,11]. In the presence of spherical aberration, the intensity distribution in the focal region becomes broad and distorted and therefore the laser power is attenuated and the fabrication performance becomes poor.

For cases of 3D micro-fabrication, 3D optical data storage, and optical trapping, a laser beam needs to be tightly focused into a sample. The refractive indices of the sample and the immersion material are not identical in most cases. Therefore, the resulting aberration affects the fabrication of micro-structures and the density of optical storage. The larger the refractive indices mismatch, the stronger the aberration. In this chapter we will discuss spherical aberration for high-NA objectives introduced by the refractive indices mismatch between the immersion medium and the sample and its compensation by a change of objective tube length. In Section 3.1 we introduce the 3D intensity point-spread function (IPSF) under the

3D Laser Microfabrication. Principles and Applications.
Edited by H. Misawa and S. Juodkazis
Copyright © 2006 WILEY-VCH Verlag GmbH & Co. KGaA, Weinheim
ISBN: 3-527-31055-X

refractive indices mismatch condition. The spherical aberration compensation method by a change of the tube length will be discussed in Section 3.2. In Sections 3.3 and 3.4, we will discuss the applications of this spherical aberration compensation method in 3D optical data storage and laser trapping, respectively, corresponding to typical focusing cases from low (high) to high (low) refractive index medium. A summary is given in Section 3.5.

3.1
Three-dimensional Indensity Point-spread Function in the Second Medium

3.1.1
Refractive Indices Mismatch-induced Spherical Aberration

As is well known, for a modern commercial objective, spherical aberration and chromatic aberration have been corrected to visually imperceptible levels, if the specified operating variables for the objective are exactly satisfied, i.e., a standard immersion media, a standard cover slip (if needed), and the correct laser light wavelength. In the fields of 3D micro-fabrication, 3D optical data storage, and laser trapping, chromatic aberration correction conditions can be easily satisfied by correctly choosing an objective with the designed wavelength range that covers the laser light wavelength. However, spherical aberration correction conditions are hard to obey in real experiments. First, in most cases the refractive index of the sample is not equal to the immersion media. Second, a laser beam must be tightly focused deeply into the sample. When a wave is focused by an objective through an interface of mismatched refractive indices n_1 and n_2, its wavefront becomes distorted. Figure 3.1 schematically shows the behaviors of a convergent ray when focused through a mismatch interface. In the case of $n_1 > n_2$ (see Fig. 3.1a), the wavefront after refraction on the interface moves faster than it does before refraction and thus leads to a larger curvature. As a result, it is focused

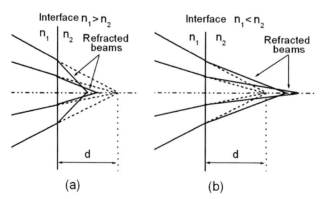

Fig. 3.1 Schematic diagram of beams being refracted on an interface between two media: (a) $n_1 > n_2$ and (b) $n_1 < n_2$.

before the geometrical focus. For $n_1 < n_2$ (see Fig. 3.1b), the traveling speed of the wavefront after refraction on the interface is slower than it is before refraction. Finally, the diffraction focus is located beyond the geometrical focus.

It can be seen that rays with different angles of incidence are focused at different positions on the axis. This feature implies that the diffraction light distribution is enlarged along the axial direction. A high-NA objective suffers from more distortion than does a low-NA one. This aberration function depends on the focus depth d, the refractive indices, the excitation wavelength and the angle of convergence of a ray, and can be written as [11, 12]

$$\Phi(\theta_1, \theta_2, d) = -kd(n_1\cos\theta_1 - n_2\cos\theta_2) \tag{1}$$

where θ_1 and θ_2 denote the angles of incidence and refraction, respectively, and are linked by Snell's law. d is the distance from the interface of the two media to the diffraction-limited focus. It is clear that the function Φ acts as a spherical aberration source because of its dependence on the angle θ_1, which leads to a distortion of the diffraction pattern. Thus the effect of the aberration becomes significant, especially at large depths. For a given value of d, the larger the difference in the refractive indices between the two media, the stronger the effect of the spherical aberration. $k = 2\pi/\lambda$ is the wave number in a vacuum. It is therefore clear that, for a given depth d, two-photon excitation experiences less aberration than does one-photon excitation, due to the longer excitation wavelength in the former case.

3.1.2
Vectorial Point-spread Function through Dielectric Interfaces

For a high-NA objective, the vectorial Debye theory has been used to describe the point-spread function at the focus [9]. In the case of refractive indices mismatch between the sample and immersion medium and plane wave incidence, the 3D vectorial IPSF can be expressed as [9, 11]

$$E(r_2, \Psi, z_2) = \frac{\pi i}{\lambda_1}\left\{\left[I_0^e + \cos(2\Psi)I_2^e\right]\boldsymbol{i} + \sin(2\Psi)I_2^e\boldsymbol{j} + 2i\cos(\Psi)I_1^e\boldsymbol{k}\right\} \tag{2}$$

where

$$I_0^e = \int_0^\alpha P(\theta_1)\sin(\theta_1)(t_s + t_p\cos\theta_2)\exp[-ik_0\Phi(\theta_1)]J_0(k_1 r_2\sin\theta_1)\exp(-ik_2 z_2\cos\theta_2)d\theta_1 \tag{3}$$

$$I_1^e = \int_0^\alpha P(\theta_1)\sin(\theta_1)(t_p\sin\theta_2)\exp[-ik_0\Phi(\theta_1)]J_1(k_1 r_2\sin\theta_1)\exp(-ik_2 z_2\cos\theta_2)d\theta_1 \tag{4}$$

$$I_2^e = \int_0^\alpha P(\theta_1)\sin(\theta_1)(t_s - t_p\cos\theta_2)\exp[-ik_0\Phi(\theta_1)]J_2(k_1 r_2\sin\theta_1)\exp(-ik_2 z_2\cos\theta_2)d\theta_1 \tag{5}$$

Here $P(\theta_1)$ is the apodization function for an objective, J_n is the Bessel function of the first kind of order n, r_2 and z_2 are radial and axial coordinates, respectively, with an origin at the focus which would occur if there were no second medium. The factor a is the maximum angle of convergence of an objective, determined by the numerical aperture of the objective ($\mathrm{NA} = n_1\sin a$), k_1 and k_2 are respectively the wave number in the immersion medium and sample, and t_s and t_p are the Fresnel transmission coefficients for the s and p polarization states, which can be expressed, respectively, as [13]

$$t_s = \frac{2\sin\theta_2\cos\theta_1}{\sin(\theta_1 + \theta_2)} \tag{6}$$

$$t_p = \frac{2\sin\theta_2\cos\theta_1}{\sin(\theta_1 + \theta_2)\cos(\theta_1 - \theta_2)} \tag{7}$$

Similarly, the magnetic field vector can be expressed as

$$B(r_2, \Psi, z_2) = \frac{\pi i}{\lambda_1}\frac{n_1}{c}\left\{\sin(2\Psi)I_2^b\,\boldsymbol{i} + \left[I_0^b - \cos(2\Psi)I_2^b\right]\boldsymbol{j} + 2i\sin\Psi\,I_1^b\boldsymbol{k}\right\} \tag{8}$$

where

$$I_0^b = \int_0^a P(\theta_1)\sin(\theta_1)(t_p + t_s\cos\theta_2)\exp[-ik_0\Phi(\theta_1)]J_0(k_1 r_2\sin\theta_1)\exp(-ik_2 z_2\cos\theta_2)d\theta_1 \tag{9}$$

$$I_1^b = \int_0^a P(\theta_1)\sin(\theta_1)(t_s\sin\theta_2)\exp[-ik_0\Phi(\theta_1)]J_1(k_1 r_2\sin\theta_1)\exp(-ik_2 z_2\cos\theta_2)d\theta_1 \tag{10}$$

$$I_2^b = \int_0^a P(\theta_1)\sin(\theta_1)(t_p - t_s\cos\theta_2)\exp[-ik_0\Phi(\theta_1)]J_2(k_1 r_2\sin\theta_1)\exp(-ik_2 z_2\cos\theta_2)d\theta_1 \tag{11}$$

For nonmagnetic materials ($\mu = 0$), only the electrical component should be considered. For a low-NA objective, I_1^e and I_2^e are very small and can be neglected. For a high-NA objective, I_1^e and I_2^e will become large and will affect the point-spread function. For a circularly polarized beam, it can be resolved into two orthogonal linearly polarized components. Each of them can be dealt with by Eqs. (2) and (8). For further information about the calculations and experiments based on a vectorial point-spread function, please refer to [14].

3.1.3
Scalar Point-spread Function through Dielectric Interfaces

Although the terms I_1^e and I_2^e will affect the point-spread function of the focus for a high-NA objective, the majority of the energy can be expressed by Eq. (3). Therefore, Eq. (3) can be used to express the IPSF under the scalar approximation, which is equivalent to neglecting the depolarization effect of the objective. This assumption holds for a maximum convergence angle of less than 45 degrees [15]. Even for an objective with an NA of 1.4, the vectorial effect does not alter the energy within the 3D IPSF appreciably.

In this chapter, we focus our discussions on objectives which satisfy a so-called "sine condition". If the projection of a ray at a radius of r and the focal length of the objective satisfy $r = f \sin\theta$, we call this objective satisfies sine condition [13]. Under this condition, the apodization function of the objective $P(\theta_1)$ in Eq. (3) equals $\sqrt{\cos\theta_1}$. In this case, if an incident plane wave is focused from the first medium of refractive index n_1 into the second medium of refractive index n_2, the scalar 3D IPSF can be expressed as [11, 16, 17]

$$I(r_2, z_2) = \left| \int_0^a \sqrt{\cos\theta_1} \sin\theta_1 (t_s + t_p \cos\theta_2) J_0(k_1 r_1 n_1 \sin\theta_1) \exp(-i\Phi - ik_2 z_2 n_2 \cos\theta_2) d\theta_1 \right|^2 \quad (12)$$

3.2
Spherical Aberration Compensation by a Tube-length Change

A commercial objective is designed to operate at a given tube length. A tube length is defined as the distance between the objective rear focal plane and the intermediate or primary image at the fixed diaphragm of the eyepiece. When this tube length is altered to deviate from design specifications, spherical aberrations are introduced into the microscope and the focal point suffers from deterioration. With a finite tube-length microscope system, whenever an accessory such as a polarizing intermediate piece is placed in the light path between the back of the objective and the eyepiece, the tube length becomes longer than designed which results in spherical aberration. To overcome the problem, almost all microscope manufacturers are now designing their microscopes to support an infinity-corrected objective. Such objectives project an image of the specimen to infinity. To make a view of the image possible, the body tube of the microscope must contain a tube lens. This lens has the formation of the image at the plane of the eyepiece diaphragm, the so-called intermediate image plane.

For a commercial objective which satisfies the sine condition, the spherical aberration caused by a change in tube length can be described as [16]

$$\Phi_t = B \sin^4(\theta_1/2) \quad (13)$$

Here

$$B = -\frac{2ks^2 \Delta l}{l^2} = -\frac{2k\Delta l}{M^2} \quad (14)$$

where l and s are the conjugate distances in image space and object space of the objective, respectively and M is the transverse magnification factor of the objective. By replacing Φ with $\Phi + \Phi_t$ in Eq. (12), we can minimize the effect of the total aberration if the sign and magnitude of B are chosen appropriately to reach a balanced condition between the two aberration sources. Eqs. (13) and (14) hold for both an infinite tube-length and a finite tube-length objective.

3.3
Effects of Refractive Indices Mismatch-induced Spherical Aberration on 3D Optical Data Storage

3D optical data storage has become an active research area because of the 3D imaging ability of confocal microscopy [15, 18]. A number of materials, including photochromic [19], photobleaching [20], photorefractive [5], and photopolymerizable [4] media, have been successfully employed to achieve 3D optical data storage. However, the currently achieved 3D recording density is far below the possible limit of Tbit cm^{-3} (assume that the volume of recorded bits is $0.5 \times 0.5 \times 1.0 \ \mu m^3$). One of the main reasons for this is the mismatch of the refractive indices between the recording material and its immersion medium, resulting in spherical aberration [11]. It has been demonstrated that this aberration source can dramatically alter the distribution of the light intensity in the focal region of a high numerical aperture objective and reduce the intensity at the focus [11]. The aim of this section is to explore the effect of the spherical aberration resulting from the refractive-index mismatch on the 3D optical data storage density in a two-photon (2p) bleaching polymer block [20]. In the case of $n_1 < n_2$, the compensation for this aberration, by changing the tube length of an objective used for recording and reading 3D data, is studied and an experimental demonstration of the aberration compensation is described.

3.3.1
Aberrated Point-spread Function Inside a Bleaching Polymer

Considering that the working distance of an objective used for recording and reading 3D data should be long enough to access a large depth of a volume-recording medium, a 0.75-NA objective is used as an example. The refractive index of the recording material is 1.48 at 798 nm [6].

A dry objective is more practically relevant to 3D data storage. The refractive index mismatch between the immersion media (air) and the polymer is huge. Based on Eq. (12), the transverse cross-section of the 3D IPSF at different depths in a 2p bleaching polymer is shown in Fig. 3.2. Clearly, the peak intensity decreases dramatically as the focal depth increases if compared with that at the surface where the effect of Eq. (1) disappears. The solid curves in Fig. 3.3 show the axial cross-section of the 3D IPSF at different depths in the 2p bleaching polymer. When $d \neq 0$, the light intensity distribution along the axial direction is no longer symmetric and exhibits a series of strong sidelobes on one side (+z direction) of the intensity peak and the peak intensity drops dramatically as the laser beam focal depth increases. Compared with the case when $d = 0$, the peak position of the light intensity is shifted forwards as illustrated in Fig. 3.3.

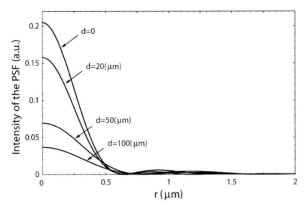

Fig. 3.2 Transverse cross-section of the 3D IPSF at different depths in a bleaching polymer. A dry 0.75-NA objective is assumed.

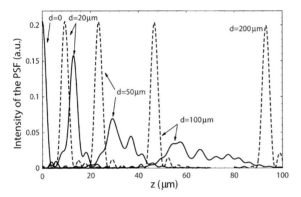

Fig. 3.3 Axial cross-section of the 3D IPSF at different depths in a bleaching polymer under unbalanced (solid curves) and balanced (dashed curves) conditions. The objective is the same as that in Fig. 3.2.

The solid curves in Fig. 3.4 show the full width half maximum (FWHM) in the transverse and axial directions of the 3D IPSF as a function of the depth *d*. It is seen that the FWHM increases appreciably with the depth when *d* is larger than 40 µm, in particular in the axial direction. This phenomenon implies that the 3D recording density decreases considerably when a laser beam is focused deeper than 40 µm into the polymer. To estimate the recording density, the volume of each recorded bit may be defined as

$$\Delta V = \frac{4\pi}{3}(\Delta r)^2 \Delta z \tag{15}$$

where Δr and Δz are the FWHMs of the 3D IPSF along the radial and axial directions, respectively. According to the relationship of the FWHMs to the depth *d*, as shown in Fig. 3.4, the recording density decreases with the depth d. Thus the average 3D recording density N is given by the integration of $1/(D\Delta V)$ over the depth of a recording material, where D is the thickness of the recording material. For the polymer used, N is approximately 0.05 Tbit cm^{-3} for a dry 0.75-NA objective.

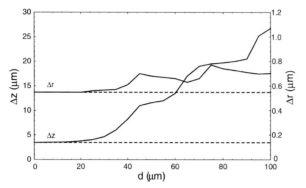

Fig. 3.4 Transverse and axial FWHMs of the 3D IPSF, Δr and Δz, as a function of the depth d of the 2p bleaching polymer under the unbalanced (solid) and balanced (dashed) conditions. The objective is the same as that in Fig. 3.2.

The maximum intensity and the focal shift of the 3D IPSF as a function of the depth d (see the solid curves) are depicted in Fig. 3.5. It is noted that the maximum intensity drops quickly with increasing depth d. At $d = 40$ μm, it is only 35% of the maximum intensity at $d = 0$. Since the fluorescence intensity under 2p excitation is proportional to the square of the incident intensity, the 2p fluorescence intensity at $d = 40$ μm is only approximately 12% of that at $d = 0$. This conclusion means that there is difficulty in recording and reading 3D data beyond $d = 40$ μm in the polymer, if the intensity of the incident laser is kept constant.

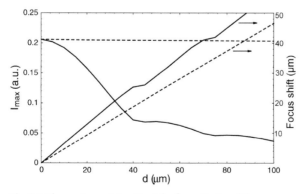

Fig. 3.5 The maximum intensity at the focus (I_{max}) and the focus shift (z_f) as a function of the depth d of the 2p bleaching polymer under unbalanced (solid) and balanced (dashed) conditions. The objective is the same as that in Fig. 3.2.

The effect of the spherical aberration on the 3D IPSF can be reduced considerably if an oil-immersion 0.75-NA objective is used, although the use of immersion oil is not a practical method in data storage. Because the refractive indices mismatch between the immersion oil (1.518) and polymer (1.48) is small, the axial FWHMs of the 3D IPSF are almost unchanged at $d = 200$ μm, as shown in Fig. 3.6. As a result, the average 3D recording density can be estimated to be 0.22 Tbit cm^{-3}, which is four times as large as that for a dry objective with the same NA. Further, the maximum intensity changes only by 0.2% from the surface to a depth of 200 μm.

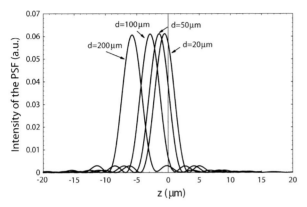

Fig. 3.6 Axial cross-sections of the 3D IPSF at different depths in the beaching polymer for an oil-immersion 0.75-NA objective.

Although the effect of the refractive-index mismatch can be reduced if an oil-immersion objective is used, the residual mismatch of the refractive indices between the oil and the polymer can still play a significant role if the numerical aperture of an objective becomes large. Figure 3.7 shows the axial cross-section of the 3D IPSF for an oil-immersion 1.4-NA objective at different depths in the polymer.

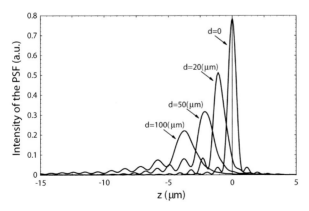

Fig. 3.7 Axial cross-sections of the 3D intensity point-spread function at different depths in the beaching polymer for an oil-immersion 1.4-NA objective.

It is clear that the axial FWHM of the 3D IPSF at $d = 100$ μm becomes three times as large as that for $d = 0$. A similar effect occurs in the transverse direction. The broadening of the FWHM accordingly results in a reduction of the 3D recording density.

3.3.2
Compensation for Spherical Aberration Based on a Variable Tube Length

By a change in the tube length, a phase shift Φ_t is introduced (Eq. (13)). Replacing Φ by $\Phi + \Phi_t$ in Eq. (12), we can minimize the effect of the total aberration if the sign and magnitude of B are chosen appropriately, to reach a balanced condition between the two aberration sources. For a dry 0.75-NA objective and a polymer with a refractive index of 1.48, the 3D IPSF simulations show that the best B value and depth d have a linear relationship that can be expressed as

$$B = -1.35kd \tag{16}$$

The dashed curves in Figs. 3.3–3.5 show the parameters of 3D IPSF under the balance condition. The negative balanced value of B means that the tube length is increased in order to compensate for the spherical aberration caused by the air–polymer interface [16]. It is seen from the dashed curves in Figs. 3.4 and 3.5, that both the intensity and the FWHMs hardly vary with depth d, under the balanced condition. Therefore, the 3D recording density in this case is approximately 0.22 Tbit cm^{-3}, as expected. Another important result is that the balanced intensity drops only 0.1% for a depth of up to 200 μm. These features clearly demonstrate that the use of a variable tube length can efficiently reduce the influence of the refractive-index mismatch between the recording material and its immersion medium.

3.3.3
Three-dimensional Data Storage in a Bleaching Polymer

To demonstrate the effect of the refractive-index mismatch on the focal point and its compensation, experimental work on 3D recording and reading in a 2p bleaching polymer has been performed. Figure 3.8 is a schematic diagram of an experimental 2p fluorescence microscope used in recording and reading 3D data. A femtosecond Ti:Sapphire laser (Tsunami, Spectra-Physics, USA) was used to provide 80 fs pulses with a repetition rate of 82 MHz at a wavelength of 798 nm. An objective O1 (NA 0.25) and a lens L1 were used to expand the laser beam to fit the back aperture of the objective, O2. A neutral density filter ND was used to control the intensity of the incident light in the recording and reading processes. The laser beam was focused onto a sample by an infinitely corrected objective O2 with a numerical aperture of 0.75 (either an Olympus dry objective or a Zeiss oil-immersion objective). A Melles Griot *x-y-z* translation stage (50 nm resolution) was used to control the position of the sample. The collected two-photon fluorescence was reflected by a dichroic beamsplitter DB, and focused by lens L2 onto a photomultiplier tube. A 540 nm short-pass edge filter was inserted in front of the detector to reject the residual signal of the excitation beam.

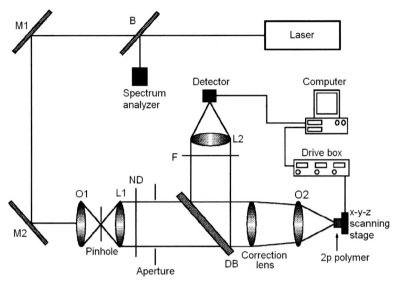

Fig. 3.8 Experimental setup for recording and reading 3D optical data in a bleaching polymer.

In the recording process, the exposure time for each point of data was approximately 2 seconds. The average laser power was 7.7 mW and 0.8 mW for recording and reading, respectively. Figure 3.9 shows a series of bleached lines recorded in the x-z plane of the polymer, with a separation of 10 μm in the z (axial) direction. For the dry objective (Fig. 3.9a), there is no visible bleaching by the 7th line, which corresponds to an axial depth of 70 μm. This result is caused by the dramatic reduction in the intensity of the focus in the presence of the spherical aberration shown in Eq. (1), which is qualitatively consistent with the theoretical prediction in Fig. 3.5. If the oil-immersion objective is used (Fig. 3.9b), the bleached lines are clearly visible at a depth of 100 μm, compared with the 2p fluorescence of the background.

According to the balanced condition given in Eq.(13), the spherical aberration caused by the air–polymer interface can be compensated for if the tube length of an objective is increased. To compensate for the effect of the spherical aberration in Fig. 3.9(a), a positive correction lens of focal length 300 mm was inserted between the objective O and the dichroic beamsplitter DB, so the effective tube length of the objective was increased. By moving it to an optimum position at which the 2p fluorescence is almost constant along the depth of the polymer, a series of bleached lines were recorded in the x-z plane of the polymer (Fig. 3.9c). The line at a depth of 100 μm is now clearly seen, showing that the spherical aberration caused by the refractive-index mismatch can be considerably reduced by increasing the tube length of the objective used in recording.

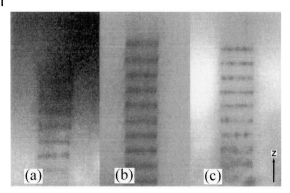

Fig. 3.9 Recorded 2p bleached lines with a separation of 10 µm along the axial direction with: (a) an uncompensated dry 0.75-NA objective; (b) an oil-immersion 0.75-NA objective; (c) an dry 0.75-NA objective and f = 300 mm correction lens.

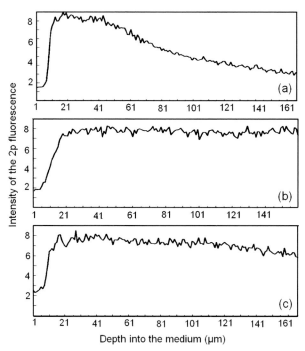

Fig. 3.10 Axial responses to a thick 2p bleaching polymer in the reading process with: (a) a dry 0.75-NA objective; (b) an oil-immersion 0.75-NA objective; (c) an dry 0.75-NA objective and f = 300 mm correction lens.

To understand the effect of the spherical aberration on the reading process, the 2p fluorescence axial responses to the thick polymer block were measured and shown in Fig. 3.10. In the case of the dry objective (Fig. 3.10a), the fluorescence intensity decreases with the depth. This behavior is a typical result in the presence of spherical aberration as shown in Eq. (1). At $d = 50$ μm, the intensity drops by 50%. This decrease in intensity can be compensated for by either using an oil-immersion objective (Fig. 3.10b) or altering the tube length of the objective (Fig. 3.10c). It is seen that the intensity of the axial response in Fig. 3.10c is slightly reduced for a depth up to 100 μm. This behavior is caused by the fact that the position of the correction lens was fixed at an optimum position. In fact, the correction lens should be moved in step with the objective along the axial direction because the balanced B value is linearly proportional to the probe depth d (see Eq. 16).

3.4
Effects of Refractive Index Mismatch Induced Spherical Aberration on the Laser Trapping Force

Laser trapping technology has now been widely used in biology [21] and scanning optical microscopy [22], where specimens (cells, bacteria, viruses and particles) are trapped and manipulated by a sharply focused laser beam, produced by a microscope objective of high numerical aperture. Due to the refractive index mismatch between the cover slip and the medium in which particles or cells are suspended, the trapping performance becomes poor when a trapping beam is focused deeply into the medium [23]. In this part, we theoretically and experimentally show the laser trapping performance in the presence of the refractive indices mismatch and the spherical aberration compensation method by a change in tube length of the objective. This example demonstrates the case $n_1 > n_2$.

3.4.1
Intensity Point-spread Function in Aqueous Solution

Based on Eq. (12), we first do some simulations using the same parameters that will be used later in experiments. The trapping laser beam is tightly focused into an aqueous sample, where the particles are suspended by a 1.25-NA oil immersion objective. Due to the refractive indices mismatch between the cover slip (1.5) and water (1.33), a spherical aberration is generated when the laser beam was focused deeply into the sample. Figure 3.11 shows the axial and transverse cross-section of the 3D IPSF for different probe depths. The probe depths 39 μm, 69 μm, and 108 μm correspond to three different sample cell thicknesses of 34 μm, 60 μm, and 94 μm, respectively. When $d \neq 0$, the light intensity distribution along the axial direction is no longer symmetric and exhibits a series of strong sidelobes on one side (minus z direction) of the intensity peak and the peak intensity drops dramatically as the laser beam focal depth increases. Compared with the case when $d = 0$, the peak position of the light intensity is shifted backwards as illustrated in Fig. 3.11(b). When the probe depth increases from 39 µm to 108 µm, the peak intensity drops dramatically from 0.1221 to 0.05656 and the FWHM of the 3D IPSF in the axial direction ($\Delta z_{1/2}$) and transverse direction ($\Delta r_{1/2}$) increases from 1.86 µm to 3.04 µm and from 0.346 µm to 0.41 µm, respectively, as shown in Fig. 3.12.

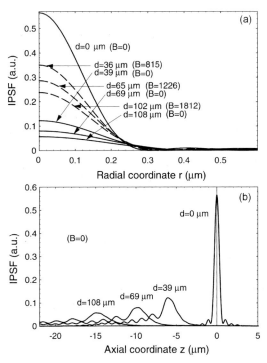

Fig. 3.11 Axial (a) and transverse (b) cross-sections of the 3D IPSF for different probe depths under the uncompensated (solid curves) and compensated (dashed curves) conditions ($\lambda = 0.633$ µm, NA′ = 1.25).

3.4.2
Compensation for Spherical Aberration Based on a Change of Tube Length

By changing the tube length of an objective, the refractive index mismatch induced spherical aberration can be compensated for. For the three sample conditions, the probe depth d should be chosen to be 36 µm, 65 µm and 102 µm to keep the focus of the laser beam in the equatorial plane of a particle. The values of B are 815, 1226 and 1812 under the compensation condition, respectively. The peak intensity of the 3D IPSF thus increases from 0.1221 to 0.3505, 0.0801 to 0.2865, and 0.0566 to 0.2390, respectively (Fig. 3.12a). Accordingly, the axial FWHMs reduce from 1.86 µm to 0.90 µm, 2.42 µm to 1.006 µm, and 3.04 µm to 1.12 µm (Fig. 3.12b). Compared with the uncompensated ($B = 0$) case (Fig. 3.11b), the positions of the intensity peak are shifted towards the positive axial direction and the sidelobes of the IPSF in the axial direction become much weaker (Fig. 3.13).

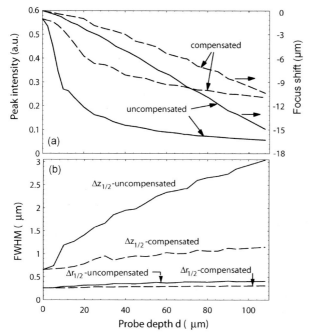

Fig. 3.12 The peak intensity and the axial focal shift (a), as well as the axial and transverse FWHMs (b), of the 3D IPSF as functions of the probe depth under the uncompensated (solid curves) and compensated (dashed curves) condition ($\lambda = 0.633$ μm, NA' = 1.25).

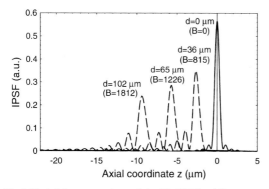

Fig. 3.13 Axial cross-sections of the 3D IPSF for different probe depths under the compensated condition ($\lambda = 0.633$ μm NA' = 1.25).

3.4.3
Transverse Trapping Efficiency and Trapping Power under Various Effective Numerical Apertures

To demonstrate the compensation of the spherical aberration by a change of tube length, a series of laser trapping experiments has been carried out. Figure 3.14 shows the experimental setup. A linearly polarized He-Ne laser with an output power of 17 mW was used as the trapping laser source. The laser beam was expanded and collimated to a size of 20 mm in diameter by objective 1 (40 ×, NA = 0.65) and lens 1 (f = 100 mm), respectively. A diaphragm is then used to control the diameter of the collimated beam. Lens 2 (f = 400 mm) is placed after the diaphragm and focuses the laser beam to a point b as shown in Fig. 3.14. A microscope objective 2 (oil immersion, 100 ×, NA = 1.25) was used to focus the laser beam into a sample. A long-pass edge filter and a relay lens were used to prevent the reflected trapping laser beam from entering a CCD camera that was used to view a real-time trapping process. A sample cell was translated by a piezo-driven scanning stage in parallel with the polarization direction of the laser beam. The power of the trapping light is determined by the power over the entrance aperture c of the microscope objective, multiplied by a factor of 0.861, which is the measured transmittance of the microscope objective.

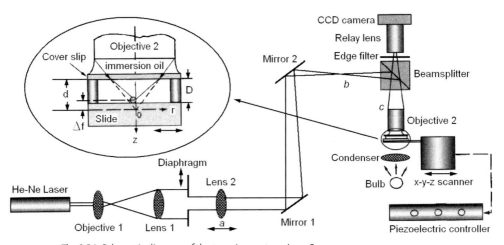

Fig. 3.14 Schematic diagram of the trapping system. Lens 2 is mounted on a translation stage, and the position of lens 2 determines the effective tube length $\overline{bc}$ of objective 2.

The transverse trapping force F on a trapped particle (ϕ = 1.893 μm, a polystyrene latex sphere suspended in water) is measured by observing the maximum translation speed of the scanning stage at which the particle falls out of the trap, and is then calculated from Stokes law

$$F = 6\pi R v \mu \tag{17}$$

where R is the radius of a trapped particle, v is the maximum translation speed, and μ is the viscosity of the surrounding medium [23]. The transverse trapping efficiency Q is calculated from the expression

$$Q = Fc/n_2 P \tag{18}$$

where c is the speed of light in vacuum, n_2 is the refractive index of water, and P is the trapping power [24]. According to the ray-optics model [25], the maximum transverse trapping force occurs when a particle is trapped near the surface in its equatorial plane. In our experiment, the transverse trapping force is measured when a particle is transversely trapped as it is just lifted.

The effect of spherical aberration induced by the refractive index mismatch between the cover slip and the water solution is not only affected by the thickness of a sample cell D, but is also determined by the NA of a microscope objective. The effective NA of the objective 2, NA′, is evaluated by the relation NA′ $= 1.25 \times \phi_2/\phi_1$, where ϕ_1 corresponds to the diameter of the diaphragm when the trapping laser beam just fills the aperture of the objective, and ϕ_2 is the reduced diameter of the diaphragm.

The measured transverse trapping efficiency as a function of the effective numerical aperture for different sample cell thickness (D) is shown in Fig. 3.15. For a given value of D, the transverse trapping efficiency decreases when the effective numerical aperture becomes large. This is due to the larger projection of the gradient force on the transverse direction for an objective of a smaller numerical aperture [25]. When the thickness of the sample cell increases from 34 μm to 94 μm, the trapping efficiency drops by 25.5% and 66% for NA $= 0.6$ and NA $= 1.25$, respectively. By a change in the tube length of the objective, this sperical aberration can be compensated to a certain extent. In the experiment, the tube length ($\overline{bc}$ in Fig. 3.14) of the microscope objective 2 is altered from the designed value of 160 mm to 140 mm by a change in the position of lens 2. Figure 3.15(b) shows

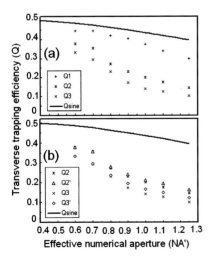

Fig. 3.15 Transverse trapping efficiency as a function of the effective numerical aperture NA′ for different values of the sample cell thickness D at a tube length of 160 mm (a) and 140 mm (b), respectively. Q_1, Q_2, and Q_3 correspond to the transverse trapping efficiencies for $D = 34$ μm, 60 μm and 94 μm, respectively. Q_2' and Q_3' correspond to the transverse trapping efficiencies for $D = 60$ μm and $D = 94$ μm at a 140 mm tube length. Q_{sine} is the theoretical prediction by the ray-optics model under the sine condition [10].

the measured transverse trapping efficiency. For $NA' = 1.25$, the improvement in the transverse trapping efficiency for $D = 34$ μm, 60 μm and 94 μm is 6%, 12% and 20%, respectively.

While the trapping force is related to the distribution of the 3D IPSF for an objective according to the wave-optics model [26], it is also directly proportional to the trapping power, according to the ray-optics model. Based on the calculated 3D IPSF above, the average trapping power P on a trapped particle can be estimated according to the following expression:

$$P = I_m \Delta r_{1/2}^2 \qquad (19)$$

where I_m is the peak light intensity at the focus of the objective, and $\Delta r_{1/2}$ is the transverse FWHM of the 3D IPSF. The average trapping power as a function of the probe depth d, for the uncompensated $(B = 0)$ and the compensated $(B \neq 0)$ conditions, is illustrated in Fig. 3.16.

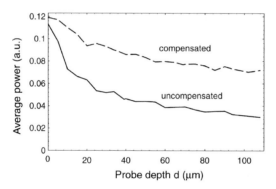

Fig. 3.16 Average trapping power as a function of the probe depth d under the uncompensated (solid curve) and compensated (dashed curve) conditions ($\lambda = 0.633$ μm, $NA' = 1.25$).

Without changing the tube length $(B = 0)$, the trapping power (a.u.) drops from 0.0460 to 0.0299 when the probe depth increases from 39 μm to 108 μm. After the laser beam penetrates through a sample cell of thickness 34 μm, 60 μm and 94 μm, the average trapping power drops by 59%, 65% and 74%, respectively. After compensation, the average trapping power drops by 21%, 32% and 39%, respectively.

It should be pointed out that the experimental trapping force improvement of 6%, 12% and 20% is much lower than the calculated improvement of 93% $(B = 815$ and $d = 36$ μm), 94.1% $(B = 1226$ and $d = 65$ μm), and 135% $(B = 1812$ and $d = 102$ μm) under the balanced condition (Fig. 3.16). The discrepancy between theory and experiment is because the objective operating at a tube length of 140 mm is not under the compensation condition. For the B value of 815, 1226 and 1812, the corresponding change in the tube length Δl should be -294 mm, -443 mm, and -654 mm, according to Eq. (14). Such a large tube length change is clearly infeasible for an objective with a designed tube length of 160 mm. In fact, the tube length change induced phase Φ_t only holds when the tube length change Δl is small compared with the tube length l. Therefore, this simple spherical aberration compensation method is more suitable for the cases where the refractive index mismatch is not too large (less than 0.2).

3.5
Summary

In this chapter we have discussed the effects of spherical aberration resulting from the refractive indices mismatching on focusing a high-NA objective, as well as a method for compensating this aberration. A vectorial form of the aberrated point-spread function is given. We use a scalar approximation to discuss two examples that demonstrate the focusing of a high-NA objective from low (high) to high (low) refractive index media. The method presented in this chapter can be applicable to the fabrication of photonic crystals using a femtosecond laser [3] where the sample refractive index n_2 may be much higher (> 0.2) than that of the immersion medium n_1. However, for the cases with even larger (more than 0.5) refractive indices mismatch, a wavefront regeneration method (also called the phase modulation) can achieve better compensation results [27, 28].

References

1 B.H. Cumpston, S.P. Ananthavel, S. Barlow, D.L. Dyer, J.E. Ehrlich, L.L. Erskine, A.A. Heikal, S.M. Kuebler, I.-Y.S. Lee, S.M. Maughon, J. Qin, H. Röckel, M. Rumi, X.L. Wu, S.R. Marder, J.W. Perry, "Two-photon polymerization initiators for three-dimensional optical data storage and microfabrication," Nature **398**, 51 (1999).

2 S. Kawata, H.-B. Sun, T. Tanaka, and K. Takadaf, "Finer features for functional microdevices," Nature **412**, 697–698 (2001).

3 M. Straub, and M. Gu, "Near-infrared photonic crystals with higher-order bandgaps generated by two-photon photopolymerization", Opt. Lett. **27**, 1824–1826 (2002).

4 J.H. Strickler and W.W. Webb, "Three-dimensional optical data storage in refractive media by two-photon pint excitation," Opt. Lett. **16**, 1780–1782 (1991).

5 H. Ueki, Y. Kawata, and S. Kawata, "Three-dimensional optical bit-memory recording and reading with photonrefractive crystal: analysis and experiment," Appl. Opt. **35**, 2457–2465 (1996).

6 D. Day, M. Gu, and A. Smallridge, "Rewritable 3D Bit Optical Data Storage in a PMMA-Based Photorefractive Polymer," Adv. Mater. **13**, 1005–1007 (2001).

7 J.D. Bhawalkar, G.S. He, and P.N. Prasad, "Nonlinear multiphoton processes in organic and polymeric materials," Rep. Prog. Phys. **59**, 1041–1070 (1996).

8 B. Richards, and E. Wolf, "Electromagnetic diffraction in optical systems, II. Structure of the image in an aplanatic system," Proc. Royal Soc. A. **253** 358–379 (1959).

9 M. Gu, *Advanced Optical Imaging Theory* (Springer, Heidelberg, 2000).

10 M.A.A. Neil, R. Juškaitis, M.J. Booth, T. Willson, T. Tanaka, and S. Kawata, "Adaptive aberration correction in a two-

photon microscope," J. Microscopy. **200**, 105–108 (2000).

11 P. Török, P. Verga, Z. Laczik, and G.R. Booker, "Electromagnetic diffraction of light focused through a planar interface between materials of mismatched refractive indices: an integration representation," J. Opt. Soc. Am. A **12**, 325–332 (1996).

12 C.J.R. Sheppard and C.J. Cogswell, "Effects of aberrating layers and tube length on confocal imaging properties," Optik, **87**, 34–38 (1991).

13 M. Born and E. Wolf, *Principles of Optics* (Pergamon, Oxford, 1980).

14 D. Ganic, X. Gan, and M. Gu, "Reduced effects of spherical aberration on penetration depth under two-photon excitation," Appl. Opt. **39**, 3945–3947 (2000).

15 T. Wilson and C.R. Sheppard, *Theory and Practive of Optical Scanning Microscopy* (Academic, London, 1984).

16 C.J.R. Sheppard and M. Gu, "Imaging by a high aperture optical system," J. Mod. Opt. **40**, 1631–1651 (1993).

17 D. Day and M. Gu, "Effects of refractive-index mismatch on three-dimensional optical data-storage density in a two-photon bleaching polymer," Appl. Opt. **37**, 6299–6304 (1998).

18 M. Gu, *Principles of Three-Dimensional Imaging in Confocal Microscopes* (World Scientific Publishing, Singapore, 1996).

19 D.A. Parthenopoulos and P.M. Rentzepis, "Three-dimensional optical storage memory," Science **245**, 843–845 (1989).

20 P.C. Cheng, J.D. Bhawalkar, S.J. Pan, J. Wiatakiewicz, J.K. Samarabandu, W.S. Liou, G.S. He, G.E. Ruland, N.D. Kumar, and P.N. Prasad, "Two-photon generated three-dimensional photon bleached patterns in polymer matrix," Scanning **18**, 129–131 (1996).

21 M. Zahn, J. Renken, S. Seeger, "Fluorimetric multiparameter cell assay at the single cell level fabricated by optical tweezers," FEBS Lett. **443**, 337–340 (1999).

22 Satoshi Kawata, Yasushi Inouye and Tadao Sugiura, "Near-Field Scanning Optical Microscope with a Laser Trapped Probe," Jpn. J. Appl. Phys. **33**, L1725–L1727 (1994).

23 P.C. Ke, and M. Gu, "Characterization of trapping force in the presence of spherical aberration," J. Mod. Opt. **45**, 2159–2168 (1998).

24 W.H. Wright, G.J. Sonek, and M.W. Berns, "Parametric study of the forces on microspheres held by optical tweezers," Appl. Opt. **33**, 1735–1748 (1994).

25 M. Gu, P.C. Ke, and X.S. Gan, "Trapping force by a high numerical-aperture microscope objective obeying the sine condition," Rev. Sci. Instrum. **68**, 3666–3668 (1997).

26 C.J.R. Sheppard and M. Gu, "Axial imaging through an aberrating layer of water in confocal microscopy" Opt. Commun. **88**, 180–190 (1992).

27 M.A.A Neil, R. Juskaitis, M.J. Booth, T. Wilson, T. Tanaka, and S. Kawata, "Adaptive aberration correrction in a two-photon microscope," J. Microscopy **200**, 105–108 (2000).

28 S.P. Kotova, M.Y. Kvashnin, M.A. Rakhmatulin, O.A. Zayakin, P. Clark, G.D. Love, A.F. Naumov, C.D. Saunter, M.Y.Loktev, G.V. Vdovin, and L.V. Toporkova, Opt. Exp. **10**, 1258–1272 (2002).

4
The Measurement of Ultrashort Light Pulses in Microfabrication Applications

Xun Gu, Selcuk Akturk, Aparna Shreenath, Qiang Cao, and Rick Trebino

Abstract

We review the state of the art of ultrashort-light-pulse measurement using Frequency Resolved Optical Gating (FROG) for micro-fabrication applications. Recent developments have extended the state of the art considerably. FROG devices for measuring the intensity and phase of ultrashort laser pulses have become so simple that almost no alignment is required. In addition, such devices not only operate single shot, but they also yield the two most important spatio-temporal distortions, spatial chirp and pulse-front tilt, which, when present, can complicate the micro-fabrication process. With other FROG variations, it is now possible to measure more general ultrashort *light* pulses (i.e., pulses much more complex than common laser pulses), with time-bandwidth products as large as several thousand and as weak as a few hundred photons, and despite other difficulties such as random absolute phase and poor spatial coherence. This latter capability should greatly enhance the study of the fundamental processes occurring during the microfabrication process.

4.1
Introduction

Since its introduction about a decade ago, Frequency Resolved Optical Gating (FROG) has become an effective and versatile way to measure ultrashort laser pulses, whether a 20 fs UV pulse or an oddly shaped IR pulse from a free-electron laser [1]. Indeed, FROG has measured the intensity and phase of ~4 fs pulses [2] and variations on it are now measuring attosecond pulses.

But now that we have achieved the ability to measure such ephemeral events reliably, it is important to go beyond the measurement of mere ultrashort *laser* pulses, whose intensity and phase are well-behaved in space, time, and frequency, and which have fairly high intensity. It is important to be able to measure ultra-short *light* pulses, whose intensity and phase are *not* well-behaved in space, time, and frequency, and which often are not very intense. It is also important to be able

3D Laser Microfabrication. Principles and Applications.
Edited by H. Misawa and S. Juodkazis
Copyright © 2006 WILEY-VCH Verlag GmbH & Co. KGaA, Weinheim
ISBN: 3-527-31055-X

to measure such pulses as broadband emission from materials illuminated by high intensities – light pulses whose measurement will lead to new technologies or teach us important things about our world, not just how well our laser is aligned. And it is important to do so with a simple device, not one so complex that it could easily introduce the same distortions it hopes to measure. In short, the goal is not a complex device that can only measure simple pulses, but a simple device that can measure complex pulses.

We have recently made significant progress in all of these areas. It is now possible to measure ultrashort light pulses [3] whose time-bandwidth product exceeds 1000, pulses with as little as a few hundred photons (and simultaneously with poor spatial coherence and random absolute phase) [4], and pulses with spatiotemporal distortions like spatial chirp and pulse-front tilt [5, 6]. It is also possible to measure pulses in a train of very different pulses [3] and it is possible to do so quite easily. Of course, measuring ultrashort *laser* pulses remains easier than measuring ultrashort *light* pulses but, recently, measuring ultrashort laser pulses became *extremely* easy. The new variation on FROG, called GRating Eliminated No-nonsense Observation of Ultrafast Incident Laser Light E-fields (GRENOUILLE) [7], has no sensitive alignment knobs, only a few elements, and a cost and size considerably less than previously available devices (including now obsolete autocorrelators). In addition, GRENOUILLE easily measures the spatiotemporal distortions, spatial chirp and pulse-front tilt [5, 6].

In this chapter, we review these developments, which are quite general and so should have many applications in a wide range of fields, including micro-fabrication. In particular, in practical micro-fabrication settings, it is important to be able to easily measure the laser pulse. On the other hand, in studies of the fundamental processes occurring during micro-fabrication, it is important to be able to measure the complex light emissions occurring on an ultrafast time scale. We first give a short review of competing techniques and show why they are not capable of achieving these goals, and then we give a detailed review of the use of FROG variations for measuring the wide range of pulses described above.

4.2
Alternatives to FROG

A few alternatives to the FROG class of techniques include simple autocorrelation [8] and interferometric autocorrelation [9, 10], various techniques that use the auto- or cross-correlation and spectrum [11, 12], and interferometric methods, such as spectral interferometry [13, 14] and Spectral Phase Interferometry for Direct Electric Field Reconstruction (SPIDER) [15]. The various autocorrelation methods only give rough estimates of the pulse length and do not attempt to determine the pulse intensity or the phase (see, for example, [16]) and so are generally considered obsolete. The two above classes of interferometric methods have been more successful, yielding the complete characterization of the spectral phase, which, along with an independently measured spectrum, yields the complete pulse intensity and phase. Indeed, spectral interferometry has measured a train of identical laser pulses with less than one photon per pulse [14]. Unfortunately, spectral interferometry requires an independent, highly time-synchronized, coherent reference pulse with the same spectrum. On the other hand, SPIDER is self-referencing and can therefore measure pulses directly from lasers. SPIDER has great difficulty in extending to more complex pulses, however, due to the need to create two monochromatic pulses of different frequency from the pulse to be measured. And, as an interferometric method, it cannot measure pulses with complex spatial behavior, shot-to-shot jitter, or random absolute phase, which are common properties of light pulses. Also, it is a highly complex technique in itself, with a Michelson interferometer and pulse stretcher in each of the two arms of a FROG, and with over ten sensitive alignment knobs that require alignment. So, in the remainder of this article, we will concentrate on simpler methods. We will see that it will be possible with FROG to reduce the number of sensitive alignment knobs to zero and still accurately measure ultrashort laser pulses, and if we are willing to allow that number to increase to three, we can measure the most complex pulses ever generated.

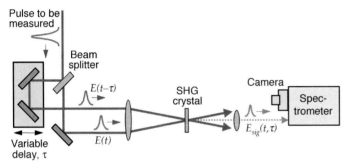

Fig. 4.1 Schematic of a FROG (a frequency-resolved autocorrelation) apparatus. A pulse is split into two, and one pulse gates the other in a nonlinear-optical medium (here a SHG crystal). The SH pulse spectrum is then measured vs. elay. XFROG involves an independent, previously measured gate pulse.

4.3
FROG and Cross-correlation FROG

FROG (see Fig. 4.1) involves time-gating the pulse with itself and measuring the spectrum vs. the delay between the two pulses [1]. When a well-characterized reference pulse is available, cross-correlation FROG (XFROG) takes advantage of this and gates the unknown pulse with this reference pulse [17]. The general expression for both FROG and XFROG traces is:

$$I_{XFOG}(\omega, \tau) = \left| \int_{-\infty}^{\infty} E_{sig}(t, \tau) \exp(-i\omega t) dt \right|^2 \qquad (1)$$

where the signal field, $E_{sig}(t,\tau)$, is a combined function of time and delay of the form $E_{sig}(t,\tau) = E(t)\,E_{gate}(t-\tau)$. In FROG, the gate function, $E_{gate}(t)$, is the unknown input pulse, $E(t)$, that we are trying to measure. In XFROG, $E_{gate}(t)$ can be any known function (i.e., pulse) acting as the reference pulse. In general, $E_{sig}(t,\tau)$ can be any function of time and delay that contains enough information to determine the pulse.

Like autocorrelation, which FROG replaces, these techniques use optical non-linearities to perform the gating: for example, second-harmonic generation (SHG) for FROG and the related process, sum-frequency generation (SFG), for XFROG. These processes allow the creation of a signal pulse whose field is proportional to the product of two input pulse fields.

The FROG and XFROG traces are spectrograms of the pulse (although the FROG trace might better be called the "auto-spectrogram" of the pulse) and as a result, are generally very intuitive displays of the pulse. It is easy to show that retrieving the intensity and phase from the FROG or XFROG trace is equivalent to a well-known solved problem: two-dimensional phase retrieval. We have thus used modified phase-retrieval routines, which have proved very robust and fast for retrieving pulses from traces. Indeed, there are currently two commercially available FROG programs (from the companies, Mesa Photonics and Femtosoft) that retrieve pulses from FROG traces at 10–30 pulses per second. FROG and XFROG yield the pulse intensity and phase vs. time and frequency, with only a few minor "trivial" ambiguities. This is a great improvement over autocorrelation, which yields, at best, a rough measure of the pulse length and little or no information about the actual pulse shape and phase.

4.4
Dithered-crystal XFROG for Measuring Ultracomplex Supercontinuum Pulses

Arguably, the most complex ultrashort pulse ever generated is ultrabroadband supercontinuum, which can now be generated easily in recently developed micro-structure and tapered optical fiber, using only nJ input pulses from a Ti:Sapphire oscillator [18]. As such, it is an excellent test case for a potential technique that purports to measure extremely complex pulses. Many applications of the super-continuum and other complex pulses require that we know the light well, espe-cially its phase.

FROG, specifically cross-correlation FROG (XFROG), is so far the only tech-nique that has been able to successfully measure this pulse [3, 19]. Not only does XFROG deliver an experimental trace that allows the retrieval of the intensity and phase of the pulse in both the time and frequency domains (and even more, as will be clear below), but the XFROG trace itself, which is a spectrogram of the pulse, also proves to be a very intuitive tool for the study of the generation and propagation of the supercontinuum. Many individual processes important in supercontinuum generation, such as soliton generation and fission, can be much more easily identified and studied by observing the XFROG trace than by considering the temporal or spectral intensity and phase.

Our XFROG apparatus is shown in Fig. 4.2. The main challenge in attempting to use XFROG (or any other potential method) to measure the continuum is obtaining sufficient bandwidth in the SFG crystal. This typically requires using an extremely thin crystal, in this case a sub-five-micron crystal, which is not practical and which would generate so few SFG photons that the measurement would not be possible were it to be used. Instead we *angle-dither* a considerably thicker (1 mm) crystal [20] to solve this problem. Because the crystal angle determines the frequencies that are phase-matched in the SFG process, varying this angle in the course of the measurement allows us to obtain as broad a range of phase-matched frequencies as desired.

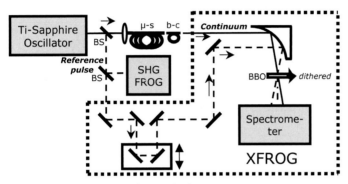

Fig. 4.2 Schematic diagram of our multi-shot XFROG measurement apparatus. BS, beam-splitter; μ-s, microstructure fiber; b-c, butt-coupling fiber.

We performed the first XFROG measurement of the microstructure-fiber supercontinuum on supercontinuum pulses generated in a 16 cm-long microstructure fiber with an effective core diameter of ~ 1.7 microns. In the measurement, SFG between the supercontinuum and the 800 nm Ti: Sapphire pump pulse acts as the nonlinear gating process. In order to phase-match all the wavelengths in the supercontinuum, the nonlinear crystal (BBO) was rapidly dithered during the measurement with a range of angles corresponding to the entire supercontinuum bandwidth. We measured an experimental trace that was parabolic in shape, in agreement with the known group-velocity dispersion of the microstructure fiber (Fig. 4.3). We found that the continuum pulses had a time-bandwidth product of ~ 4000, easily the most complicated pulses ever characterized. Despite the general agreement between the measured and retrieved traces, the results from the intensity-and-phase retrieval were somewhat unexpected. The retrieved trace contained an array of fine structure not present in the measured trace, and the retrieved spectrum also contained ~ 1 nm-scale fine structure, contrary to the smooth spectrum previous experiments had shown. However, single-shot spectrum measurements confirmed our findings, that is, the ~ 1 nm-scale fine features do exist in

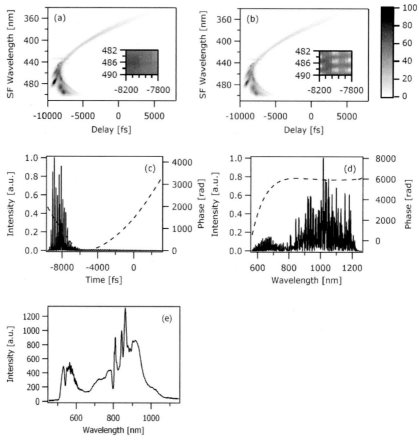

Fig. 4.3 XFROG measurement of microctructure-fiber continuum with an 88 nm 30 fs precharacterized reference pulse: (a) measured trace, (b) retrieved trace, (c) temporal intensity (solid) and phase (dash), and (e) independently measured spectrum. The XFROG error was 0.012. The insets in plots (a) and (b) are higher-resolution sections in the traces. Traces are 8096 × 8096 in dimension.

the supercontinuum spectrum, but only on a single-shot basis, as wild shot-to-shot fluctuations wash them out completely in multi-shot measurements in spectrometers. These fine spectral features agreed with theoretical calculations very well.

The reason that XFROG recovered the unstable fine spectral features lies in the intrinsic information *redundancy* of FROG traces. Indeed, any FROG trace is a two-dimensional temporal-spectral representation of a complex field, and the two axes are two sides of the same coin. The same information is present in both axes. In our case, the unstable fine spectral features, also correspond to slow temporal modulations, which are detectable in a multi-shot measurement. Although the experimental XFROG trace that we measured lacked the fine spectral features because our measurement was made on a multi-shot basis (10^{11} shots!), the long temporal features in the traces, however, were sufficient to assist the retrieval algorithm to find a result with fine spectral features, as such a trace is closest to the measured trace among all possible solutions. This is a great advantage of FROG: lost frequency resolution is recoverable from the FROG measurement via redundant temporal information.

These results have been instrumental in helping us to understand the underlying spectral broadening mechanisms and in confirming recent advances in numerical simulations of supercontinuum generation in the microstructure fiber. Simulations using the extended nonlinear Schrödinger equation (NLSE) model have matched experiments amazingly well [21, 22]. Although most microstructure-fiber supercontinuum experiments have used 10–100 cm of fiber, simulations have revealed that most of the spectral broadening occurs in the first few mm of fiber. Further propagation, which still slowly broadens the spectrum through less important nonlinear processes, such as Raman self-frequency shift, yields only increasingly unstable and fine spectral structure due to the interference of multiple solitons in the continuum spectrum.

This observation suggests that it would be better to use a short (< 1 cm) length of microstructure fiber for supercontinuum generation, as the resulting continuum will still be broad, but short, more stable, and with less fine spectral structure. Indeed, we generated supercontinuum in an 8 mm-long microstructure fiber with 40 fs Ti: Sapphire oscillator pulses, and we performed a similar dithered-crystal-angle XFROG measurement [19].

We see from Fig. 4.4 that the retrieved trace is in good agreement with the measured one, reproducing all the major features. The additional structure that appears in the retrieved trace can be attributed to shot-to-shot instability of spectral fine structure in the continuum spectrum as discussed in detail above. The retrieved continuum intensity and phase vs. time and frequency are shown in Fig. 4.4 (b) and (c). The most obvious feature in this figure is that the continuum from the 8 mm-long fiber is significantly shorter than the picosecond continuum generated in the 16 cm-long fiber and, indeed, consists of a series of sub-pulses that are shorter than the input 40 fs pulse. At the same time, the short fiber continuum has less complex temporal and spectral features than the continuum pulses previously measured from longer fibers. The spectral phase of the short-fiber continuum varies only in the range of 25 rad, which is relatively flat compared with the spectral phase of the long-fiber continuum, which is dominated by cubic phase spanning over 1000 π rad.

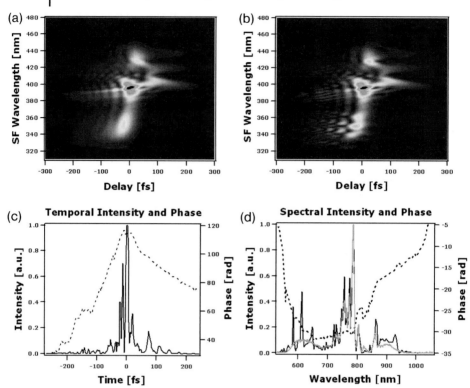

Fig. 4.4 XFROG measurement (a) and retrieval (b) of the 8 mm-long microstructure-fiber continuum with an 800 nm 40 fs pre-characterized reference pulse. (c): retrieved temporal intensity (solid) and phase (dashed). (d): retrieved spectral intensity (solid) and phase (dashed). Light gray: independently measured multi-shot spectrum using a spectrometer. Note that this spectrum agrees with the FROG-measured spectrum but, unlike the FROG measured spectrum, is a bit smoothed due to its multi-shot nature.

4.5
OPA XFROG for Measuring Ultraweak Broadband Emission

Whereas measuring continuum is challenging, due to its extreme complexity and instability, continuum is nevertheless a relatively intense (nJ), spatially coherent beam, which vastly simplifies its measurement. Unfortunately, this cannot be said of ultrashort emitted light pulses from a medium undergoing micro-fabrication. Such pulses can also be spatially incoherent, and they have random absolute phase. While their measurement would yield important insight into the dynamics of the micro-fabrication process [23–26], their measurement proves even more challenging.

Indeed, spectral interferometry, which is well-known for its high sensitivity, proves inadequate for such measurements due both to the light's spatial incoherence and random absolute phase.

In this section we present a noninterferometric technique capable of measuring trains of few-photon spatially incoherent light pulses with random absolute phase [27]. This technique is a variation on the XFROG method and hence involves spectrally resolving a time-gated pulse and measuring its spectrum as a function of delay to yield an XFROG trace or a spectrogram of the pulse. The nonlinearity used in this technique, however, is Optical Parametric Amplification (OPA) or Difference Frequency Generation (DFG), both of which involve not only gating, but also *gain* in the process. The weak pulses are amplified exponentially by up to $\sim 10^5$ by an intense, bluer, shorter, synchronized gate pulse and then spectrally resolved to generate an OPA XFROG trace. We then use a modified FROG retrieval algorithm to retrieve the intensity and phase of the ultraweak pulse measured from the OPA XFROG trace.

In addition to the above complexities, such ultrafast light emissions are also usually broadband. We use a Noncollinear OPA (NOPA) geometry in order to phase-match the broad bandwidth while scanning the delay and generating the OPA XFROG trace. Group Velocity Mismatch (GVM) becomes an important issue in time-gating such broadband pulses with the much shorter gate pulse [28–33]. But GVM can be minimized in the OPA XFROG measurement by using the NOPA geometry as well. A suitable crossing angle can be chosen so that the GVM is minimized while simultaneously maximizing the phase-matched bandwidth. This allows the use of thicker OPA crystals to improve the gain.

We will first discuss the basic theory behind OPA XFROG. We shall then demonstrate OPA XFROG measurements of trains of pulses as weak as 50 aJ, that is, having ~ 150 photons per pulse. These pulses have average powers of 50 fW. Finally we also demonstrate NOPA XFROG for pulses having large bandwidths of ~ 100 nm.

In both OPA and DFG, a strong bluer "pump" pulse is coincident in time in a nonlinear-optical crystal with another pulse (which, in the OPA literature, is usually called the "signal" pulse, but we will avoid this terminology as it conflicts with our use of the term "signal," and call it "seed" instead). If the pump pulse is strong, it exponentially amplifies both the seed pulse (OPA) and also noise photons at the same frequency (usually referred to as the optical parametric generation, or OPG, process), and simultaneously generating difference-frequency (DFG, often called the "idler") photons [34, 35]. Either the OPA or the DFG pulse can be spectrally resolved to generate an XFROG trace.

From the coupled-wave OPA equations, the electric field of the OPA XFROG signal from the crystal has the form:

$$E_{sig}^{OPA}(t, \tau) = E(t) E_{gate}^{OPA}(t - \tau) \tag{2}$$

where, as before, $E(t)$ is the unknown input pulse and we have assumed that the pump pulse intensity remains unaffected by the process, which should be valid when the pulse to be measured is weak and we only need to amplify it enough to measure it. The OPA gate pulse is given by

$$E_{gate}^{OPA}(t-\tau) = \cosh\left(g\left|E_{ref}(t-\tau)\right|z\right) \tag{3}$$

where the gain parameter, g, is given by the expression:

$$g = \frac{4\pi d_{eff}}{\sqrt{n_{OPA}\lambda_{OPA}}\sqrt{n_{DFG}\lambda_{DFG}}} \tag{4}$$

Thus the unknown pulse undergoes exponential gain during OPA. And very importantly, the gating and gain processes do not alter the pulse phase.

It must be pointed out that, in OPA XFROG, unlike other FROG methods, the input pulse is present as a background, even at large delays in the OPA XFROG trace. The equation and the corresponding XFROG algorithm take this into account while retrieving the intensity and phase of the pulse. For high gain, this background becomes negligible.

In the case of DFG XFROG, the idler is spectrally resolved to yield the DFG XFROG trace. Although it has been known that DFG can be used to measure fairly weak pulses [36], the method has never been demonstrated for cases with gain. Including the effect of gain the DFG electric field is given by:

$$E_{sig}^{DFG}(t,\tau) = E(t)E_{gate}^{DFG}(t-\tau)^* \tag{5}$$

The unknown input pulse here is the same as in the case of OPA. The gate function now has the form:

$$E_{gate}^{DFG}(t-\tau) = \exp\left(i\phi_{ref}(t-\tau)\right)\sinh\left(g\left|E_{ref}(t-\tau)\right|z\right) \tag{6}$$

where $\phi_{ref}(t-\tau)$ is the phase of the reference pulse. If the reference pulse is weak, the net gain is small and the above expression reduces to the form $E_{gate}^{DFG}(t-\tau) = E_{ref}(t-\tau)$.

The unknown pulse can thus be easily retrieved from the measured trace using the iterative XFROG algorithm, modified for the appropriate gate pulse. For high gains, the reference-gate pulse experiences gain-shortening in time, a generally desirable effect.

GVM between the pump/gate pulse and the unknown/seed pulse can distort measurement of phase by affecting the gain experienced by the unknown pulse. Thus the interaction length between the pump and unknown pulse during parametric amplification is limited by GVM. The larger the GVM, the shorter will be the interaction length. Therefore, in order to obtain simultaneous gain over the entire bandwidth, it is necessary to choose a crystal whose length is of the order – but less than – the interaction length.

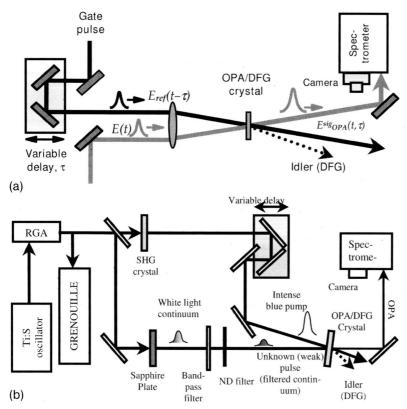

Fig. 4.5 (a) Schematic of the OPA XFROG beam geometry.
(b) Schematic of experimental apparatus for OPA XFROG.

It is also possible to eliminate GVM in OPA XFROG by crossing the pump and unknown pulse at a crossing angle that can be calculated for specific wavelengths using a public domain computer program "GVM" within nonlinear optics software SNLO [37]. The noncollinear geometry is particularly useful in working with broadband pulses, since it is possible to choose an optimal crossing angle which will minimize the GVM over the entire bandwidth range, while simultaneously allowing the entire bandwidth to be phase-matched.

Our experimental set-up for OPA/DFG XFROG is shown in Fig. 4.5. In our experiments, the output from a femtosecond KM Labs Ti:Sapphire (Ti:S) oscillator was amplified by a kilohertz repetition rate Quantronix 4800 series Ti:S regenerative amplifier. The amplified 800 nm pulse was first characterized using a Swamp Optics GRENOUILLE. The pulse was then split into two. One pulse generated a white-light continuum (with poor spatial coherence) in a 2 mm thick sapphire plate, which was then spectrally filtered using a band-pass filter to yield a narrow spectrum. This pulse was attenuated using neutral density filters to act as the weak unknown pulse.

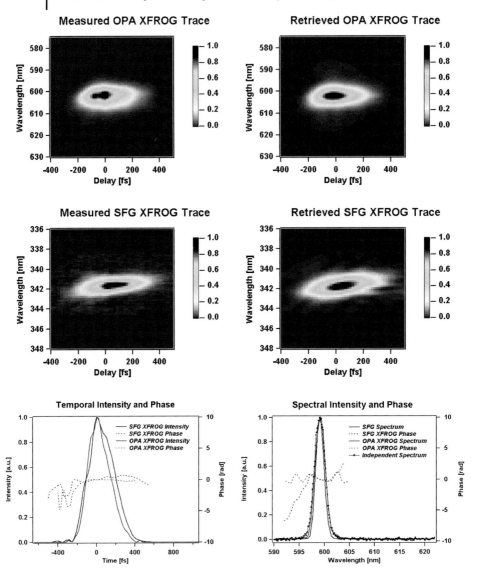

Fig. 4.6 The measured and retrieved traces and retrieved intensity and phase vs. time and the spectrum and spectral phase vs. wavelength of a spectrally filtered continuum from a sapphire plate. The retrieved intensity and phase from the OPA XFROG measurement of 80 fJ pulses agree well with the retrieved intensity and phase of unattenuated continuum of 80 pJ using the established technique, SFG XFROG.

The other pulse was frequency-doubled using a 1 mm thick Type I BBO crystal and passed through a variable delay line to act as the gate (pump) pulse for the OPA process. The two pulses were focused at ~ 3° crossing angle using a 75 mm spherical mirror into a 1 mm BBO Type I crystal. The resulting OPA signal was spectrally resolved and imaged onto a CCD camera integrated over a few seconds.

In the first case, we attenuated the filtered white light continuum to 80 fJ and measured its OPA XFROG trace. The pulse in this case experienced an average gain (G) of about cosh(5.75) ~ 150. Its intensity and phase retrieved using the OPA XFROG algorithm are shown in Fig. 4.6. A comparison of the intensity and phase of the same pulse, unattenuated at 80 pJ, is also shown. This was made using the less sensitive, but well established technique of SFG XFROG. Both techniques yielded identical pulses and the independently measured spectrum of the filtered white light matched well with the OPA XFROG retrieved spectrum. This established OPA XFROG as a legitimate pulse measurement technique which could measure pulses ~ 10^3 weaker than those measured by SFG XFROG.

Next, we pushed the technique to the limit by attenuating the filtered white light continuum down to 50 aJ and retrieved its intensity and phase using the OPA XFROG technique. Shown in Fig. 4.7 are the measured traces with their

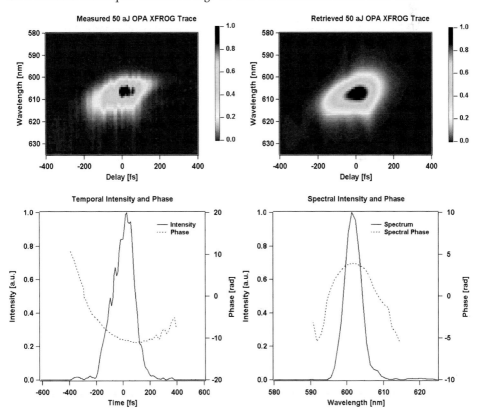

Fig. 4.7 OPA XFROG measurement of a 50 aJ attenuated and filtered continuum generated using a sapphire plate.

intensity and phase retrieved for an average gain of G ~ 10^5. The OPA signal was only about five times more intense than the background caused by OPG in the nonlinear crystal. This background could prove to be the lower limit on how weak the unknown pulse can be and still be measured accurately using the OPA XFROG technique. Despite this, OPA XFROG is still the most sensitive ultrafast pulse measurement technique, able to measure pulses in the range of tens of fW, as opposed to interferometric techniques such as spectral interferometry, which have been demonstrated in measuring pulses with zJ (10^{-21} J) of energy, but having average power of hundreds of fW.

Finally, going to the NOPA geometry, we crossed the pump pulse and white light continuum at an angle of ~ 6.5° (internal in the crystal), chosen in order to minimize GVM. Using filters once again, we spectrally filtered the white light continuum, this time with a bandwidth of ~ 100 nm. This broadband pulse was phase-matched and the OPA XFROG measurement was made for two cases, as shown in Fig. 4.8. For the first OPA XFROG trace, the energy of the pulse was measured to be 500 pJ. The gain experienced in this instance was ~ 50, which we considered the low gain condition. This pulse was then attenuated by four orders of magnitude to 50 fJ and its OPA XFROG trace measured again. This condition had a higher gain of ~1000. The intensity and phase from the two cases compared well, showing that higher gain did not distort the spectral phase during the OPA XFROG measurement process.

The group-delay mismatch (GDM) over the broad bandwidth was minimized and calculated to be ~100 fs over the nearly 60 nm spectral envelope FWHM of a 860 fs long (temporal envelope FWHM) pulse. A thinner crystal would further reduce the GDM, but at the same time a compromise would be made on the gain that could be achieved. This sets a limitation on how weak a pulse can be measured. In this demonstration, we used a 2 mm-thick Type I crystal which was able to measure 50 fJ weak broadband pulses. Geometrical smearing effects in both the longitudinal and transverse directions were calculated to be 56 fs and 34 fs respectively for the noncollinear geometry.

As an aside, it must be pointed out that the structure in the white light continuum is real, as was shown in the previous section. This structure would not be observed in spectral measurements using spectrometers, for two reasons. The continuum generation process is extremely sensitive to the intensity fluctuations from the amplifier used to generate the continuum. Since the amplifier output is not very stable, the continuum itself varies from shot-to-shot. The time averaging performed over a multi-shot spectral measurement would wash this structure out. If however, a single-shot measurement were to be performed, the structure in the spectrum would still be absent. This is because the white light continuum from the sapphire plate was collected from multiple filaments, in order to duplicate the spatial behavior of broadband material emission. So the spatial incoherence would wash out the structure. Our OPA XFROG measurements retrieved a typical spectrum of the broadband continuum. This robust behavior of the XFROG algorithm has been demonstrated in other broadband continuum measurements [3].

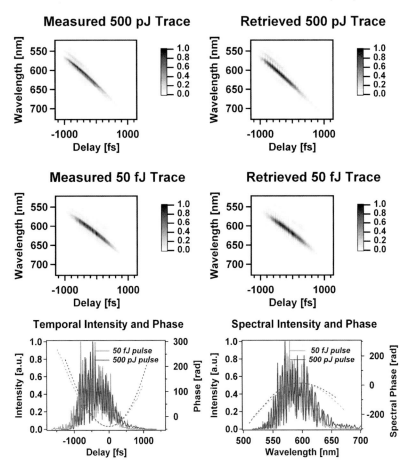

Fig. 4.8 OPA XFROG measurements of broadband white light continuum for cases of low gain in a 500 pJ strong pulse and high gain in a 50 fJ weak pulse.

The experiments discussed above have all been performed using the OPA XFROG geometry. DFG XFROG should yield similar results with the same gain.

Thus OPA/DFG XFROG promises to be a powerful new technique which opens up the field of pulse measurement to ultrafast and ultraweak, complex and broadband, arbitrary light pulses.

4.6
Extremely Simple FROG Device

While the above methods can measure very complex light pulses, they do not involve complex devices. However, if the pulse to be measured is a fairly simple laser pulse, then we might expect the device to be very simple. In fact, we recently showed that it is possible to create a SHG FROG device for measuring ultrashort laser pulses that is so simple that it contains only a few simple elements and, once set up, it never requires realignment, and it consists entirely of only four or five optical elements.

We call this simple variation GRENOUILLE [7]. GRENOUILLE involves two innovations. First, a Fresnel biprism replaces the beam splitter and delay line in a FROG, and second a thick crystal replaces the thin crystal and spectrometer in a FROG, yielding a very simple device (Fig. 4.9).

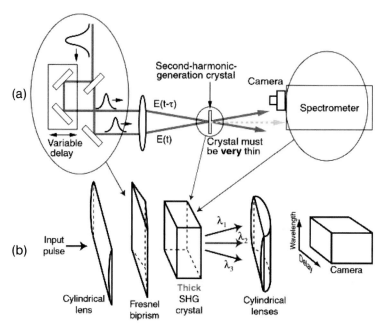

Fig. 4.9 FROG device (a) and the much simpler GRENOUILLE (b), which involves replacing the more complex components with simpler ones.

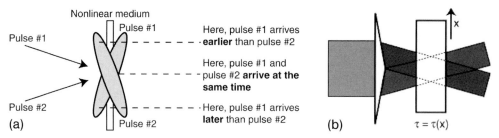

Fig. 4.10 Single-shot FROG measurements involve crossing large beams at a large angle, so that the relative delay between the two beams varies transversely across the crystal (a). This can be accomplished more easily and without the need for alignment using a prism with a large apex angle (b).

Specifically, when a Fresnel biprism (a prism with an apex angle close to 180°) is illuminated by a wide beam, it splits the beam into two and crosses these beam-lets at an angle as in conventional single-shot autocorrelator and FROG beam geometries, in which the relative beam delay is mapped onto horizontal position at the crystal (see Fig. 4.10). But, better than conventional single-shot geometries, the beams here are *automatically aligned* in space and in time – a significant simplification. Then, as in standard single-shot geometries, the crystal is imaged onto a CCD camera, where the signal is detected vs. position (i.e., delay) in the horizontal direction.

FROG also involves spectrally resolving a pulse that has been time-gated by itself. GRENOUILLE (see Fig. 4.11) combines both of these operations in a single *thick* SHG crystal. As usual, the SHG crystal performs the self-gating process: the two pulses cross in the crystal with variable delay. But, in addition, the thick crystal has a very small phase-matching bandwidth, so the phase-matched wavelength produced by it varies with angle. Thus, the thick crystal also acts as a *spectrometer*. The first cylindrical lens must focus the beam into the thick crystal tightly enough to yield a range of crystal incidence (and hence exit) angles large enough to include the entire spectrum of the pulse. After the crystal, a cylindrical lens then maps the crystal exit angle onto the position at the camera, with wavelength a near-linear function of (vertical) position.

The resulting signal at the camera will be an SHG FROG trace with delay running horizontally and wavelength running vertically.

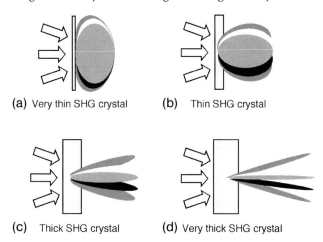

(a) Very thin SHG crystal   **(b)**   Thin SHG crystal

(c)   Thick SHG crystal   **(d)** Very thick SHG crystal

Fig. 4.11 Polar plots of SHG efficiency vs. output angle for various colors of a broadband beam impinging on a SHG crystal. Different shades of gray indicate different colors. Note that, for a thin crystal (a), the SHG efficiency varies slowly with angle for all colors, leading to large a phase-matching bandwidth for a given angle. As the crystal thickness increases, the polar plots become narrower, leading to very small phase-matching bandwidths. The thinnest crystal shown here would be required for all pulse-measurement techniques. GRENOUILLE, however, uses a thick crystal (d) to create and spectrally resolve the autocorrelation signal, yielding a FROG trace – without the need for a spectrometer.

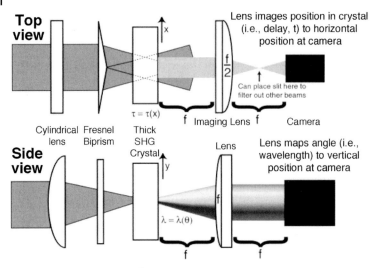

Fig. 4.12 Top and side views of GRENOUILLE.

The key issue in GRENOUILLE is the crystal thickness. Ordinarily, achieving sufficient phase-matching bandwidth requires *minimizing* the group-velocity mismatch, GVM: the fundamental and the second harmonic must overlap for the entire SHG crystal length, L. This condition is: $GVM \cdot L \ll \tau_p$, where τ_p is the pulse length, $GVM \equiv 1/v_g(\lambda_0/2) - 1/v_g(\lambda_0)$, $v_g(\lambda)$ is the group velocity at wavelength λ, and λ_0 is the fundamental wavelength. For GRENOUILLE, however, the opposite is true; the phase-matching bandwidth must be *much less than* that of the pulse:

$$GVM \cdot L \gg \tau_p \tag{7}$$

which ensures that the fundamental and the second harmonic *cease* to overlap well before exiting the crystal, which then acts as a frequency filter.

On the other hand, the crystal must not be too thick, or group-velocity *dispersion* (GVD) will cause the pulse to spread in time, distorting it:

$$GVD \cdot L \ll \tau_c \tag{8}$$

where $GVD \equiv 1/v_g(\lambda_0 - \delta\lambda/2) - 1/v_g(\lambda_0 + \delta\lambda/2)$, $\delta\lambda$ is the pulse bandwidth, and τ_c is the pulse coherence time (~ the reciprocal bandwidth, $1/\Delta v$), a measure of the smallest temporal feature of the pulse. Since $GVD < GVM$, this condition is ordinarily already satisfied by the usual GVM condition. But here it will not necessarily be satisfied, so it must be considered.

Combining these two constraints, we have:

$$GVD \left(\tau_p / \tau_c \right) \ll \tau_p / L \ll GVM \tag{9}$$

There exists a crystal length L that satisfies these conditions simultaneously if:

$$GVM / GVD \gg TBP \tag{10}$$

where we have taken advantage of the fact that τ_p / τ_c is the time-bandwidth product (TBP) of the pulse. Equation (10) is the fundamental equation of GRENOUILLE.

For a near-transform-limited pulse ($TBP \sim 1$), this condition is easily met because $GVM \gg GVD$ for all but near-single-cycle pulses. Consider typical near-transform-limited (i.e., $\tau_p \sim \tau_c$) Ti:Sapphire oscillator pulses of ~100 fs duration, where $\lambda_0 \sim 800$ nm, and $\delta\lambda \sim 10$ nm. Also, consider a 5 mm BBO crystal – about 30 times thicker than is ordinarily appropriate. In this case, Eq. (9) is satisfied: 20 fs cm$^{-1} \ll$ 100 fs/0.5 cm = 200 fs cm$^{-1} \ll$ 2000 fs cm^{-1}. Note that, for GVD considerations, shorter pulses require a thinner, less dispersive crystal, but shorter pulses also generally have broader spectra, so the same crystal will provide sufficient spectral resolution. For a given crystal, simply focusing near its front face yields an effectively shorter crystal, allowing a change of lens or a more expanded beam to "tune" the device for shorter, broader-band pulses. Less dispersive crystals, such as KDP, minimize GVD, providing enough temporal resolution to accurately measure pulses as short as 50 fs. Measurements of somewhat complex ~100 fs pulses are shown in Fig. 4.13. Conversely, more dispersive crystals, such as LiIO$_3$, maximize GVM, allowing for sufficient spectral resolution to measure pulses as narrowband as 4.5 nm (~200 fs transform-limited pulse length at 800 nm). Still longer or shorter pulses are also measurable, but with less accuracy (although the FROG algorithm can incorporate these effects and extend GRENOUILLE's range). Note that the temporal-blurring effect found in thick nonlinear media [5] does not occur in the single-shot SHG geometry.

The main factor limiting GRENOUILLE's accurate measurement of shorter pulses is material-induced dispersion in the transmissive optics, including the necessarily thick crystal. Since shorter pulses have broader spectra, material dispersion is more significant and problematic. Another factor is that, for GRENOUILLE to work properly, the entire pulse spectrum must be phase-matched for some beam angle, requiring a large range of angles in the nonlinear crystal. This can be accomplished using a tighter focus, but then the resulting shorter confocal parameter of the beam reduces the effective crystal length that can be used, thus reducing spectral resolution.

Fortunately, these problems can be solved by designing a tighter focused, nearly-all-reflective GRENOUILLE, which can measure 800 nm laser pulses as short as 20 fs [38]. We convert almost all the transmissive components to reflective ones, except the Fresnel biprism (~1.3 mm of fused silica). This eliminates most of the material dispersion that would be introduced by the device. Moreover, the "thick" crystal required to spectrally resolve (using phase-matching) a 20 fs pulse

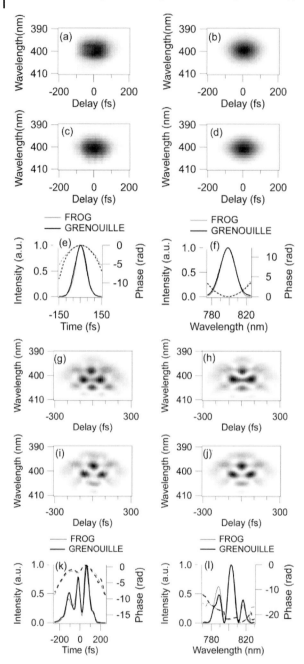

Fig. 4.13 GRENOUILLE (and, for comparison, FROG) measurements of a pulse.
(a) Measured GRENOUILLE trace. (b) Measured FROG trace. (c) Retrieved
GRENOUILLE trace. (d) Retrieved FROG trace. (e, f) Measured intensity and
phase vs. time and frequency for the above traces. (g–l) Analogous traces and
retrieved intensities and phases for a more complex pulse. Note the good
agreement among all the traces and retrieved pulses.

is also thinner: only 1.5 mm. This not only allows us to eliminate dispersion induced by a crystal, but also allows us to focus more tightly (this yields a shorter beam confocal parameter, decreasing the effective nonlinear interaction length), covering the spectra of short pulses. This is important because the device must be able to measure pulses with bandwidths of ~50 nm, that is, the device should have ~100 nm of bandwidth itself. A short interaction length in the crystal reduces the device spectral resolution, but fortunately, due to their broadband nature, shorter pulses require less spectral resolution. With these improvements, a GRENOUILLE can be made that is as simple and as elegant as the previously reported device (Fig. 4.14), but which is capable of accurately measuring much shorter pulses: 20 fs or shorter.

Transmissive GRENOUILLE

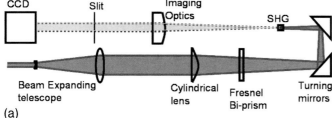

(a)

Reflective GRENOUILLE

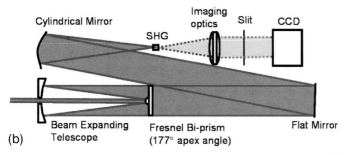

(b)

Fig. 4.14 Compact GRENOUILLE geometries. Previous transmissive design for measuring pulses as short as 50 fs (a) and reflective GRENOUILLE design for measuring ~20 fs pulses. In the reflective design, the primary mirror of the Cassegrain telescope is conveniently cemented to the back of the Fresnel biprism (the only transmissive optic).

To test the reliability of our short-pulse GRENOUILLE, we used a KM Labs Ti:Sapphire oscillator operating with ~ 60 nm (FWHM) of bandwidth, we used an external prism pulse compressor to compress the pulse as much as possible. Then we measured the output pulse with conventional multi-shot FROG and with our GRENOUILLE. We then used the Femtosoft FROG code to retrieve the intensity and phase for both measurements. Figure 4.15 shows measured and retrieved traces as well as the retrieved intensity and phase for multi-shot FROG and

GRENOUILLE measurements, all in excellent agreement with each other. The pulse that GRENOUILLE retrieved in these measurements is 19.73 fs FWHM. This is the shortest pulse ever measured with GRENOUILLE to the best of our knowledge.

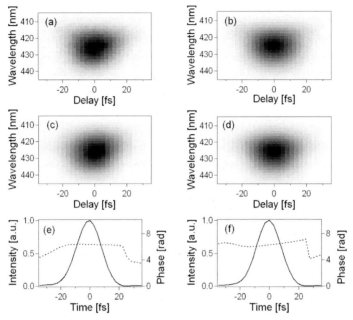

Fig. 4.15 Comparisons of short-pulse GRENOUILLE and multi-shot FROG measurements. (a) measured GRENOUILLE trace; (b) measured multi-shot FROG trace; (c) retrieved GRENOUILLE trace; (d) retrieved multi-shot FROG trace; (e) retrieved intensity and phase vs. time for GRENOUILLE measurements (temporal pulse width 19.73 fs FWHM); (f) retrieved intensity and phase vs. time for multi-shot FROG measurements (temporal pulse width 19.41 fs FWHM).

Because ultrashort laser pulses are routinely dispersed, stretched, and (hopefully) compressed, it is common for them to contain spatio-temporal distortions, especially spatial chirp (in which the average wavelength of the pulse varies spatially across the beam) and pulse-front tilt (in which the pulse intensity fronts are not perpendicular to the propagation vector). Unfortunately, convenient measures of these distortions have not been available. Fortunately, we have recently shown that GRENOUILLE and some other single-shot SHG FROG devices automatically measure both of these spatio-temporal distortions [5, 6]. And they do so without a single alteration in their setup.

Specifically, spatial chirp introduces a shear in the SHG FROG trace, and pulse-front tilt displaces the trace along the delay axis. Indeed, the single-shot FROG or GRENOUILLE trace shear is approximately *twice* the spatial chirp when plotted vs. frequency and *one half* when plotted vs. wavelength (Fig. 4.16). Pulse-front tilt measurement involves simply measuring the GRENOUILLE trace displacement

(Fig. 4.16). These trace distortions can then be removed and the pulse retrieved using the usual algorithm, and the spatio-temporal distortions can be included in the resulting pulse intensity and phase.

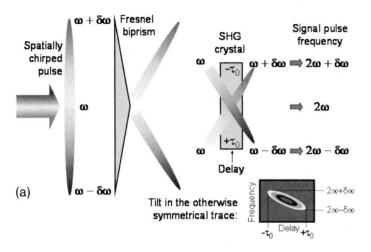

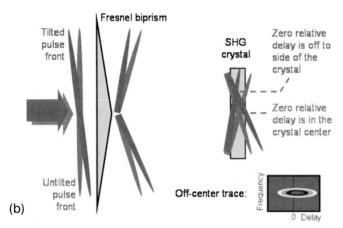

Fig. 4.16 Spatial chirp tilts (shears) the trace (a), and pulse-front tilt translates the trace (b) in GRENOUILLE measurements. This allows GRENOUILLE to measure these distortions easily and without modification to the apparatus.

We have also made independent measurements of spatial chirp by measuring spatio-spectral plots (that is, spatially resolved spectra), obtained by sending the beam through a carefully aligned imaging spectrometer (ordinary spectrometers are not usually good diagnostics for spatial chirp due to aberrations in them that mimic the effect) and spatially resolving the output on a 2D camera, which yields a tilted image (spectrum vs. position) in the presence of spatial chirp. We find very good agreement between this measurement of spatial chirp and that from GRENOUILLE measurements (Fig. 4.17).

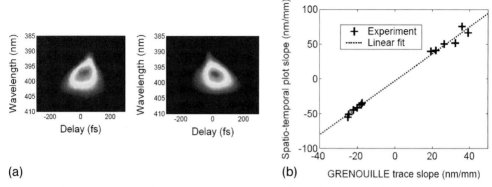

(a) (b)

Fig. 4.17 Experimental GRENOUILLE traces (a) for pulses with positive and negative spatial chirp. The tilt in GRENOUILLE traces reveals the magnitude and sign of spatial chirp. (b) Slopes of GRENOUILLE traces and corresponding spectrum vs. position slopes for various amounts of spatial chirp.

To vary the pulse-front tilt of a pulse, we placed the last prism of a pulse compressor on a rotary stage. By rotating the stage we were able to align and misalign the compressor, obtaining positive, zero, or negative pulse-front tilt. Figure 4.18(b) shows theoretical and experimental values of pulse-front tilt in our experiments and some experimental GRENOUILLE traces for different amounts of pulse-front tilt (a). We find very good agreement between theoretical values of pulse-front tilt and that from GRENOUILLE measurements.

4.7
Other Progress

There has been a tremendous amount of additional progress, by many groups, on FROG techniques for measuring a wide range of pulses. It is not possible to review them all here. Indeed, progress up to 2001 is described in detail in an entire book on FROG [1]. However, some recent progress is worth mentioning here.

A very fast FROG algorithm has been developed by Kane [39, 40], which now achieves pulse retrieval from a FROG trace in less than 30 ms, and a user-friendly commercial version of it is available (from Mesa Photonics). Indeed, the well-known generalized-projections FROG algorithm, which was fairly slow in its original implementation, has been significantly optimized, and it is now very fast. The commercial version of it (from Femtosoft) now retrieves approximately ten pulses per second.

It has become possible to place error bars on pulse intensities and phases vs. time and frequency when measured by FROG [41, 42]. This approach uses the bootstrap method of statistics, which has the nice advantage that it does not require tedious accounting of the error in each component measurement, as is

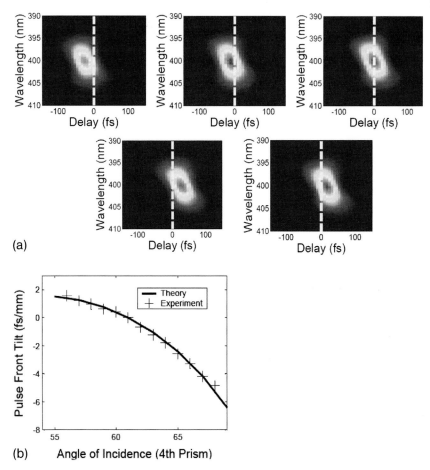

(a)

(b)

Fig. 4.18 Measured GRENOUILLE traces for pulses with very negative, slightly negative, zero, and slightly positive, and very positive pulse-front tilt (note that spatial chirp is also present in all these traces). The horizontal trace displacement is proportional to the pulse-front tilt (a). Theoretically predicted pulse-front tilt and the experimentally measured pulse-front tilt using GRENOUILLE (b).

usually necessary. Instead, it takes advantage of the over-determination of the pulse by the FROG trace and simply requires the running of the algorithm many times, but with some points removed each time.

FROG has been extended into the UV [43–45] and even XUV [46–48]. Variations on FROG have been used to measure the XUV pulses that happen to be only atto-seconds long [49].

On the long-wavelength side, simple GRENOUILLE devices have been developed and few-cycle pulses have been measured in the IR [50, 51]. Reid et al., have measured 4 μm pulses [52]. If required, FROG and GRENOUILLE could easily measure pulses out as far as 10–20 μm. The only reason that such measurements have not yet been made is the expense, availability, and sensitivity of cameras at long wavelengths, although scanning a single-element detector is an option.

Also worth mentioning is that FROG has also been used in automated procedures for pulse shaping for coherent control [53, 54].

Finally, we leave a summary of applications of FROG and its relatives for another time, as they are too numerous to mention.

4.8
Conclusions

In short, GRENOUILLE provides not only the pulse intensity and phase vs. time and frequency, but also the otherwise difficult-to-measure spatio-temporal distortions, spatial chirp and pulse-front tilt. Indeed, we have found that GRENOUILLE is the most sensitive measure of pulse-front tilt available. And other FROG variations, XFROG and OPA XFROG, can measure extremely complex light pulses with as little as ~100 photons. In addition, these latter methods can measure these pulses despite such additional pulse complexities as poor spatial coherence, random absolute phase, and massive shot-to-shot jitter – jitter not in the pulse energy, but in the pulse *shape*. FROG and its relatives are now in use in hundreds of laboratories throughout the world. We look forward to their application in the study of many exotic new phenomena!

Acknowledgment

This work was supported by the National Science Foundation, grants #ECS-9988706, ECS-0200223, and DBI0116564. Much of this chapter has been reprinted from a review article in the *Optical Review*.

References

1 R. Trebino: Frequency-Resolved Optical Gating: The Measurement of Ultrashort Laser Pulses (2002) Kluwer Academic Publishers, Boston.

2 A. Baltuska, M. S. Pshenichnikov and D. Wiersma: Opt. Lett. 23 (1998) 1474.

3 X. Gu, L. Xu, M. Kimmel, E. Zeek, P. O'Shea, A. P. Shreenath, R. Trebino and R. S. Windeler: Opt. Lett. 27 (2002) 1174.

4 J. Zhang, A. P. Shreenath, M. Kimmel, E. Zeek, R. Trebino and S. Link: Opt. Expr. 11 (2003) 601.

5 S. Akturk, M. Kimmel, P. O'Shea and R. Trebino: Opt. Expr. 11 (2003) 68.

6 S. Akturk, M. Kimmel, P. O'Shea and R. Trebino: Opt. Expr. 11 (2003) 491.

7 P. O'Shea, M. Kimmel, X. Gu and R. Trebino: Opt. Lett. 26 (2001) 932.

8 M. Maier, W. Kaiser and J. A. Giordmaine: Phys. Rev. Lett.17 (1966) 1275.

9 J. C. Diels, J. J. Fontaine and F. Simoni: Proc. Int'l Conf. Lasers, STS Press, McLean, VA (1983) 348.

10 J. C. M. Diels, J. J. Fontaine, I. C. Mcmichael and F. Simoni: Appl. Opt. 24 (1985) 1270.

11 J. Peatross and A. Rundquist: J. Opt. Soc. Amer. B 15 (1998) 216.

12 J. W. Nicholson, J. Jasapara, W. Rudolph, F. G. Omenetto and A. J. Taylor: Opt. Lett. 24 (1999) 1774.

13 C. Froehly, A. Lacourt and J. C. Vienot: J. Opt. Paris 4 (1973) 183.

14 D. N. Fittinghoff, J. L. Bowie, J. N. Sweetser, R. T. Jennings, M. A. Krumbügel, K. W. Delong, R. Trebino and I. A. Walmsley: Opt. Lett. 21 (1996) 884.

15 C. Iaconis and I. A. Walmsley: IEEE J. Quant. Electron. 35 (1999) 501.

16 J.-H. Chung and A. M. Weiner: IEEE J. Sel. Top. Quant. Electron. 7 (2001) 656.

17 S. Linden, H. Giessen and J. Kuhl: Physica Status Solidi B Conference (Germany) 206 (1998) 119.

18 J. K. Ranka, R. S. Windeler and A. J. Stentz: Opt. Lett. 25 (1999) 25.

19 Q. Cao, X. Gu, E. Zeek, M. Kimmel, R. Trebino, J. Dudley and R. S. Windeler: Appl. Phys. B 77 (2003) 239.

20 P. O'Shea, X. Gu, M. Kimmel and R. Trebino: Opt. Expr. 7 (2000) 342.

21 A. L. Gaeta: Opt. Lett. 27 (2002) 924.

22 J. Dudley, X. Gu, X. Lin, M. Kimmel, E. Zeek, P. O'Shea, R. Trebino, S. Coen and R. S. Windeler: Opt. Expr.10 (2002) 1215.

23 S. Haacke, S. Schenkl, S. Vinzani and M. Chergui: Biopolymers 67 (2002) 306.

24 N. Hampp: Chemical Reviews 100 (2000) 1755.

25 T. Kobayashi, T. Saito and H. Ohtani: Nature 414(2001) 531.

26 S. Schenkl, E. Portuondo, G. Zgrablic, M. Chergui, S. Haacke, N. Friedman and M. Sheves: Physical Chemistry Chemical Physics 4 (2002) 5020.

27 J. Y. Zhang, A. P. Shreenath, M. Kimmel, E. Zeek, R. Trebino and S. Link: Optics Express 11 (2003) 601.

28 G. Cerullo and S. De Silvestri: Review of Scientific Instruments 74 (2003) 1.

29 A. V. Smith: Opt. Lett. 26 (2001) 719.

30 A. Andreoni, M. Bondani and M. A. C. Potenza: Optics Communications 154 (1998) 376.

31 R. Danielius, A. Piskarskas, A. Stabinis, G. P. Banfi, P. Ditrapani and R. Righini: J. of the Opt. Soc.of Am. B 10 (1993) 2222.

32 C. Radzewicz, Y. B. Band, G. W. Pearson and J. S. Krasinski: Optics Communications 117 (1995) 295.

33 P. Ditrapani, A. Andreoni, G. P. Banfi, C. Solcia, R. Danielius, A. Piskarskas, P. Foggi, M. Monguzzi and C. Sozzi: Physical Review A 51 (1995) 3164.

34 R. W. Boyd: Nonlinear Optics (2002) Second ed, Academic Press.

35 R. L. Sutherland: Handbook of Nonlinear Optics (1996) Marcel Dekker, Inc.

36 S. Linden, J. Kuhl and H. Giessen: Opt. Lett. 24 (1999) 569.

37 A. V. Smith: Proceedings of the SPIE 3928 (2000) 62.

38 S. Akturk, M. Kimmel, P. O'Shea and R. Trebino: Opt. Lett. (2004) submitted.

39 D. J. Kane: IEEE Journal of Selected Topics in Quantum Electronics 4 (1998) 278.

40 D. J. Kane: IEEE J. Quant. Electron.35 (1999) 421.

41 Z. Wang, E. Zeek, R. Trebino and P. Kvam: J. Opt. Soc. Amer. B 20 (2003) 2400.

42 Z. Wang, E. C. Zeek, R. Trebino and P. Kvam: Opt. Expr.11 (2003) 3518.

43 D. J. Kane, A. J. Taylor, R. Trebino and K. W. Delong: Opt. Lett.19 (1994) 1061.

44 K. Michelmann, T. Feurer, R. Fernsler and R. Sauerbrey: Appl. Phys. B 63 (1996) 485.

45 C. G. Durfee, S. Backus, H. C. Kapteyn and M. M. Murnane: Opt. Lett. 24 (1999) 697.

46 K. Ohno, T. Tanabe and F. Kannari: 13th International Meeting on Ultrafast Phenomena (2002) Vancouver, BC, Canada.

47 T. Sekikawa, T. Katsura, S. Miura and S. Watanabe: Phys. Rev. Lett. 88 (2002) 193902.

48 T. Sekikawa, T. Katsura and S. Watanabe: Review of Laser Engineering 30 (2002) 503.

49 M. Hentschel, R. Klenberger, C. Spielmann, G. A. Reider, N. Milosevic, T. Brabec, P. B. Corkum, U. Heinzmann, M. Drescher and F. Krause: Nature 414 (2001) 509.

50 S. Akturk, M. Kimmel and R. Trebino: Opt. Expr. 12 (2004) 4483.

51 S. Akturk, M. Kimmel, R. Trebino, S. Naumov, E. Sorokin and I. Sorokina: Opt. Expr. 11 (2003) 3461.

52 D. T. Reid, P. Loza-Alvarez, C. T. A. Brown, T. Beddard and W. Sibbett: Opt. Lett., 25 (2000) 1478.

53 T. Brixner, M. Strehle and G. Gerber: Applied Physics B B68 (1999) 281.

54 T. Brixner, A. Oehrlein, M. Strehle and G. Gerber: Applied Physics B B70 (2000), (suppl. issue) 119.

5
Nonlinear Optics

John Buck and Rick Trebino

Abstract

We describe the basic physics of nonlinear optics. Starting from simple pictures of this class of processes, we then solve Maxwell's Equations in the Slowly Varying Envelope Approximation (SVEA) – a good approximation even for extremely short pulses – to find simple expressions for the effects of essentially arbitrary nonlinear-optical processes. The phenomenon of phase-matching emerges, providing a potentially very strict constraint that severely limits some nonlinear-optical processes, but not others. We consider in detail the particular second-order process, second-harmonic generation (SHG). Finally, we discuss several third-order nonlinear-optical processes and describe the approximate strength of nonlinear-optical processes.

5.1
Linear versus Nonlinear Optics

The great advantage of ultrashort laser pulses is that all their energy is crammed into a very short time, so they have very high power and intensity. A typical ultrashort pulse from a Ti:Sapphire laser oscillator has a paltry nanojoule of energy, but it is crammed into 100 fs, so its peak power is 10 000 Watts. And it can be focused to a micron or so, yielding an intensity of 10^{12} W cm^{-2}. And it is easy to amplify such pulses by a factor of 10^6.

What this means is that ultrashort laser pulses easily experience *high-intensity effects* – those not ordinarily seen because even sunlight on the brightest day does not approach the above intensities. And all high-intensity effects fall under the heading of nonlinear optics. Some of these effects are undesirable, such as optical damage. Others are desirable, such as *second-harmonic generation*, which allows us to generate light at a new frequency, twice that of the input light. Or like *four-wave mixing*, which allows us to generate light with an electric field proportional to $E_1(t) \, E_2{}^*(t) \, E_3(t)$, where $E_1(t)$, $E_2(t)$, and $E_3(t)$ are the complex electric-field amplitudes of three different light waves. Whereas linear optics requires that light

3D Laser Microfabrication. Principles and Applications.
Edited by H. Misawa and S. Juodkazis
Copyright © 2006 WILEY-VCH Verlag GmbH & Co. KGaA, Weinheim
ISBN: 3-527-31055-X

beams should pass through each other without affecting each other, nonlinear optics allows the opposite. This chapter will describe the basics of nonlinear optics for anyone who has not experienced this field, so you can understand the basics of pulse measurement (the next chapter) and micromachining, which are both inherently nonlinear-optical phenomena.

The fundamental equation of optics – whether linear or nonlinear – is the wave equation:

$$\frac{\partial^2 \mathcal{E}}{\partial z^2} - \frac{1}{c^2}\frac{\partial^2 \mathcal{E}}{\partial t^2} = \mu_0 \frac{\partial^2 \mathcal{P}}{\partial t^2} \tag{1}$$

where μ_0 is the magnetic permeability of free space, c is the speed of light in vacuum, $\mathcal{E}$ is the real electric field, and $\mathcal{P}$ is the real induced polarization. The induced polarization contains the effects of the light on the medium and the effect of the medium back on the light wave. It drives the wave equation.

The induced polarization contains linear-optical effects (the absorption coefficient and refractive index) and also nonlinear-optical effects. At low intensity (or low field strength), the induced polarization is proportional to the electric field that is already present. In the frequency domain, this is

$$\mathcal{P} = \varepsilon_0 \chi^{(1)} \mathcal{E} \tag{2}$$

where ε_0 is the electric permittivity of free space, and the linear susceptibility, $\chi^{(1)}$, describes the linear-optical effects. This expression follows from the fact that the light electric field, $\mathcal{E}$, forces electric dipoles in the medium into oscillation at the frequency of the field; the dipole oscillators then emit an additional electric field at the same frequency. The total electric field (incident plus emitted) is what appears as $\mathcal{E}$ in Eqs. (1) and (2). If we assume a lossless medium, for example, we find that the electric and polarization field expressions, $E(z, t) \propto E_0 \cos(\omega t - kz)$ and, $\mathcal{P} = \varepsilon_0 \chi^{(1)} E_0 \cos(\omega t - kz)$ will solve the wave equation under the condition that $k = n\omega/c$, in which the refractive index is defined as $n = (1+\chi^{(1)})^{1/2}$.

In linear optics, (where Eq. (.2) applies), the wave equation is linear, so $\mathcal{E}$ is a sum of more than one beam (field), then so is $\mathcal{P}$. As a result, $\mathcal{P}$ drives the wave equation to produce light with *only* those frequencies present in $\mathcal{P}$, and these arise from the original input beams. In other words, light does not change color (see Fig. 5.1).

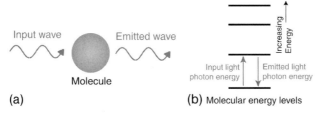

(a)   (b) Molecular energy levels

Fig. 5.1 Linear optics. (a) A molecule excited by a light wave oscillates at that frequency and emits only that frequency. (b) This process can be diagrammed by showing the input light wave as exciting ground-state molecules up to an excited level, which re-emits the same frequency.

Also, with a linear wave equation, the principle of superposition holds, and beams of light can pass through each other but do not affect each other.

Life at low intensity is dull.

5.2
Nonlinear-optical Effects

At high intensity, the induced polarization ceases to be a simple linear function of the electric field. Put simply, like a cheap stereophonic amplifier driven at too high a volume, the medium does not follow the field perfectly (see Figs. 5.2–5.4), and higher-order terms must be included [1–3]:

$$\mathcal{P} = \varepsilon_0 \left[\chi^{(1)} \mathcal{E} + \chi^{(2)} \mathcal{E}^2 + \chi^{(3)} \mathcal{E}^3 + \ldots \right) \tag{3}$$

where $\chi^{(2)}$ and $\chi^{(3)}$ are the frequency-dependent second and third-order susceptibilities. $\chi^{(n)}$ is called the nth-order susceptibility.

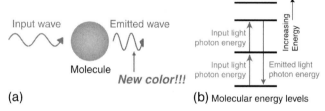

(a)   (b) Molecular energy levels

Fig. 5.2 Nonlinear optics. (a) A molecule excited by a light wave oscillates at other frequencies and emits those new frequencies. (b) This process can be diagrammed by showing the input light wave as exciting ground-state molecules up to highly excited levels, which re-emit the new frequencies.

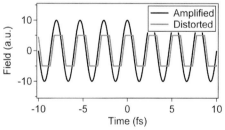 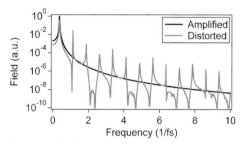

Fig. 5.3 Nonlinear electronic effects in a cheap audio amplifier. The input wave from the audio source is taken here to be a sine wave. In an expensive amplifier, the sine wave is accurately reproduced at higher volume, but, because the cheap amplifier cannot achieve the desired volume, the output wave saturates and begins to look more like a square wave. This produces new frequency components at harmonics of the input wave.

Nonlinear-optical effects are analogous: a sine-wave electric wave drives a molecular system, which also does not reproduce the input sine wave accurately, producing new frequencies at harmonics of the input wave. Whereas audiophiles spend a great deal of money to avoid the above nonlinear electronic effects, optical scientists spend a great deal of money to achieve nonlinear-optical effects.

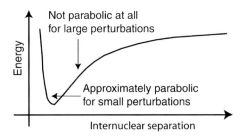

Fig. 5.4 Potential surface of a molecule, showing the energy vs. separation between nuclei. Note that the potential is nearly parabolic near the bottom, but it is far from parabolic for excitations that hit the molecule harder, forcing it to vibrate with larger ranges of nuclear separations. This molecule will emit frequencies other than the one driving it.

What do nonlinear-optical effects look like? They are easy to calculate. Recall that the real field, $\mathcal{E}$, is given by:

$$\mathcal{E}(t) = \frac{1}{2} E(t)\exp(i\omega t) + \frac{1}{2} E^*(t)\exp(-i\omega t) \tag{4}$$

where we have temporarily suppressed the space dependence. $E(t)$ is the complex field amplitude, which we assume to be slowly-varying. So squaring this field yields:

$$\mathcal{E}^2(t) = \frac{1}{4} E^2(t)\exp(2i\omega t) + \frac{1}{2} E(t) E^*(t) + \frac{1}{4} E^{*2}(t)\exp(-2i\omega t) \tag{5}$$

Notice that this expression includes terms that oscillate at 2ω, the *second harmonic* of the input light frequency. These terms then drive the wave equation to yield light at this new frequency. This process is very important; it is called *second-harmonic generation* (SHG). Optical scientists, especially ultrafast scientists, make great use of SHG to create new frequencies. Figure 5.5 shows a schematic of SHG.

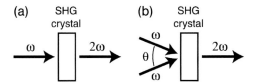

Fig. 5.5 Second-harmonic generation. (a) Collinear beam geometry. (b) Noncollinear beam geometry with an angle, θ, between the two input beams. Such noncollinear beam geometries are possible in nonlinear optics because more than one field is required at the input.

The above expression also contains a zero-frequency term, so light can induce a dc electric field. This effect is called optical rectification; it is generally quite weak, so we will not say much more about it.

If we consider the presence of two beams and this time do not suppress the spatial dependence, $\mathcal{E}(\vec{r}, t) = \frac{1}{2} E_1(\vec{r}, t)\exp[i(\omega_1 t - k_1\,\vec{r})] + \frac{1}{2} E_2(\vec{r}, t)\exp[i(\omega_2 t - k_2\,\vec{r})] + c.c.$ In this case, we have:

$$
\begin{aligned}
\mathcal{E}^2(\vec{r}, t) = &\frac{1}{4} E_1^2 \exp[2i(\omega_1 t - \vec{k}_1 \cdot \vec{r})] + \frac{1}{2} E_1 E_1^* + \frac{1}{4} E_1^{*2}\exp[-2i(\omega_1 t - \vec{k}_1 \cdot \vec{r})] \\
&+ \frac{1}{4} E_2^2 \exp[2i(\omega_2 t - \vec{k}_2 \cdot \vec{r})] + \frac{1}{2} E_2 E_2^* + \frac{1}{4} E_2^{*2}\exp[-2i(\omega_2 t - \vec{k}_2 \cdot \vec{r})] \\
&+ \frac{1}{2} E_1 E_2 \exp\{i[(\omega_1 + \omega_2)t - (\vec{k}_1 + \vec{k}_2) \cdot \vec{r}\,]\} \\
&+ \frac{1}{2} E_1^* E_2^* \exp\{-i[(\omega_1 + \omega_2)t - (\vec{k}_1 + \vec{k}_2) \cdot \vec{r})]\} \\
&+ \frac{1}{2} E_1 E_2^* \exp\{i[(\omega_1 - \omega_2)t - (\vec{k}_1 - \vec{k}_2) \cdot \vec{r})]\} \\
&+ \frac{1}{2} E_1^* E_2 \exp\{-i[(\omega_1 - \omega_2)t - (\vec{k}_1 - \vec{k}_2) \cdot \vec{r})]\}
\end{aligned}
\tag{6}
$$

The first two lines are already familiar: they are the SHG and optical-rectification terms for the individual fields. The next line is new: it yields light at the frequency, $\omega_1 + \omega_2$, the sum frequency, and hence is called *sum-frequency generation* (SFG). The last line is also new: it yields light at the frequency, $\omega_1 - \omega_2$, the difference frequency, and hence is called *difference-frequency generation* (DFG). These two processes are also quite important.

Notice something else. The new beams are created in new directions, $k_1 + k_2$ and $k_1 - k_2$. This can be very convenient if we desire to see these new – potentially weak – beams in the presence of the intense input beams that create them.

Many third-order effects are collectively referred to as *four-wave-mixing (4WM)* effects because three waves enter the nonlinear medium, and an additional one is created in the process, for a total of four. We will not write out the entire third-order induced polarization, but, in third order, as you can probably guess, we see effects including *third-harmonic generation* (THG) and a variety of terms like:

$$
\mathcal{P}_i = \frac{3}{4}\varepsilon_0\, \chi^{(3)}\ E_1 E_2^* E_3 \exp\{i[(\omega_1 - \omega_2 + \omega_3)t - (\vec{k}_1 - \vec{k}_2 + \vec{k}_3) \cdot \vec{r}\,]\}
\tag{7}
$$

Notice that, if the factor of the electric field envelope is complex-conjugated, its corresponding frequency and k-vector are both negative, while, if the field is not complex-conjugated, the corresponding frequency and k-vector are both positive. Such third-order effects, in which one k-vector is subtracted, are often called *induced grating* effects because the intensity arising from the interference of two of the beams, say, E_1 and E_2, has a sinusoidal spatial dependence (see Fig. 5.6). The sinusoidal intensity pattern affects the medium in some way, creating a sinusoidal modulation of its properties, analogous to those of a diffraction grating.

The process can then be modeled as diffraction of the third beam off the induced grating.

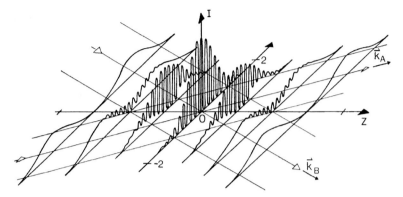

Fig. 5.6 Intensity pattern produced when two beams cross. When the beams cross in a medium, the medium is changed more at the intensity peaks than at the troughs, producing a laser-induced grating.

Third-order effects include a broad range of interesting phenomena (some useful, some irritating), many beyond the scope of this book. But we will consider a few that are important for many applications, including ultrafast laser spectroscopy and the measurement of ultrashort pulses. For example, suppose that the second and third beams in the above expression are the same: $E_2 = E_3$ and $k_2 = k_3$. In this case, the above induced polarization becomes:

$$\mathcal{P}_i = \frac{3}{4} \varepsilon_0 \, \chi^{(3)} \, E_1 |E_2|^2 \exp\{i[(\omega_1 t - \vec{k}_1 \cdot \vec{r}]\} + c.c. \tag{8}$$

This yields a beam that has the same frequency and direction as beam #1, but allows it to be affected by beam #2 through its mag-squared factor. So beams that pass through each other can affect each other! Of course, the strength of all such effects is zero in empty space ($\chi^{(3)}$ of empty space is zero), but the strength can be quite high in a solid, liquid, or gas. It is often called *two-beam coupling* (see Fig. 5.7).

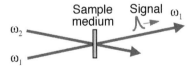

Fig. 5.7 Two-beam coupling. One beam can affect the other in passing through a sample medium. The pulse at the output indicates the signal beam, here collinear with one of the beams and at the same frequency. This idea is the source of a variety of techniques for measuring the properties of the sample medium.

A particularly useful implementation of the above third-order effect is *polarization gating*, (see Fig. 5.8), which involves the input of two co-propagating waves whose fields are polarized at 45° to each other. One of these is usually a weak field, such that by itself it would not induce any appreciable nonlinear response in the medium. The other wave (known as the pump, and represented as E_2 in Eq. (3.8)) is intense enough to induce a nonlinear response, which will affect the weak field. The weak wave can be decomposed into orthogonal field components that are aligned parallel and perpendicular to the strong field. This is represented by two versions of Eq. (8), in which the role of E_1 is played by the weak field components that are parallel and perpendicular to E_2; the values of $\chi^{(3)}$ will differ for the two cases. The effect of the pump field is to induce *nonlinear refractive index* changes in the medium that will affect the weak field components. The index changes differ, however, for waves that are polarized parallel and perpendicular to the strong field, and so a *nonlinear birefringence* is set up. The weak wave components now accumulate a phase difference as they propagate, which leads to a change in the polarization state of the original weak field by the time it reaches the output. The medium acts as a wave-plate, and – with sufficient pump intensity and medium length – a full 90° rotation of the weak wave polarization can occur. In any event, some field component that is orthogonal to the original weak wave input will be generated, which can be isolated by using crossed polarizers. This beam geometry is convenient and easy to set up, and it is much more sensitive than two-beam coupling.

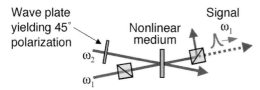

Fig. 5.8 Polarization gating. If the polarizers are oriented at 0° and 90°, respectively, the 45° polarized beam (at frequency ω_2) induces polarization rotation of the 0° polarized beam (at frequency ω_1), which can them leak through the second 90° polarizer. The pulse at the output indicates the signal pulse, again collinear with one of the input beams, but here with the orthogonal polarization.

By the way, another process is simultaneously occurring in polarization gating called *induced birefringence*, in which the electrons in the medium oscillate along with the incident field at +45°, which stretches the formerly spherical electron cloud into an ellipsoid elongated along the +45° direction. This introduces anisotropy into the medium, typically increasing the refractive index for the +45° direction and decreasing it for the –45° direction. The medium then acts like a wave plate, slightly rotating the polarization of the field, E_1, allowing some of it to leak through the crossed polarizers.

However you look at it, you obtain the same answer when the medium responds rapidly.

Another type of induced-grating process is *self-diffraction* (see Fig. 5.9). It involves beams #1 and #2 inducing a grating, but beam #1 also diffracting off it. Thus beams #1 and #3 are the same beam. This process has the induced-polarization term:

$$P_i = \frac{3}{8} \varepsilon_0 \, \chi^{(3)} \; E_1^2 E_2^* \exp\{i[(2\omega_1 - \omega_2)t - (2\vec{k}_1 - \vec{k}_2) \cdot \vec{r}]\} + c.c. \tag{9}$$

It produces a beam with frequency $2\omega_1 - \omega_2$ and k-vector $2\vec{k}_1 - \vec{k}_2$. This beam geometry is also convenient because only two input beams are required.

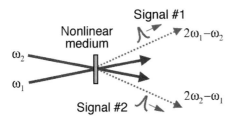

Fig. 5.9 Self-diffraction. The two beams yield a sinusoidal intensity pattern, which induces a grating in the medium. Then each beam diffracts off the grating. The pulses at the output indicate the signal pulses, here in the $2k_1 - k_2$ and $2k_2 - k_1$ directions.

And it is also possible to perform third-harmonic generation using more than one beam (or as many as three). An example beam geometry is shown in Fig. 5.10, using two input beams.

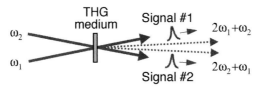

Fig. 5.10 Third-harmonic generation. While each beam individually can produce a hird harmonic, it can also be produced by two factors of one field and one of the other. These latter two effects are illustrated here.

5.3
Some General Observations about Nonlinear Optics

Nonlinear-optical effects are usually shown as in Fig. 5.11. Upward-pointing arrows indicate fields without complex conjugates and with frequency and k-vector contributions with plus signs. Downward-pointing arrows indicate complex-conjugated fields in the polarization and negative signs in the contributions to the frequency and k-vector of the light created. Unless otherwise specified, ω_0 and k_0 denote the signal frequency and k-vector.

Notice that, in all of these nonlinear-optical processes, the polarization propagates through the medium just as the light wave does. It has a frequency and k-vector. For a given process of Nth order, the frequency is usually denoted by ω_0 and is given by:

$$\omega_0 = \pm \omega_1 \pm \omega_2 \pm \ldots \pm \omega_N \tag{10}$$

where the signs obey the above complex-conjugate convention.

The polarization has a k-vector with an analogous expression:

$$\vec{k}_0 = \pm \vec{k}_1 \pm \vec{k}_2 \pm \ldots \pm \vec{k}_N \tag{11}$$

where the same signs occur in both Eqs. (10) and (11).

In all of these nonlinear-optical processes, terms with products of the E-field complex envelopes, such as $E_1{}^2 E_2{}^*$, are created.

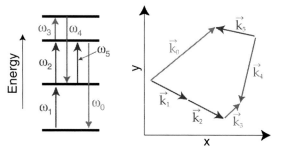

Fig. 5.11 Sample complex nonlinear-optical process, $\mathcal{P} \propto E_1 E_2 E_3 E_4{}^* E_5$. Here, $\omega_0 = \omega_1 + \omega_2 + \omega_3 - \omega_4 + \omega_5$ and $k_0 = k_1 + k_2 + k_3 - k_4 + k_5$. The k-vectors are shown adding in two-dimensional space, but, in third- and higher-order processes, the third space dimension is potentially also involved. The different frequencies (colors) of the beams are shown as different shades of gray.

5.4
The Mathematics of Nonlinear Optics

5.4.1
The Slowly Varying Envelope Approximation

How do we calculate the effects of these induced polarizations? We must substitute into the wave equation, Eq. (1), and solve the nonlinear differential equation that results. While this is hard to do exactly, a few tricks and approximations make it quite easy in most cases of practical interest.

The first approximation is that we consider only a range of frequencies near one frequency at a time. We will write the wave equation for one particular signal frequency, ω_0, and only consider a small range of nearby frequencies. Anything happening at distant frequencies will alternately be in phase and then out of phase with the fields and polarizations in this range and so should have little effect. We will also assume that the nonlinear optical process is fairly weak, so it will not affect the input beams. Thus we will only consider the one signal field of interest. If you are interested in more complex situations, check out a full text on nonlinear optics (see, for example, the list at the end of this chapter).

The second is the *Slowly Varying Envelope Approximation* (SVEA), which, despite its name, remains a remarkably good approximation for all but the shortest pulses. It takes advantage of the fact that, as short as they are, most ultrashort laser pulses are still not as short as an optical cycle (about 2 fs for visible wavelengths). Thus the pulse electric field can be written as the product of the carrier sine wave and a relatively slowly varying envelope function. This is what we have been doing, but we have not explicitly used this fact; now we will. Since the measure of the change in anything is the derivative, we will now neglect second derivatives of the slowly varying envelope compared to those of the more rapidly varying carrier sine wave. And the wave equation, which is what we must solve to understand any optics problem, is drowning in derivatives.

Assume that the driving polarization propagates along the z-axis, and write the electric field and polarization in terms of slowly varying envelopes:

$$\mathcal{E}(\vec{r},t) = \tfrac{1}{2} E(\vec{r},t) \exp[i(\omega_0 t - k_0 z)] + c.c. \tag{12}$$

$$\mathcal{P}(\vec{r},t) = \tfrac{1}{2} P(\vec{r},t) \exp[i(\omega_0 t - k_0 z)] + c.c. \tag{13}$$

where we have chosen to consider the creation of light at the same frequency as that of the induced polarization, ω_0. But we have also assumed that the light field and polarization have the same k-vectors, k_0, which is a big – and often unjustified – assumption, as discussed above. But bear with us for now, and we will explain later.

Recall that the wave equation calls for taking second derivatives of E and P with respect to t and/or z. Let us calculate them:

$$\frac{\partial^2 \mathcal{E}}{\partial t^2} = \frac{1}{2}\left[\frac{\partial^2 E}{\partial t^2} + 2i\omega_0 \frac{\partial E}{\partial t} - \omega_0^2 E\right] \exp[i(\omega_0 t - k_0 z)] + c.c. \tag{14}$$

$$\frac{\partial^2 \mathcal{E}}{\partial z^2} = \frac{1}{2}\left[\frac{\partial^2 E}{\partial z^2} + 2i k_0 \frac{\partial E}{\partial z} - k_0^2 E\right] \exp[i(\omega_0 t - k_0 z)] + c.c. \tag{15}$$

$$\frac{\partial^2 \mathcal{P}}{\partial t^2} = \frac{1}{2}\left[\frac{\partial^2 P}{\partial t^2} + 2i\omega_0 \frac{\partial P}{\partial t} - \omega_0^2 P\right] \exp[i(\omega_0 t - k_0 z)] + c.c. \tag{16}$$

As we mentioned above, we will assume that derivatives are small and that derivatives of derivatives are even smaller:

$$\left|\frac{\partial^2 E}{\partial t^2}\right| \ll \left|2i\omega_0 \frac{\partial E}{\partial t}\right| \ll |\omega_0^2 E| \tag{17}$$

Letting $\omega_0 = 2\pi/T$, we find that this condition will be true as long as:

$$\left|\frac{\partial^2 E}{\partial t^2}\right| \ll \left|2\frac{2\pi}{T}\frac{\partial E}{\partial t}\right| \ll \left|\frac{4\pi^2}{T^2}E\right| \tag{18}$$

where T is the optical period of the light, again about 2 fs for visible light. These conditions hold if the field envelope is not changing on a time scale of a single cycle, which is nearly always true. So we can neglect the smallest term and keep the larger two.

The same is true for the spatial derivatives. We will also neglect the second spatial derivative of the electric field envelope.

And the same derivatives arise for the polarization. But since the polarization is small to begin with, we will neglect both the first and second derivatives.

The wave equation becomes:

$$\left[-2ik_0 \frac{\partial E}{\partial z} - \frac{2i\,n^2\omega_0}{c^2}\frac{\partial E}{\partial t} - k_0^2 E + \frac{\omega_0^2}{c^2}E\right]\exp[i(\omega_0 t - k_0 z)] =$$
$$-\mu_0\omega_0^2 P\exp[i(\omega_0 t - k_0 z)] \tag{19}$$

since we can factor out the complex exponentials.

We can also cancel the exponentials. Recalling that E satisfies the wave equation by itself, $k_0^2 E = n^2(\omega_0^2/c^2)E$, and those two terms can also be canceled. Then dividing through by $-2ik_0$ yields:

$$\frac{\partial E}{\partial z} + \frac{n}{c}\frac{\partial E}{\partial t} = -i\frac{\mu_0\omega_0^2}{2k_0}P \tag{20}$$

This expression is actually rather oversimplified. A more accurate inclusion of dispersion (see [4]) yields the same equation, but with the phase velocity of light in the medium, c/n, replaced by the group velocity, v_g:

$$\frac{\partial E}{\partial z} + \frac{1}{v_g}\frac{\partial E}{\partial t} = -i\frac{\mu_0\omega_0^2}{2k_0}P \tag{21}$$

We can now simplify this equation further by transforming the time coordinate to be centered on the pulse. This involves new space and time coordinates, z_v and t_v, given by: $z_v = z$ and $t_v = t - z/v_g$. To transform to these new coordinates requires replacing the derivatives:

$$\frac{\partial E}{\partial z} = \frac{\partial E}{\partial z_v}\frac{\partial z_v}{\partial z} + \frac{\partial E}{\partial t_v}\frac{\partial t_v}{\partial z} \tag{22}$$

$$\frac{\partial E}{\partial t} = \frac{\partial E}{\partial z_\nu}\frac{\partial z_\nu}{\partial t} + \frac{\partial E}{\partial t_\nu}\frac{\partial t_\nu}{\partial t} \tag{23}$$

Computing the simple derivatives and substituting, we find:

$$\frac{\partial E}{\partial z} = \frac{\partial E}{\partial z_\nu} + \frac{\partial E}{\partial t_\nu}\left[-\frac{1}{v_g}\right] \tag{24}$$

$$\frac{\partial E}{\partial t} = 0 + \frac{\partial E}{\partial t_\nu} \tag{25}$$

The time derivative of the polarization is also easily computed. This yields:

$$\frac{\partial E}{\partial z_\nu} + \frac{\partial E}{\partial t_\nu}\left[-\frac{1}{v_g}\right] + \frac{1}{v_g}\left[\frac{\partial E}{\partial t_\nu}\right] = -i\frac{\mu_0\omega_0^2}{2k_0}P \tag{26}$$

Canceling the identical terms leaves:

$$\frac{\partial E}{\partial t} = -i\frac{\mu_0\omega_0^2}{2k_0}P \tag{27}$$

where we have dropped the subscripts on t and z for simplicity. This nice simple equation is the SVEA equation for most nonlinear-optical processes in the simplest case. Assumptions that we have made to arrive here include that: 1) the nonlinear effects are weak; 2) the input beams are not affected by the fact that they are creating new beams (okay, so we are violating Conservation of Energy here, but only by a little); 3) the group velocity is the same for all frequencies in the beams; 4) the beams are uniform spatially; 5) there is no diffraction; and 6) pulse variations occur only on time scales longer than a few cycles in both space and time. And we have assumed that the electric field and the polarization have the same frequency and k-vector. While the other assumptions mentioned above are probably reasonable in practical situations, this last assumption will be wrong in many cases – in fact it is actually difficult to satisfy, and we usually have to go to some trouble in order to satisfy it. This effect is called "phase-matching," and we will consider it in detail in the next section. But the rest of these assumptions are quite reasonable in most pulse-measurement situations.

5.4.2
Solving the Wave Equation in the Slowly Varying Envelope Approximation

If the polarization envelope is constant, then the wave equation in the SVEA is the world's easiest differential equation to solve, and here is the solution:

$$E(z,t) = -i\frac{\mu_0\omega_0^2}{2k_0}P\,z \tag{28}$$

and we can see that the new field grows linearly with distance. Since the intensity is proportional to the mag-squared of the field, the intensity then simply grows quadratically with distance:

$$I(z,t) = \frac{c\mu_0\omega_0^2}{8n}|P|^2 z^2 \tag{29}$$

5.5
Phase-matching

There is an ubiquitous effect that must always be considered when we perform nonlinear optics and it is another reason why nonlinear optics is not part of our everyday lives. This is *phase-matching*. What it refers to is the tendency, when propagating through a nonlinear-optical medium, of the generated wave to become out of phase with the induced polarization after some distance. If this happens, then the induced polarization will create new light that is out of phase with the light it created earlier, and, instead of making more such light, the two contributions will *cancel out*. The way to avoid this is for the induced polarization and the light it creates to have the same phase velocities. Since they necessarily have the same frequencies, this corresponds to having the same k-vectors, the issue which we discussed a couple of sections ago. Then the two waves are always in phase, and the process is orders of magnitude more efficient. In this case, we say that the process is *phase-matched*.

We have been implicitly assuming phase-matching so far by using the variable k_0 for both k-vectors. But because they can be different, let us reserve the variable, k_0, for the k-vector of the light at frequency ω_0 [$k_0 = \omega_0\, n(\omega_0)\ /\ c$, where c is the speed of light in a vacuum], and we will now refer to the induced polarization's k-vector, as given by Eq. (11), as. We must recognize that k_P will not necessarily equal, the k-vector of light with the polarization's frequency ω_0 – light that the induced polarization itself creates. Indeed, there is no reason whatsoever for the sum of the k-vectors above, all at different frequencies with their own refractive indices and directions, to equal $\omega_0 n(\omega_0)/c$.

Equation (27) now becomes:

$$2ik_0\frac{\partial E}{\partial z}\exp[i(\omega_0 t - k_0 z)] = \mu_0\omega_0^2 P\exp[i(\omega_0 t - k_P z)] \tag{30}$$

Simplifying:

$$\frac{\partial E}{\partial z} = -i\frac{\mu_0\omega_0^2}{2k}P\exp(i\Delta k\, z) \tag{31}$$

where:

$$\Delta k \equiv k_0 - k_P \tag{32}$$

We can also solve this differential equation simply:

$$E(L,t) = -i\frac{\mu_0\omega_0^2}{2k_0}P\left[\frac{\exp(i\,\Delta k\,z)}{i\,\Delta k}\right]_0^L \tag{33}$$

$$= -i\frac{\mu_0\omega_0^2}{2k_0}P\left[\frac{\exp(i\,\Delta k\,L)-1}{i\,\Delta k}\right] \tag{34}$$

$$= -i\frac{\mu_0\omega_0^2 L}{2k_0}P\exp(i\,\Delta k\,L/2)\left[\frac{\exp(i\,\Delta k\,L/2)-\exp(-i\,\Delta k\,L/2)}{2i\,\Delta k\,L/2}\right] \tag{35}$$

The expression in the brackets is $\sin(\Delta kL/2)/(\Delta kL/2)$, which is just the function called $\mathrm{sinc}(\Delta kL/2)$. Ignoring the phase factor, the light electric field after the non-linear medium will be:

$$E(L,t) = -i\frac{\mu_0\omega_0^2}{2k_0}P\,L\,\mathrm{sinc}(\Delta k\,L/2) \tag{36}$$

Mag-squaring to obtain the light irradiance or intensity, I, we have:

$$I(L,t) = \frac{c\mu_0\omega_0^2}{8n}|P|^2\,L^2\,\mathrm{sinc}^2(\Delta k\,L/2) \tag{37}$$

Since the function, $\mathrm{sinc}^2(x)$, is maximal at $x = 0$, and also highly peaked there (see Fig. 5.12), the nonlinear-optical effect of interest will experience much greater efficiency if $\Delta k = 0$. This confirms what we said earlier, that the nonlinear-optical efficiency will be maximized when the polarization and the light it creates remain in phase throughout the nonlinear medium, that is, when the process is *phase-matched*.

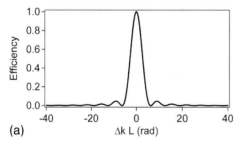
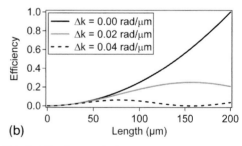

(a)

(b)

Fig. 5.12 (a) Plot of $\mathrm{sinc}^2(\Delta k\,L/2)$ vs. $\Delta k\,L$. Note that the sharp peak at $\Delta k\,L = 0$. (b) Plot of the generated intensity vs. L, the nonlinear-medium thickness for various values of Δk. Note that, when $\Delta k \neq 0$, the efficiency oscillates sinusoidally with distance and remains minimal for all values of L.

Phase-matching is crucial for creating more than just a few photons in a nonlinear-optical process. To summarize, the phase-matching conditions for an N-wave-mixing process are (see Fig. 5.11):

$$\omega_0 = \pm\omega_1 \pm \omega_2 \pm \ldots \pm \omega_N \tag{38}$$

$$\vec{k}_0 = \pm\vec{k}_1 \pm \vec{k}_2 \pm \ldots \pm \vec{k}_N \tag{39}$$

where k_0 is the k-vector of the beam at frequency, ω_0, which may or may not naturally equal the sum of the other k-vectors, and it is our job to make it so.

Note that if we were to multiply these equations by $\hbar$, they would correspond to energy and momentum conservation within the material for the photons involved in the nonlinear-optical interaction.

Let us consider phase-matching in collinear SHG. Let the input beam (often called the *fundamental beam*) have frequency ω_1 and k-vector, $k_1 = \omega_1\, n(\omega_1)/c$. The second harmonic occurs at $\omega_0 = 2\omega_1$, which has the k-vector, $k_0 = 2\omega_1\, n(2\omega_1)/c$. But the induced polarization's k-vector has magnitude, $k_p = 2k_1 = 2\omega_1\, n(\omega_1)/c$. The phase-matching condition becomes:

$$k_0 = 2\, k_1 \tag{40}$$

which, after canceling common factors $(2\omega_1/c)$ simplifies to:

$$n(\omega_1) = n(2\omega_1) \tag{41}$$

Thus, in order to phase-match SHG, it is necessary to find a nonlinear medium whose refractive indices at ω and 2ω are the same (to several decimal places). Unfortunately – and this is another reason why you do not see this type of thing every day – all media have dispersion, the tendency of the refractive index to vary with wavelength (see Fig. 5.13). This effect quite effectively prevents seeing SHG in nearly all everyday situations.

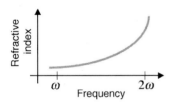

Fig. 5.13 Refractive index vs. wavelength for a typical medium. Because phase-matching SHG requires the refractive indices of the medium to be equal for both ω and 2ω, it is not possible to generate much second harmonic in normal media.

It turns out to be possible to achieve phase-matching for birefringent crystals, whose refractive-index curves are different for the two orthogonal polarizations (see Fig. 5.14).

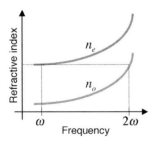

Fig. 5.14 Refractive index vs. wavelength for a typical *birefringent* medium. For the two polarizations (say, vertical and horizontal, corresponding to the *ordinary* and *extraordinary* polarizations) see different refractive index curves. As a result, phase-matching of SHG is possible. This is the most common method for achieving phase-matching in SHG. The extraordinary refractive index curve depends on the beam propagation angle (and temperature), and thus can be shifted by varying the crystal angle in order to achieve the phase-matching condition.

Far from phase-matching   Closer to phase-matching

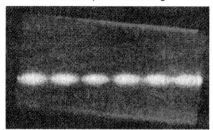 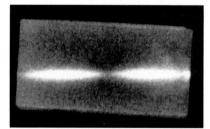

Six coherence lengths   Two coherence lengths

Fig. 5.15 Light inside a SHG crystal for two different amounts of phase-mismatch (i.e., for two different crystal angle orientations). Note that, as the crystal angle approaches the phase-matching condition, the periodicity of the intensity with position decreases, and the intensity increases. At phase-matching, the intensity increases quadratically along the crystal, achieving nearly 100% conversion efficiency, in practice.

In *noncollinear* SHG, we must consider that there is an angle, θ, between the two beams (see Fig. 5.15). The input vectors have longitudinal and transverse components, but, by symmetry, the transverse components cancel out, leaving only the longitudinal component of the phase-matching equation:

$$k_1 \cos(\theta/2) + k_1 \cos(\theta/2) = k_0 \tag{42}$$

Simplifying, we have as our phase-matching condition. Substituting for the k-vectors, the phase-matching becomes:

$$n(\omega_1) \cos(\theta/2) = n(2\omega_1) \tag{43}$$

Figure 5.16 shows a nice display of noncollinear SHG phase-matching processes involving one intense beam and scattered light in essentially all directions. This picture does not yield any particular insights for any particular application of nonlinear optics, but it is very attractive, and we thought you might like to see it. By the way, the star is not really nonlinear-optical; this is just due to the high intensity of the spot at its center (and the "star filter" on the camera lens when the picture was taken). The ring is real, however, and there can be as many as three of them.

Finally, whether for a collinear or non-collinear beam geometry, it is also possible to achieve phase-matching using two orthogonal polarizations for the (two) input beams. In other words, the input beam is polarized at a 45° angle to the output SH beam. This is referred to as *Type II phase-matching*, while the above process is called *Type I phase-matching*. Type II phase-matching is more complex than Type I because the two input beams have different refractive indices, phase velocities, and group velocities, which must be kept in mind when performing measurements.

Fig. 5.16 Interesting non-collinear phase-matching effects in second-harmonic generation. (Picture taken by Rick Trebino.)

Phase-matching is easier to achieve in third order, largely because we have an extra k-vector to play with. In fact, it can be so easy that it happens automatically. In two-beam coupling and polarization gating, the phase-matching equations become:

$$\omega_0 = \omega_1 - \omega_2 + \omega_2 \tag{44}$$

$$\vec{k}_0 = \vec{k}_1 - \vec{k}_2 + \vec{k}_2 \tag{45}$$

These equations are *automatically satisfied* when the signal beam has the same frequency and k-vector as beam 1: ω_1 and k_1, respectively.

For other third-order processes, phase-matching is not automatic, but it can be achieved with a little patience. For some processes, however, it can be impossible, as is the case for self-diffraction. In the latter case, sufficient efficiency can be achieved for most purposes, provided that the medium is kept thin to minimize the phase-mismatch.

5.6
Phase-matching Bandwidth

5.6.1
Direct Calculation

While, at most one frequency can be exactly phase-matched at any one time, some nonlinear-optical processes are more forgiving about this condition than others. Since it will turn out to be important for most applications to achieve efficient SHG (or another nonlinear-optical process) for all frequencies in the pulse, phase-matching bandwidth is an important issue. Figures 5.17(a) and (b) show the SHG efficiency vs. wavelength for two different crystals and for different incidence angles. Notice the huge variations in phase-matching efficiency for different crystal angles and thicknesses.

We can easily calculate the range of frequencies that will be approximately phase-matched in, for example, SHG. Assuming that the SHG process is exactly phase-matched at the wavelength, λ_0, the phase-mismatch, Δk, will be a function of wavelength:

$$\Delta k\,(\lambda) = 2k_1 - k_2 \tag{46}$$

$$\Delta k(\lambda) = 2\left[2\pi\frac{n(\lambda)}{\lambda}\right] - \left[2\pi\frac{n(\lambda/2)}{\lambda/2}\right] \tag{47}$$

$$\Delta k(\lambda) = \frac{4\pi}{\lambda}[n(\lambda) - n(\lambda/2)] \tag{48}$$

Expanding $1/\lambda$ and the material dispersion to first order in the wavelength,

$$\Delta k(\lambda) = \frac{4\pi}{\lambda_0}\left[1 - \frac{\delta\lambda}{\lambda_0}\right]\left[n(\lambda_0) + \delta\lambda n'(\lambda_0) - n(\lambda_0/2) - \frac{\delta\lambda}{2}n'(\lambda_0/2)\right] \tag{49}$$

where $\delta\lambda = \lambda - \lambda_0$, $n'(\lambda) = dn\,/\,d\lambda$, and we have taken into account the fact that, when the input wavelength changes by $\delta\lambda$, the second-harmonic wavelength changes by only $\delta\lambda/2$.

Recalling that the process is phase-matched for the input wavelength, λ_0, we note that $n(\lambda_0/2) - n(\lambda_0) = 0$, and we can simplify this expression:

$$\Delta k(\lambda) = \frac{4\pi}{\lambda_0}\left[\delta\lambda n'(\lambda_0) - \frac{\delta\lambda}{2}n'(\lambda_0/2)\right] \tag{50}$$

where we have neglected second-order terms.

The sinc^2 curve will decrease by a factor of 2 when $\Delta k\,L/2 = \pm 1.39$. So solving for the wavelength range that yields $|\Delta k| < 2.78/L$, we find that the phase-matching bandwidth will be:

$$\delta\lambda = \frac{0.44\,\lambda_0/L}{|n'(\lambda_0) - \frac{1}{2}n'(\lambda_0/2)|} \tag{51}$$

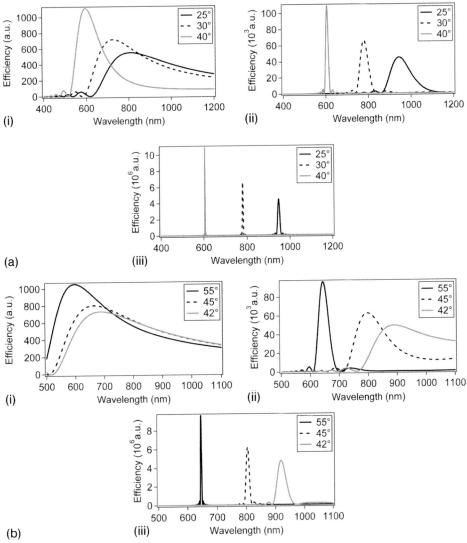

Fig. 5.17 (a) Phase-matching efficiency vs. wavelength for the nonlinear-optical crystal, beta-barium borate (BBO). (i): a 10 μm thick crystal. (ii): a 100 μm thick crystal. (iii): a 1000 μm thick crystal. These curves also take into account the ω_0^2 and L^2 factors in Eq. (37). While the curves are scaled in arbitrary units, the relative magnitudes can be compared among the three plots. (These curves do not, however, include the nonlinear susceptibility, $\chi^{(2)}$, so comparison of the efficiency curves in Figs. 5.17(a) and (b) requires inclusion of this factor.)

(b) The same as Fig. 5.17(a), except for the nonlinear-optical crystal, potassium di-hydrogen phosphate (KDP). (i): a 10 μm thick crystal. (ii): a 100 μm thick crystal. (iii): a 1000 μm thick crystal. The curves for the thin crystals (top row) do not fall to zero at long wavelengths because KDP simultaneously phase-matches for two wavelengths, that shown and a longer (IR) wavelength, whose phase-matching ranges begin to overlap when the crystal is thin.

Notice that the $\delta\lambda$ is inversely proportional to the thickness of the nonlinear medium. Thus, in order to increase the phase-matching bandwidth, we must use a medium with dispersion such that $n(\lambda_0 = \frac{1}{2}n'(\lambda_0/2))$, or more commonly decrease the thickness of the medium (see Fig. 5.18).

Finally, note the factor of 1/2 multiplying the second-harmonic refractive index derivative in Eq. (51). This factor does not occur in results appearing in some journal articles. These articles use a different derivative definition for the second harmonic [that is, $dn/d(\lambda/2)$] because the second harmonic necessarily varies by only one-half as much as the fundamental wavelength. We, on the other hand, have used the same definition – the standard one, $dn/d\lambda$ – for both derivatives, which, we think, is less confusing, but it yields the factor of 1/2. It is easy to see that the factor of 1/2 is correct: assuming that the process is phase-matched at λ_0, maintaining a phase-matched process [i.e., $n(\lambda/2) = n(\lambda)$] requires that the variation in refractive index per unit wavelength near $\lambda_0/2$ be twice as great as that near λ_0, since the second harmonic wavelength changes only half as fast as the fundamental wavelength.

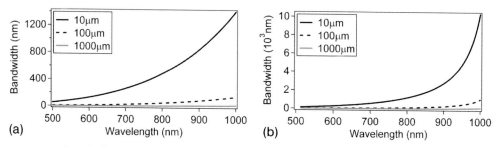

Fig 5.18 Phase matching bandwidth vs. wavelength for BBO (a) and KDP (b).

5.6.2
Group-velocity Mismatch

There is an alternative approach for calculating the phase-matching bandwidth, which seems like a completely different effect until you realize that you get the same answer, and that it is just a time-domain approach, while the previous approach was in the frequency domain. Consider that the pulse entering the SHG crystal and the SH it creates may have the same phase velocities (they are phase-matched), but they could have different group velocities. This is called *group-velocity mismatch* (GVM). If so, then the two pulses could cease to overlap after propagating some distance into the crystal; in this case, the efficiency will be reduced because SH light created at the back of the crystal will not coherently combine with SH light created in the front. This effect is illustrated in Fig. 5.19.

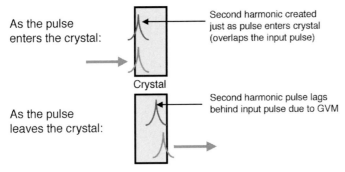

As the pulse enters the crystal:

Second harmonic created just as pulse enters crystal (overlaps the input pulse)

Crystal

As the pulse leaves the crystal:

Second harmonic pulse lags behind input pulse due to GVM

Fig. 5.19 Group-velocity mismatch. The pulse entering the crystal creates SH at the entrance, but this light travels at a different group velocity from that of the fundamental light, and light created at the exit does not coherently add to it.

We can calculate the bandwidth of the light created when significant GVM occurs. Assuming that a very short pulse enters the crystal, the length of the SH pulse, δt, will be determined by the difference in light-travel times through the crystal:

$$\delta t = \frac{L}{v_g(\lambda_0/2)} - \frac{L}{v_g(\lambda_0)} = L\ GVM \qquad (52)$$

where $GVM \equiv 1/v_g(\lambda_0/2) - 1/v_g(\lambda_0)$. This expression can be rewritten using expressions for the group velocity:

$$v_g(\lambda) = \frac{c/n(\lambda)}{1 - \frac{\lambda}{n(\lambda)}n'(\lambda)} \qquad (53)$$

Substituting for the group velocities in Eq. (50), we find:

$$\delta t = \frac{L\,n(\lambda_0/2)}{c}\left[1 - \frac{\lambda_0/2}{n(\lambda_0/2)}n'(\lambda_0/2)\right] - \frac{L\,n(\lambda_0)}{c}\left[1 - \frac{\lambda_0}{n(\lambda_0)}n'(\lambda_0)\right] \qquad (54)$$

Now, recall that we would not be doing this calculation for a process that was not phase-matched, so we can take advantage of the fact that $n(\lambda_0/2) = n(\lambda_0)$. Things then simplify considerably:

$$\delta t = \frac{L\,\lambda_0}{c}\left[n'(\lambda_0) - \frac{1}{2}n'(\lambda_0/2)\right] \qquad (55)$$

Take the second-harmonic pulse to have a Gaussian intensity, for which $\delta t\,\delta v = 0.44$. Rewriting in terms of the wavelength, $\delta t\,\delta \lambda = \delta t\,\delta v\,[dv/d\lambda]^{-1} = 0.44\,[dv/d\lambda]^{-1} = 0.44\,\lambda^2/c$, where we have neglected the minus sign, since we are computing the bandwidth which is inherently positive. So the bandwidth is:

$$\delta\lambda \approx \frac{0.44\,\lambda_0/L}{\left|n'(\lambda_0) - \frac{1}{2}n'(\lambda_0/2)\right|} \tag{56}$$

Note that the bandwidth calculated from GVM considerations precisely matches that calculated from phase-matching bandwidth considerations.

5.6.3
Phase-matching Bandwidth Conclusions

As we mentioned, it is usually important to achieve efficient (or at least uniform) phase-matching for the entire bandwidth of the pulse. Since ultrashort laser pulses can have extremely large bandwidths (a 10 fs pulse at 800 nm has a bandwidth of over a hundred nm), it will be necessary to use extremely thin SHG crystals. Crystals as thin as 5 μm have been used for few-fs pulses.

But also recall that the intensity of the phase-matched SH produced is proportional to L^2. So a very thin crystal yields very little SHG efficiency. Thus there is a nasty trade-off between efficiency and bandwidth.

5.7
Nonlinear-optical Strengths

Just how strong are nonlinear-optical effects? Clearly they are not so strong that sunlight, even on the brightest day, efficiently produces enough of them for us to see. Of course, phase-matching is also not occuring.

But what sort of laser intensities are necessary to see these effects? We start with Eq. (36), which can be rewritten (with $\omega_0 = 2\omega$) in the form:

$$E^{2\omega}(L,t) = -i\frac{\mu_0\omega^2 L}{k}\,P\exp(i\,\Delta k\,L/2)\mathrm{sinc}(\Delta k\,L/2) \tag{57}$$

where $P = \frac{1}{2}\varepsilon_0\chi^{(2)}(E^\omega)^2$. Then, we relate intensity to electric field strength by $I = (n/2\eta_0)|E|^2$, where $\eta_0 = \sqrt{\mu_0/\varepsilon_0}$. With these, we re-write Eq. (57) in terms of intensities to find:

$$I^{2\omega} = \frac{\eta_0\omega^2(\chi^{(2)})^2(I^\omega)^2 L^2}{2c_0^2 n^3}\,\mathrm{sinc}^2(\Delta k\,L/2) \tag{58}$$

Next, suppose we consider the best case, in which the process is phase-matched ($\mathrm{sinc}^2(0)$) and re-write Eq. (58) in terms of an *SHG efficiency*.

$$\frac{I^{2\omega}}{I^\omega} = \frac{2\eta_0\omega^2 d^2 I^\omega L^2}{c_0^2 n^3} \tag{59}$$

where we define the *d-coefficient* as $d = \frac{1}{2}\chi^{(2)}$. d is what we usually find quoted in handbooks. It will depend not only on the material, but also on the field configuration – how the fields are polarized with respect to the crystal orientation.

Again, we refer you to a more detailed treatment of nonlinear optics to fully understand these issues. Our concern now is just to get a feeling for the numbers involved and what we can hope to achieve in SHG efficiency in the lab. As a quick calculation, suppose we use beta-barium borate (BBO) as our nonlinear crystal, in which $d \approx 2 \times 10^{-12}$ mV^{-1}, and where $n \approx 1.6$ (note that we can get away with approximate values for n when it appears in an amplitude calculation, but we must have *very* accurate values for n when computing phase – or phase mismatch). If we wish to frequency-double an input beam of wavelength, $\lambda = 0.8\,\mu$m, we find from Eq. (59):

$$\frac{I^{2\omega}}{I^{\omega}} \approx 5 \times 10^{-8} I^{\omega} L^2 \tag{60}$$

where I is in W m^{-2} and L is in m.

From the small coefficient in front, some pretty high intensities are needed for modest crystal lengths in order to achieve anything like a decent efficiency! Suppose we consider an ultrafast laser. Basically, if you have an unamplified Ti:Sapphire laser, which produces nanojoule (nJ) pulses, 100 fs long, you have pulses with intensities on the order of 10^{14} W m^{-2} (when focusing to a about a 10 μm spot diameter). But, of course, when focusing this tightly, the beam does not stay focused for long, which limits the crystal length we can use. Additionally, because ultrashort pulses are broadband, the requirement of phase matching the entire bandwidth limits the SHG crystal thickness to considerably less than 1 mm, and usually less than 100 μm. Choosing a crystal length of 100 μm, and using the other numbers, we would achieve an efficiency of about 5%. This again is best-case for this configuration because: 1) the beam does not stay focused to its minimum size throughout the entire length (as the above calculation assumes); and 2) d is reduced somewhat below its maximum value; this is because the fields are not necessarily at the best orientation within the crystal to most effectively excite the anharmonic oscillators. Phase matching decides the field orientation, and the price is paid through a slightly reduced nonlinear coefficient (known as d_{eff}). So we are trying to optimize all of these parameters until we are satisfied with the SHG power that we receive. Then we stop.

This brings us to $\chi^{(3)}$. To have an idea of its order of magnitude for non-resonant materials, consider glass. Single-mode optical fibers, made of glass, guide light with a cross-sectional beam diameter of slightly less than 10 μm. So we can achieve similar intensities to those we saw before in our SHG example, but over much longer distances. In silica glass, $\chi^{(3)} \approx 2.4 \times 10^{-22}$ m^2 V^{-2}. One can make a comparison with a second-order process by calculating the second and third-order polarizations that result at a given light intensity. In our 100 fs 1 nJ pulse, focused to 10 μm diameter, the field strength is $E \approx 2.5 \times 10^8$ V m^{-1}. Then $\chi^{(3)} E \approx 6 \times 10^{-14}$ m V^{-1}. Compare this to $\chi^{(2)} = 2d \approx 4 \times 10^{-12}$ m V^{-1} for BBO. From here, the nonlinear polarizations for both processes are found by multiplying these results by the light intensity. As this example demonstrates, third-order processes in non-resonant materials are substantially weaker than second-order processes. But this can be made up for sometimes by: 1) tuning the frequency of one or more of the interacting waves near a material resonance (but at some cost in higher losses for those waves that are near resonance); or 2) taking advantage of long interaction lengths that may be possible in phase-matched situations (such as in optical fibers). Turning up the intensity will also help. Microjoule pulses can yield more than adequate signal energies from most of the third-order nonlinear optical effects mentioned in this chapter. Third order bulk media typically used are fused silica and any glass for the various induced grating effects.

The above illustrations assumed 100 fs pulse intensities on the order of 10^{12} W cm^{-2}. However, with the less tight focusing that is practical in the lab, intensities more like 10^9 W cm^{-2} are typically available. While this seems high, it is only enough to create barely detectable amounts of second harmonic. How about performing third-order nonlinear optics with such pulses? One can just barely do this in some cases, and it is difficult. It is better to have a stage of amplification, especially from a regenerative amplifier ("regen"). Microjoule pulses can yield more than adequate signal energies from most of the third-order nonlinear-optical effects mentioned in this chapter. Third-order media typically used are fused silica and any glass for the various induced-grating effects. These media are actually not known for their high nonlinearities, but they are optically very clean and hence are the media of choice for most nonlinear-optical applications.

References and Further Reading

1 N. Bloembergen, *Nonlinear Optics,* World Scientific Pub. Co. 1996 (original edition: 1965).

2 R.W. Boyd, *Nonlinear Optics,* Academic Press, 1992.

3 P. Butcher and D. Cotter, *The Elements of Nonlinear Optics* Cambridge University Press, 1991.

4 J.-C. Diels and W. Rudolph, *Ultrashort Laser Pulse Phenomena,* Academic Press, 1996.

5 F.A. Hopf and G.I. Stegeman, *Applied Classical Electrodynamics: Nonlinear Optics,* Krieger Pub. Co., reprinted 1992.

6 D.L. Mills, *Nonlinear Optics: Basic Concepts,* 2nd ed., Springer Verlag, 1998.

7 G.G. Gurzadian et al., *Handbook of Nonlinear Optical Crystals, 3rd ed.,* Springer Verlag, 1999.

8 K.-S. Ho, S.H. Liu, and G.S. He, *Physics of Nonlinear Optics,* World Scientific Pub. Co., 2000.

9 E.G. Sauter, *Nonlinear Optics,* Wiley-Interscience, 1996.

10 Y.R. Shen, *The Principles of Nonlinear Optics,* Wiley-Interscience, 1984.

11 A. Yariv, *Quantum Electronics,* 3rd ed. Wiley, 1989.

12 F. Zernike and J. Midwinter, *Applied Nonlinear Optics,* Wiley-Interscience, 1973 (out of print).

6
Filamentation versus Optical Breakdown
in Bulk Transparent Media

Eugenijus Gaižauskas

Abstract

We present theoretical treatment of 3D optical pulse evolution in normal-dispersion Kerr media. The case when beam collapse is arrested by nonlinear losses leading to beam filamentation is high-lighted. The self-transformation of femtosecond wavepackets from Gaussian toward Gauss–Bessel shape as well as X-wave formation by propagating in borosilicate glass are demonstrated by numerical simulations of the nonlinear Schrödinger equation with axial symmetry. We consider formation the multiple filaments resulted from both beam elipticity and noise by using a cw approximation. Finally, we discuss filamentation from Bessel–Gauss pulses, which is found to produce lines consisting of discrete, equidistant high intensity spots providing a potentially useful tool for laser microfabrication of transparent materials.

6.1
Introduction

Since its discovery by Braun et al. [1] in 1995, the spontaneous filament formation accompanying the intense fs-pulse propagation in transparent media received rapidly increasing attention, which was motivated primarily by the possibility of spatial (and perhaps temporal) localization of pulse energy while keeping the extremely high intensities over a long range of propagation. The potential applications of such a light channeling would be extremely promising, ranging from material processing, optical microscopy to the remote sensing and LIDAR. Despite the huge amount of theoretical and experimental studies devoted to the subject, the physics underlying the spontaneous formation of light filaments in transparent media is still under debate. A number of approaches have been proposed for its description which are widely used for the modeling of the process. The first one interprets the filament only as an optical illusion related to the use of time-integrated detection, based on a moving focus [2]. The second approach considers the filament as a genuine soliton-like, self-guided beam, whose statio-

3D Laser Microfabrication. Principles and Applications.
Edited by H. Misawa and S. Juodkazis
Copyright © 2006 WILEY-VCH Verlag GmbH & Co. KGaA, Weinheim
ISBN: 3-527-31055-X

narity is supported by the dynamical balance between Kerr-induced self-focusing and plasma-induced de-focusing, where the plasma is formed owing to multi-photon absorption (MPA) [1, 3, 4]. In the dynamic spatial replenishment model, [5] the light filament is interpreted as continuously absorbed and regenerated by subsequent focusing of different *temporal* slices of the the pulse. It should be stressed, that experimental observation of the light filaments from femtosecond pulses, in spite of different length and power scales, exhibits substantially the same features in all the investigated media (gases [1, 3], fused silica [6, 7, 8] and water [9, 10]). Specifically, filaments emerge for a power well exceeding the critical one for cw self-focusing and propagate in the absence of diffraction for a few to many diffraction lengths. Additionally, the core (central lobe) of the filament contains only a small fraction of the total beam power, limited by the so called "intensity clamping effect" [8, 11], while the remaining excess stays in the filament periphery, without apparent trapping.

The independence of the filamentation features on the media (which have to be, nevertheless, transparent for light propagation at the chosen wavelength) hints at the very simple mechanism underlying the basic physics of the filamentation processes on the femtosecond scale. In contrast to the widely accepted, self-guiding mechanism, provided by the balance between self-focusing, electron plasma de-focusing and diffraction, recently "filamentation without self-channeling" was proposed [8, 9]. According to this model, light filamentation appears mainly due to multi-photon ionization-induced nonlinear losses, rather than electron plasma induced changes in the refractive index. Note that, in contrast to the filamentation scenario, provided by balance between self-focusing, electron plasma de-focusing and diffraction, the role of the nontrapped radiation which appeared for the (see [8, 9]) "filamentation without self-channeling" mechanism, is essential. The presence of nonlinear losses during the self-focusing of pico-/femto-second pulses lead to spatial-temporal transformation of the wavepacket toward the conical waves (Besel-type beam), as well as on-axis splitting of the temporal shape of the pulse. In fact, in this model, the filament propagates as a high intensity spot of light radiation, supported by the surrounding radiation. The latter behaves as a reservoir, containing the main part of the beam energy, continually refuelling the central lobe.

It is dificult to overestimate the advantages of the such "light drilling structure" for light-based technologies. Nevertheless, promising exploitation possibilities of the nonlinear response of the media to the high intensity region of conical waves (the high-intensity part of the wave), requires a comprehensive study of the of conical waves in the nonlinear (extreme) regime for these reasons. First, it should be mentioned that, in the linear case, it is impossible to exploit the key feature of the conical beam (the highly localized nondiffracting peak) because the main part of the energy of the conical wave is localized far from this point. Thus, for example, if a Bessel beam is used for linear microscopy, very small objects deep inside a thick transparent sample can indeed be detected, which is impossible with gaussian-like beams. However, the resolution of the image is quenched to a great extent, due to the presence of the larger part of the energy in the side lobes. This explains why conical waves have not made a breakthrough in field applications, until now. On the contrary, exploitation of the conical waves in the nonlinear regime applies to conditions when the matter responds to the high-intensity part of the wave only, being transparent to the weak intensity part, even if it contains the main part of beam energy.

In this chapter we discuss the effects of diffraction-free propagation (filamentation) and the splitting of the intense femtosecond laser pulse in transparent media with Kerr nonlinearity. In contrast to the, widely accepted, self-guiding mechanism, provided by the balance between self-focusing, electron plasma de-focusing and diffraction, the recently proposed [8, 9] *filamentation without self-guiding* will be exploited in the understanding of the underlying physical processes. In the Section 6.2 the current understanding of the key questions of conical waves will be briefly summarized. In the following sections the specific effects of spatial-temporal transformation in the axially-symmetric case are analyzed, Section 6.5 is devoted to problems of the formation of multiple filaments in focusing Kerr media. In the final Section we consider the special case of the filamentation, when a Gauss–Bessel (instead of a pure Gaussian) beam is launched into the material.

6.2
Conical Waves: Tilted Pulses, Bessel Beams and X-type Waves

Conical waves are unique and peculiar wavepackets, in which the energy flow is not directed along the beam axis, as in conventional wavepackets. In contrast, here the energy arrives from a cone-shaped surface, forming a very intense and localized interference peak at the cone vertex. Significant examples of conical waves are Bessel (or Durnin) beams, discovered by James Durnin, as a family of nonspreading solutions to the Helmholtz wave equation. The simplest family member consists of a monochromatic beam whose radial profile is the zeroth-order Bessel function in the continuous-wave limit.

The peculiar spatiotemporal profile of an optical realization of the nonspreading axis-symmetric X-wave has been recorded in [12, 13, 14]. Here a simple representation of X-type waves and their correlation with plane waves has been introduced. The same group from the University of Tartu discovered how to counter chromatic dispersion and create optical X-wave pulses. The propagation features of Bessel X-waves in dispersive media were analyzed in details recently [15] by considering X-waves as cylindrically symmetric versions of the so-called tilted pulses (TP).

In order to disclose the physical nature of the X-type localized conical waves, it is instructive to start with the most simple example of the X-type wave as a result of interference between two tilted plane wave bursts. Specifically, let us consider a pair of plane wave bursts having identical temporal dependencies $A(t) = \exp(-t^2/\tau_p^2)\exp[\mathbf{k}_{1,2} \cdot \mathbf{r} - \omega t]$ the wave vectors $\mathbf{k}_1 = [k_z , k_T]$ and $\mathbf{k}_1 = [k_z, -k_T]$ with projections k_z and k_T onto the axis z and x, respectively. Consequently, the propagation directions of these waves packets make an angle $\theta = \pm\arccos(k_T/k_z)$, (i.e., they are tilted) with respect to the axis z. Figure 6.1 gives an example of the

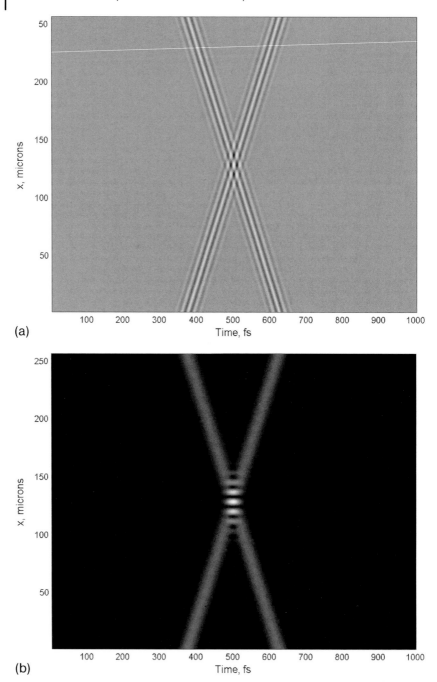

(a)

(b)

Fig. 6.1 Gray-scale images of X-type scalar wave formed by tilted plane wave pulses having durations $\tau_p = 15$ fs: (a) the field of the wave (real part of the X-wave); (b) intensity of the X-wave (amplitude squared).

interference patterns formed by the tilted plane wavepackets of 800 nm wavelengths, specified above, having durations as 15 fs. In the spatio-temporal regions where the pulses do not overlap their field is given simply by the burst profile as $A(t) = \exp[i(kz \pm k_T x - \omega t)]$. In the overlap region, where the field is given by a superposition of plane wave bursts, it forms a well-known two-wave interference pattern with doubled amplitudes as shown in Fig. 6.1(a). The points of completely constructive interference (maximum intensity) lie here on the z axis, forming highly localized energy strips in the center, for the intensity distribution shown in Fig. 6.1(b). Altogether, the superposition of the pulse pair makes up an X-shaped propagation-invariant interference pattern possessing wave-number $k_z = k_0 \cos \theta$ along the propagation axis z. Phrased differently, although separate plane wave bursts move with speed c, the interference pattern as a whole moves with superluminal speed $c/\cos \theta$ in the direction of the propagation axis z. It should be stressed, nevertheless, that here we do not encounter any problem due to the superluminal velocity of the formed interference pattern, because the energy flux along the propagation axis z does not possesses superluminal speed. Consequently, physical signals cannot be transferred in this way.

Let us proceed with the natural generalization of the simple X-wave example above described towards localized (both in time and space) wavepackets and consider the superposition of monochromatic plane wave bursts, propagating in different directions specified by wavevectors $\mathbf{k}(\omega)$. Confining (as above) with two spatial $[x, z]$ dimensions one arrives at the expression:

$$E(\mathbf{r}, t) = \frac{1}{\pi} \int_0^\infty d\omega q(\omega) \exp[i\mathbf{k}(\omega) \cdot \mathbf{r} - i\omega t] \tag{1}$$

where the $q(\omega)$ is the spectral function obtained by the Fourier transformed amplitude $A(x, z, t)$ of the plane wave burst. $\mathbf{r}(x, z)$ is the radius vector and wave vector $\mathbf{k}(x, z)(\omega)$ determines the propagation direction of each plane wave burst. We are not going to discuss dispersion effects of the TP (see [15]) here and will confine our consideration to the dispersionless (free) medium. Therefore the modulus of the vector $\mathbf{k}$ simple reads $|k(\omega)| = \omega/c$.

Gray-scale images of the wavepacket resulting from the interference of the plane waves bursts given by expression (1) are shown in Fig. 6.2. The opposite signs of the tilt angles $\pm \theta$ were taken by creating a) and b) images of the Fig. 6.2.

Figure 6.3 gives two examples of the interference patterns formed by the above, tilted, pulses of 800 nm wavelengths, having durations 15 fs. 3D plots and gray scale intensity images are depicted here for different tilt angles. The highly localized energy "bullet" in the center of the X-waves is formed here by constructive interference, as in Fig. 6.3.

Finally, going to the (3 + 1) dimension case and superimposing the axis-symmetrical pairs of TP, the propagation direction of which form a cone around the z axis with the top angle 2θ, one has an X-pulse localized in space and time. The intensity distribution of the X-pulse in the planes, containing the axis z, are analogous to those, showed in Fig. 6.3.

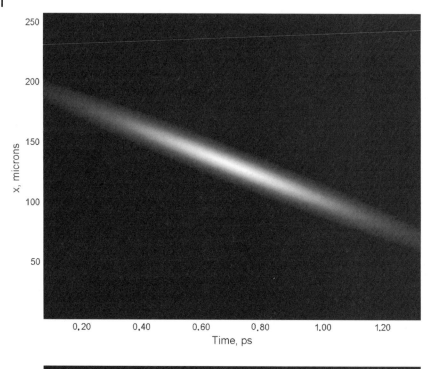

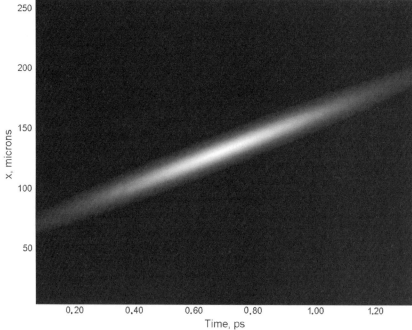

Fig. 6.2 Gray scale plots of the time variation of the intensity $E(x, t)|^2$ of the TP, titled in opposite directions.

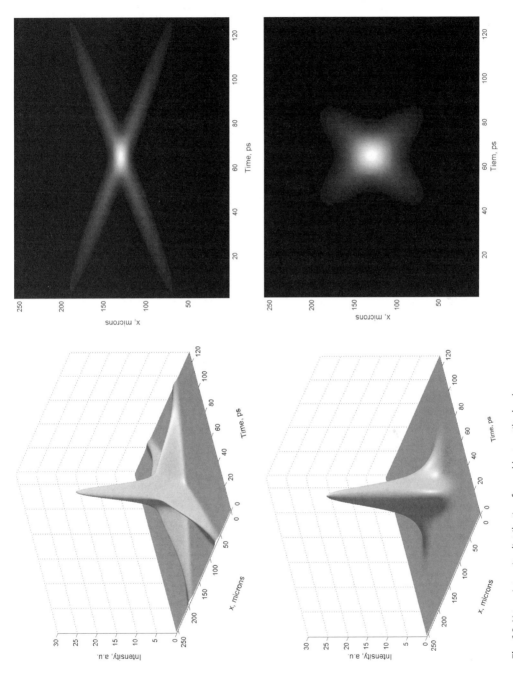

Fig. 6.3 X-type intensity distribution formed by two tilted pulses.

Proceeding to nonlinear X-waves we recall that the nonlinear response of condensed matter can compensate for the diffractive spreading of optical beams, or the dispersive broadening of pulses due to group-velocity dispersion, forming spatial or temporal solitons, respectively. Recent experimental and theoretical results concerning the behavior of intense ultrashort pulses performed by Di Trapani [16] and his collaborators indicate, however, that the spatial and temporal degrees of freedom cannot be treated separately. Moreover, it was found that different colors seemed to find the correct dispersion-defeating angle by themselves, leading to the existence of localized X-waves in the nonlinear regime.

Computer simulations of the self-focusing of the ultrashort pulses in Kerr media with nonlinear losses, provided in the next section, supports this prediction.

6.3
Dynamics of Short-pulse Splitting in Nonlinear Media with Normal Dispersion: Effects of Nonlinear Losses

One of the most significant predictions of theoretical analysis based on the three-dimensional nonlinear Schrödinger equation is that the femtosecond pulse undergoes temporal splitting as it self-focuses in a nonlinear medium with normal group-velocity dispersion. Many recent studies [6, 17, 18, 19, 20]. have shown that pulse splitting occurs in Kerr media for pulses with powers beyond the threshold for self-focusing.

The modeling of the ultrashort pulse dynamics in focusing Kerr media is one of the most challenging physical and computational tasks in nonlinear optics, which requires the use of the full Maxwell equations with plenty of nonlinear terms. As our aim in this work is to prove the relevance of the "filamentation without self-guiding" model for the description of time dependent effects accompanying femtosecond pulse propagation, we will neglect the role of the electron plasma density (created due to multi-photon ionization) and the related defocusing effect and present the theoretical results on splitting (in space and time) of intense short optical pulses in nonlinear media with normal dispersion. Our analysis is based on the numerical integration of the nonlinear envelope equation in (3+1) dimensions (containing axial symmetry) in the case where nonlinear beam collapse appears mainly due to nonlinear losses (such a possibility was predicted theoretically in [21, 22]).

Our results show that, in the normal dispersion regime, above the threshold for self-focusing, pulse-splitting dynamics follow those predicted in [23]. Specifically, it is found that, above the critical self-focusing power, the onset of pulse splitting into pulselets separated in time occurs for lengths comparable to the diffraction length $L_{DF} = kw_0^2$ and smaller than the dispersion length $L_{DS} = \tau_p^2 |k''|^2$. Results are found to be in quite good agreement with experiments which were performed in water [24].

The envelope $A(r, z, t)$ of an optical pulse centered at frequency ν propagating along the z axis in a dispersive medium with an intensity-dependent refractive index, is modeled by the NLSE. In the frame of the adopted paraxial approximation moving at the group velocity $v_g = [\partial k(\omega)/\partial \omega]^{-1}$, the resulting equation for the field amplitude reads:

$$\frac{\partial A}{\partial z} = \frac{ik''}{2}\frac{\partial^2 A}{\partial t^2} + \frac{k'''}{6}\frac{\partial^3 A}{\partial t^3} - \frac{i}{2k}\left(\frac{\partial^2 A}{\partial r^2} + \frac{1}{r}\frac{\partial A}{\partial r}\right) + \frac{i\omega_0 n_2}{c}|A|^2 A - \frac{\beta^{(K)}}{2}|A|^{2K-2}A \qquad (2)$$

where $k''(''') = \partial^{2(3)} k(\omega)/\partial \omega^{2(3)}$ are second- and third-order dispersion coefficients, z is the propagating distance, n_2 is the nonlinear index of refraction, $k = \omega n_0/c$ is the wave-vector amplitude and $\beta^{(K)}$ the MPI absorption coefficient. The energy gap of ≈ 5.8 eV in borosilicate glass suggests that multi-photon absorption starts with five photons accounting for a single photon energy of 1.4 eV at $\lambda = 800$ nm. Therefore, nonlinear losses with K = 5 given by $\beta^{(5)} = 10^{-47} \text{cm}^7 \text{ W}^{-4}$ were accounted for. Note that, among material parameters used for the modelling, the nonlinear coefficient of multiphoton ionization is the most uncertain. Several theories resembling Keldysh's original proposal have been discussed in the literature [25, 26]. It was demonstrated, in particular, that the prediction of conventional Keldysh-type theories based on the assumption of the stationarity of the initial state, differ from the predictions of the above threshold ionization (ATI) electron spectra excited by high-intensity, femtosecond, light pulses. Owning to this uncertainty of the MPA coefficient, we took for the coefficient of the nonlinear losses, the value which best fitted the experimental results [27] and [28].

Since we are going to concentrate on space-time effects in this section (and will not involve multi-filamentation phenomena, which are to be considered in next sections) we study Eq. (2) for the case in which the beam has radial symmetry and assume that the input pulse is Gaussian in space and time such that:

$$A(r, \tau, 0) = A_0 \exp\left[-\frac{r^2}{w_0^2} - \frac{t^2}{\tau^2} - \frac{ikr}{2R_0}\right] \qquad (3)$$

We numerically integrate Eq. (2) with the initial beam taken as a plane wave ($R_0 = \infty$) with the waist $w_0 = 75$ µm, temporal half-width duration $\tau = 0.13$ ps and amplitude $A_0 = \sqrt{2P_0/\pi w_0^2}$, where P_0 stands for the peak power of the incident pulse. Our technique for integration is based on a split-step procedure which separates nonlinear equations into dispersive, diffractive and nonlinear parts and solves them inside each step, separately. A fast Fourier transform (FFT) to wavenumbers-frequency space was used for the dispersive step. After an inverse transform back to a space-time representation, the fourth-order Runge–Kutta procedure was used to evaluate the nonlinear step, whereas the finite difference method was used for the diffractive step.

From now on, we examine the in-space resolved temporal profiles and angular spectra of the transmitted pulses for the input peak power of the pulse, beyond the self-focusing threshold. At the chosen wavelength 800 nm PTR glass (like fused silica) possesses normal dispersion characterized by the second and third-order dispersion parameters $k'' = 1.8 \times 10^{-4} \text{ ps}^2 \text{ cm}^{-1}$ and $k''' = 2.6 \times 10^{-7} \text{ ps}^3 \text{ cm}^{-1}$. We start with moderate beam power of 10 MW at the pulse maximum, which corresponds to a power 3.5 times above the self-focusing threshold. The following three length scales characterize the pulse interaction: diffraction length $L_{DF} = kw_0^2$; dispersion length $\tau_p^2/|k''|^2$; and nonlinear length $L_{NL} = c/\omega_0 n_2 |A_0|^2$. For the above mentioned pulse and beam parameters, we get: $L_{nl} = 0.12$ cm $\ll L_{df} = 2.2$ cm $\ll L_{ds} = 90$ cm. Hence, nonlinear effects (self-focusing and nonlinear losses) as well as diffraction prevail over dispersion for the first millimeters of propagation. Nevertheless, pulse transformation in space and time after the propagation of these first millimeters, generates complex field structure and the length scales mentioned (associated with diffraction, dispersion and nonlinearity) become comparable. Consequently, it would be a mistake to treat space and time phenomena, as well as linear and nonlinear regimes, separately, because this will fail to capture X-waves [16].

We shall display both three-dimensional surfaces of the intensity $I(r, z, t) = |A(r, z, t)|^2$ versus t and r for various propagation distances of z in the frame travelling with the group velocity and corresponding gray scale images of the far-field (power) spectra $|A(k_\perp, \omega)|^2$ versus wavelength and transverse wavenumber $k_\perp$. $A(k_\perp, \omega)$ stands for the Fourier (time-frequency space) and Bessel (the lateral coordinate-wavenumber) transforms of the field amplitude.

The left column of the Fig. 6.4 shows the progression in the pulse propagation dynamics with increasing distance z. As can be seen from Fig. 6.4(a) initially the gaussian (intime and inspace) pulse is affected mainly by nonlinearities and diffraction. Self-focusing and self-phase modulation due to third-order nonlinearity is strongest at the center of the pulse. This narrows the beam diameter at maximum intensity and widens the spectral distribution, as a result of the combined action of the third-order nonlinearity and diffraction profiles of the beam and pulse after propagation of the first millimeters. This, in turn, increases the influence of the dispersion as well as nonlinear losses. Specifically, group velocity dispersion spatially separates the different colors in the pulse. Hence, the pulse becomes prolonged (in time) and, finally, temporal pulse splitting occurs. Third-order dispersion, as it is known, breaks the time symmetry of the modeling non-linear Schrödinger equation, deforming pulses and leading to the emission of continuous radiation. In our case, third-order dispersion causes asymmetry between the leading and trailing sub-pulses with the leading peak being more intense and considerably shorter Figs. 6.4(b) and (c). In the normal dispersion regime, the leading sub-pulse is red shifted, therefore broadening of the red-shifted on-axis spectrum is clearly seen on the corresponding images (right column of Fig. 6.4).

The right column of Fig. 6.4 shows the corresponding Fourier–Bessel transform images. The dominant effect in Fig. 6.4(a) is spectral broadening, whereas starting from 6.4(b) the development of the modulation of frequencies due to pulse splitting in time as well as the appearance of the lateral rings of the radiation, should be noted.

Note that in Fig. 6.4 we have plotted only part of the r and t space region over which the computation was performed (hence the actual boundary is considerable larger than the boundary shown in the figure). This is true for all of the figures presented in this section.

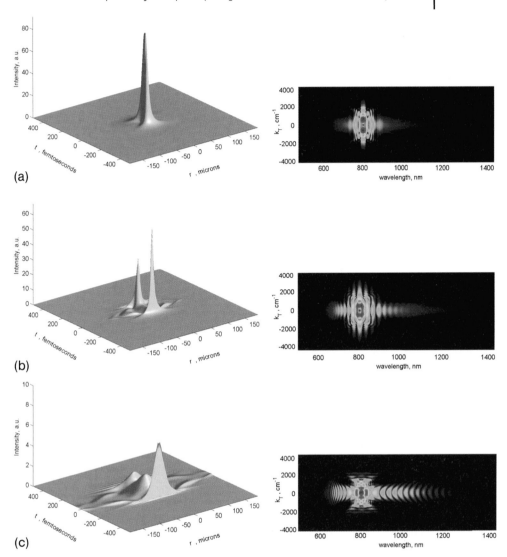

Fig. 6.4 Left: calculated space-time intensity profiles of a pulse in the moving coordinate frame at propagation distances $z = 10$ mm (a); 20 mm (b); 40 mm (c). Right: corresponding images of the angular spectra. Pump power $P = 3.5 \times P_{cr}$, beam waist $w_0 = 75$ µm at the entrance; X-wave shapes should be noted both in the physical as well as the transformed domain.

Analogous modeling for a beam power which is twice as large as above $(7\ P_{cr})$ shows results similar to Fig. 6.4. It should be noted, nevertheless that, due to shorter nonlinear length, $L_{nl} = 0.6$ mm, combined action of the self-focusing, diffraction and dispersion prolongs the pulse in time at the distance of 1 cm. The temporal pulse spitting appears after propagating 2 cm. However, relative amplitudes of the sub-pulses do not possess such a clear asymmetry as in the previous cases, the central pulse becoming more prominent. Finally, as the power is increased, the nonlinearities cause self-focusing of the pulse and flatten the beam profile (due to multi-photon absorption). Both effects are strongest near the peak of the pulse and develop annular rings around the central peak. By comparing our results with the experimental one [24] we find quite a good qualitative agreement. This supports the view that propagation of the femtosecond pulses in self-focusing Kerr media with nonlinear losses, is indeed dominated by an interplay between dispersion and nonlinearity, in which the nonlinear responses of plasma do not play a crucial role. Indeed, nonlinear space-time coupling dominates the propagation dynamics and the X nature of the pulse can be seen both in the space-time profiles and the far-field spectra of the pulse at propagation.

We have described above, the (3+1)D propagation of short intense optical pulses in glass in the presence of axial symmetry. We have shown that, above the critical self-focusing power, both pulse and beam splitting into sub-pulses occurs. Characteristic of this case, a beam structure in the form of a narrow filament and annular rings around the central peak, should be noted. Let us conclude, that the apparent filamentation regime predicted in [8, 9] by a simplified model which appeared to be capable only of describing the asymptotic (cw) case, is verified here in the transient regime. The numerics indicate the occurrence of a filament formation for pump intensities above critical, which is supported be conical radiation in the form of Bessel-type rings. The number of rings increases as the pump intensity increases and the rings propagate at larger angles.

6.4
On the Physics of Self-channeling: Beam Reconstruction from Conial Waves

Wavepackets focused by using spatial masks undergo nontrivial dynamics associated primarily with complex diffraction and interference effects. In nonlinear media with nonlinear losses such a focusing regime brings the system into the, still not well understood, realm of nonlinear propagation. In this section we will formulate physical picture of the self-channelling in focusing Kerr media with nonlinear losses. To this end we will adopt the *cw* model, describing the spatial dynamics without accounting for the (plasma induced) de-focusing effect which, nevertheless, may be an essential element for the establishment of a self-guiding regime. In this context, the only nonlinear terms that we took in the model are the self-focusing and the MPA ones. Recent experiments [29, 30] indicated that filaments created in femtosecond high-power pulses propagating in air are surprisingly robust. For example in [29] the robustness of the beam when interacting with microscopic water droplets, was demonstrated. Below, we provide closer insight into the interplay between the filaments core and periphery of the beam.

Consider the focusing nonlinear Schrödinger equation (NLS) in the presence of axial symmetry and nonlinear dissipation due to the multi-photon ionization:

$$\frac{\partial A}{\partial z} = \frac{i}{2k}\left(\frac{\partial^2 A}{\partial r^2} + \frac{1}{r}\frac{\partial A}{\partial r}\right) + i\sigma|A|^2 A - \frac{\beta^{(K)}}{2}|A|^{2K-2}A \qquad (4)$$

where $\sigma = \omega_0 n_2/c$ is the nonlinear coupling coefficient. Let us not pay attention (despite it not being strictly correct mathematically) to the asymptotic of the field, and aim instead for the possibility of narrow filament formation. In other words, let us analyze, whether the stationary pattern of the $A(r)$ field survives near the axis $r = 0$. The equation for the amplitude of the stationary solution, taken in the form $A(r)\exp(-i\alpha z)$, reads:

$$\frac{d^2 A}{dr^2} + \frac{1}{r}A + aA + \sigma A^3 + i\beta^{(K)}A^{2K-1} = 0 \qquad (5)$$

Let us linearize Eq. (2) in the vicinity of $r = 0$, i.e., in the region where $\sqrt{ar} \ll 1$. To second order in r, Eq. (5) can be approximated as follows:

$$\frac{d^2 A}{dr^2} + \frac{1}{r}A + aA + \sigma A_0^2 A + i\beta^{(K)}A_0^{2K-2}A = 0 \qquad (6)$$

In the absence of the nonlinear terms, the solution of the Eq. (6) can be written in the form of the Bessel beam:

$$A(r) = A_0 J(\sqrt{a}r), \qquad (7)$$

Note that, physically, the coefficient a represents the transverse projection of the wave vector. In the presence of the nonlinear terms in Eq. (6) this coefficient becomes complex: $\tilde{a} = a + \sigma A_0^2 + i\beta^{(K)}A^{2K-2}$. It should be be stressed, that nonlinear Bessel beams with real (positive $\tilde{a}$) tend to be focused. This means that the amplitude will be growing at the position of interest ($r = 0$). On the other hand, solution of Eq. (6) with complex $\tilde{a}$ can be written in the form of the linear combination of modified Bessel functions with different behavior both at ($r = 0$) and in the asymptotic and offers a possible physical mechanism of the filamentation without self-guiding. Specifically, a self-focused Bessel beam, affected by nonlinear losses in the main lobe, tends to arrange its transverse patterns in order to store (keep) energy in the periphery, in the case when nonlinear losses in the central part prevail over the focusing. This peripheral energy, on the other hand, is provided (due to focusing) back to the main lobe in the opposite case, when nonlinear focusing prevails over the nonlinear losses in the central lobe.

The arrest of the collapse and stationarity of the central lobe in the presence of nonlinear losses, as claimed above, suggests that a crucial role is played by conical waves (in the form of modified Bessel functions). However, nonlinear losses play a fundamental role in determining beam evolution and spontaneously induce qualitative changes in the geometry of the process. This fact opens up a new perspective in technological applications for the femtosecond pulses and creates new tasks for the physics of nonlinear waves.

In order to deepen our knowledge of the physics of beam filamentation we follow these considerations by two instructive cases of beam self-focusing. First, we will consider more precisely the energy balance mechanism during beam filamentation. Second, the effects of the two different masks (hole and diaphragm) on the beam diffraction, will be considered.

In Fig. 6.5 we plot the beam diameter as well as the calculated fractional power losses $(dP/P)/dz$ versus z, obtained by numerical integration of the modified NLSE (Occurrence of two different regimes is evident from this picture: for $z < 10$ mm, $(dP/P)/dz$ increases due to self-focusing, whereas the beam diameter decreases down to 20 µm. However, a sharp drop in $(dP/P)/dz$ occurs at the position where the filament becomes stabilized. This indicates that the beam has undertaken an important transformation which has allowed it to exhibit a relevant localization while minimizing the nonlinear losses (MPA) by transforming toward a Bessel-like profile.

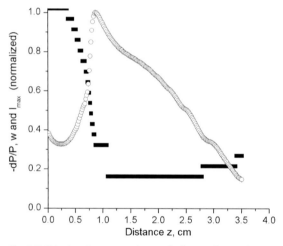

Fig. 6.5 Calculated net power losses (hollow circles) and beam diameter (black squares) versus z. $P = 10\ P_{cr}$.

Owning to this natural tendency of the wavepacket (to transform into a robust shape), it is instructive, therefore, to find the shape of the waveform in the presence of the MPA. For this purpose we have investigated nonlinear losses given by the transformation (8) and show the results in Fig. 6.6, where the MPA-induced $(dP/P)/dz$ losses are plotted versus the parameter b which describes the transformation of a beam from Gaussian ($b = 0$) to Bessel–Gauss to Bessel profile ($b = b_{max}$), according to the expression:

$$A(x,y,0) = A_0 \exp\left[-\frac{r}{r_0^2}\left(1 - \frac{b_{max}}{b}\right)\right] J_0(br), \tag{8}$$

where $r = \sqrt{x^2 + y^2}$ and $J_0(x)$ is the zero-order Bessel function and the same peak intensity is ensured for all b values.

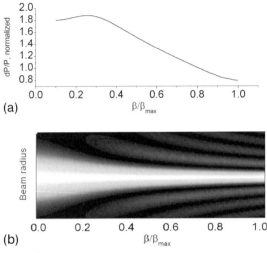

(a)

(b)

Fig. 6.6 (a) Objective loss function dP/P as the beam transforms from Gaussian to Bessel (net power losses vs beam shape parameter) for $P_0 = 50\ P_{cr}$. (b) beam shape transformation from Gaussian to Bessel according Eq. (8).

As expected, the nonlinear losses $\to 0$ in the limit $b \to b_{max}$, owing to the negligible fractional power that the Bessel beam contains in the intense spike. If the robustness of the Bessel profile under Kerr-induced spatial phase modulation is accounted for, one could foresee that indeed a Bessel-like, infinite-power exact solution of the Eq. (2) might exist, which behaves as a strong attractor for the whole nonlinear beam transformation. This test supports the physical interpretation of the filamentation as being addressed to the spontaneous transformation of a Gaussian beam into a Bessel-type one, driven by minimum (nonlinear) losses, maximum stationarity and maximum localization requirements.

The second question, which is the concern of this section, looks at the possible role of *linear replenishment* in filament reconstruction. Indeed, it is well known that the self-reconstruction property is an inherent feature of conical, nondiffracting, waves, which has been experimentally demonstrated with Bessel beams in [31, 32].

Figure 6.7 proposes the calculated evolution (versus the propagation coordinate z) of the beam-fluence profile (the time-integrated intensity in a transverse dimension) of both the free propagating (a) and stopped (c) filaments in nonlinear Kerr media. The result shows that a central spike of nearly original dimensions reappears just after a few millimeters of propagation beyond the stopper. Note how, at even higher intensities, smaller diameters and longer stationarity is reached when the stopper is inserted. The results clearly outline the self-reconstruction dynamics from the conical wave. Additionally, we have investigated linear propagation and self-reconstruction of the filament in free space, under identical input conditions. Figure 6.7(b,d) shows the calculated fluence profiles for the free (d)

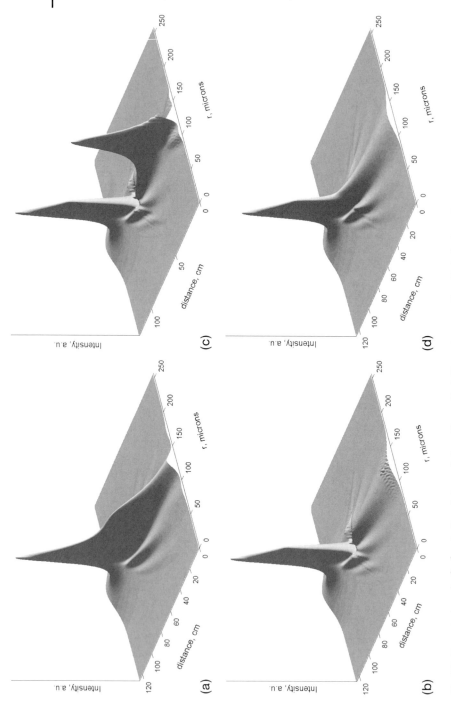

Fig. 6.7 Numerical results for free (a,d) and stopped at z = 18 mm (b,c) filament propagation. Note the boundary of water-air at z = 18 mm (b,d).

and the blocked (c) filaments. Both nonlinearity and dispersion were switched off for $z > 18$ mm. Although apparent diffraction of the central spike takes place, self-reconstruction is evident beyond the stopper even in this case.

The comparison of the free filament propagation (Fig. 6.7(b,d)) and its propagation in Kerr media (Fig. 6.7(a,c)) makes evident the presence of the two expected contributions to the filament reconstruction. One contribution is a linear, geometrical, effect, which we attribute to the conical structure of the wave. It leads to the fast (i.e. within one Rayleigh range) reconstruction of the central spot in air and to its further spreading, as expected, in the case of a finite-power linear conical wave. The other is, in contrast, a nonlinear effect, which further contributes in limiting (due to the Kerr nonlinearity) the central-spot spreading.

In conclusion, we have stressed the occurrence of two (combined) mechanisms of the filament replenishment in Kerr media with nonlinear losses. The first (linear) one arises from the conical structure of the wave, whereas the second one was addressed to the focusing of the Kerr nonlinearity.

6.5
Multi-filaments and Multi-focuses

The appearance of multiple filaments when focusing high intensity radiation, was previously thought to be associated with spontaneous noise amplification, i.e., as a natural consequence of symmetry breaking. In general, our understanding of these effects are hompered both by the complexity of the associated nonlinear model and by the limited availability of experimental results.

To study complicated behaviour characterized by multiple filamentation processes, the (3+1)D case in all three spatial dimensions has to be treated. This is a formidable task for numerical analysis. It is instructive, therefore, to make a comparison of the beam structure as obtained by modeling the modified nonlinear Schrödinger equation in the form given by Eq. (2) and in the *cw* case, as follows:

$$\frac{\partial A}{\partial z} = -\frac{i}{2k}\left(\frac{\partial^2 A}{\partial r^2} + \frac{1}{r}\frac{\partial A}{\partial r}\right) + \frac{i\omega_0 n_2}{c}|A|^2 A - \frac{\beta^{(K)}}{2}|A|^{2K-2}A \tag{9}$$

Figure 6.8 shows the results of the numerical simulations of transverse intensity images. Time-integrated fluency distributions versus lateral coordinate r and propagation distance z are plotted here. Figure 6.8(a–c) provides a comparison between the three distinct plots of intensity versus lateral coordinate r and propagation distance z, for the considered (axially symmetric) case of 130 fs pulse propagation (input power $P_{in} = 3.5\ P_{cr}$ at the pulse maximum). In Fig. 6.8(a) the intensity of the pulse at $t = 0$ in the moving coordinate frame is shown, whereas in Fig. 6.8(b) and Fig. 6.8(c) images integrated over time and calculated in the *cw* case, are depicted, respectively. Simulation results reveal that the high-power pulses are split spatially, forming Bessel-type lateral structure during propagation in all cases. (The number of cones and their angle with respect to the propagation axis increases with incident pulse energy.) Moderate (quantitative) differences in the distribution of the peripheral radiation should be noted here. On the other hand, the corresponding beam diameters (FWHM) versus the propagation length, (not shown here) clearly suggests that temporal effects play only a minor role in filament formation dynamics and can be excluded to a large extent. This statement is supported also by perfect agreement between the experimental data and the *cw* model-based numerics, as was found previously [9]. Owning to these findings and the fact that beam characteristics are of primary interest for multiple filamentation and the effects of optical breakdown, we will confine our analysis to the *cw* case throughout this section.

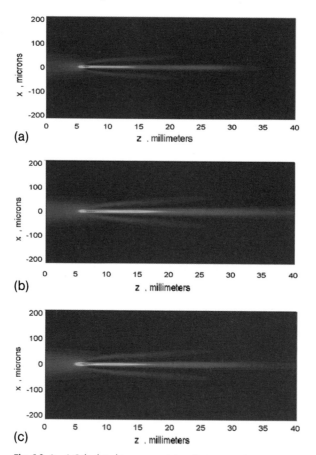

Fig. 6.8 (a–c) Calculated transverse intensity images at $t = 0$ of the moving coordinate frame (a), after integration over time (b) and the *cw* case (c); Pump power: $P = 3.5\ P_{cr}$, beam waist $w_0 = 75\ \mu m$.

6.5.1
Multiple Flamentation in Bulk Transparent Media

Above, we have considered the initial Gaussian beam profiles. On the other hand, the electromagnetic field of the ideal laser beam obeys the Helmholz equation:

$$\Delta A(\vec{r}) + k^2 A(\vec{r}) = 0 \tag{10}$$

and has an infinite set of solutions. One complete and orthogonal set of solutions is called the Hermitian-Gaussian beam, each mode being characterized by integer numbers n and m. In this, more general, case the amplitude A_0 (see Eq. (3)) of the particular Hermitian-Gaussian beam becomes:

$$A_0^{m(n)}(x, y, z) = \frac{A_{mn}}{w_0} H_m\left[\sqrt{2}\frac{x}{w_0}\right] H_n\left[\sqrt{2}\frac{y}{w_0}\right] \tag{11}$$

where $H_{m,n}(x)$ stands for the Hermite polynomial of order $m(n)$ in x.

Increasing the energy always leads to the appearance of multiple filaments. The first (historical) explanation for MF was that it is initiated by input beam noise modulation instability [33]. If it is the case, the MF pattern appears as unpredictable (different from shot to shot) in the number as well as in the location of the filaments. Recently, it was pointed out that vectorial-induced symmetry breaking can lead to MF even for cylindrically symmetric input beams [34]. Multi-filament formation was demonstrated by using saturating nonlinearity in the presence of the beam inhomogeneities or elipticity [35]. Moreover, in ensuing investigations [36, 37] it was predicted that the random nature of complete beam breakup could be regularized both by the particular focusing geometry of *cw* beams and modulation of the envelope of the beam. The possibility of organizing regular filamentation patterns in air by imposing either strong field gradients or phase distortions in the input-beam profile of an intense femtosecond laser pulse, was experimentally demonstrated in [38].

Figure 6.9 proposes a comprehensive summary of the space-time structure of the output field by plotting the intensity iso-surfaces when propagating the beam in the 4 cm sample of the PTR glass, corresponding to the incident pump power $P = 15\ P_{cr}$, supplemented with 2% noise radiation conditioned by the higher Hermite-Gaussian modes (Eq. (11)). Here we demonstrate that the effect of the multiple filament formation can be well described also in the model under consideration. Since multiple filamentation involves a complete break up of the beam cylindrical symmetry, the complex scalar envelope $A(x, y, z)$ of the beam in the nonlinear medium (including diffraction self-focusing and nonlinear losses due to MPI) evolves according to the modified nonlinear Schrödinger equation. In the frame of the adopted paraxial approximation moving at the group velocity $v_g = [\partial k(\omega)/\partial\omega]^{-1}$, the resulting equation for the field amplitude, A reads:

$$\frac{\partial A}{\partial z} = \frac{i}{2k}\left(\frac{\partial^2 A}{\partial x^2} + \frac{\partial^2 A}{\partial y^2}\right) + \frac{i\omega_0 n_2}{c}|A|^2 A - \frac{\beta^{(K)}}{2}|A|^{2K-2} A \tag{12}$$

where z is the propagating distance, n_2 is the nonlinear index of refraction and $\beta^{(K)}$ is the MPI absorption coefficient.

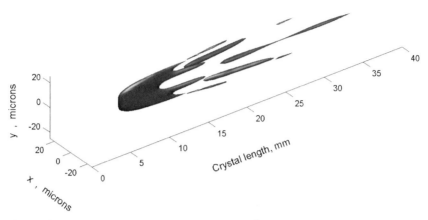

Fig. 6.9 The calculated intensity iso-surfaces at $I = 20|A_{cr}|^2$ (see text for the details). $P = 15\ P_{cr}$, $w_0 = 100\ \mu m$, $R = 2\ cm$.

For the elementary case of an optical system having ordinary astigmatism, one can simply make a separation of variables along the transverse principal axes of the system, and apply ordinary (complex-valued) Gaussian beam arguments separately along the resulting orthogonal x and y axes. Specifically, the intensity distribution of the incident radiation was taken in the form of the elliptical beam:

$$A(x, y, 0) = A_0 \exp\left[-\frac{x^2}{w_y^2} - \frac{y^2}{w_y^2} - \frac{2ikx^2}{R_x} - \frac{2iky^2}{R_y} \right]$$

(13)

with initial beam waist $w_0 = 100\ \mu m$, and amplitude $A_0 = \sqrt{2P_0/\pi w_0^2}$. The nonlinear differential equation (12) was solved numerically, following the split-step procedure described above.

As it was pointed out, a recently disclosed particular MF scenario in air [38] and water [30], in which MF starts as a nucleation of an annular ring at moderate (above the threshold for self-focusing) power, was found to be characteristic for MF in fused silica as well (Fig. 6.10). The filamentation in this case starts with occurrence of the pair of filaments along the major axis of the ellipse, in addition to the central one. By increasing the incident beam power the MF picture "rotates" by 90 degrees and filamentation along the minor axis is observable. Note that the effect of the rotation of filaments was not captured in [30] by the simplified model based on the NLS equation with saturating nonlinearity. On the contrary, the model used below, for the description of the MF, reveals the main features of the beam ellipticity induced MF scenario demonstrated previously for water [30], as well as in this work for fused silica, i.e., new filaments in the case under consideration emerge through the nucleation of annular rings, with a characteristic rotation of the nuclear structure by 90 degrees. In the images of Fig. 6.11 we show the

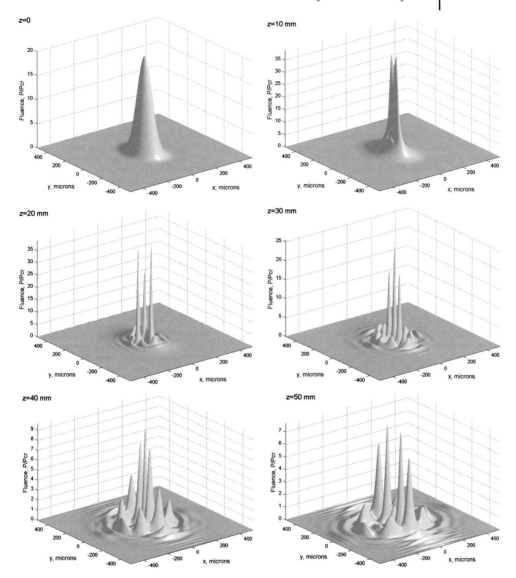

Fig. 6.10 Calculated 3D views of the intensity patterns at z = 0; 10; 20; 30; 40; 50 mm for the following incident beam parameters: intensity P = 15 P_{cr}. w_0 = 100 μ, R = 2 cm. Elipticity of the incident beam was taken to be as large as 10% and the major axis of the ellipse was taken along the x axis.

filamentation patterns of the beam at the propagation distance $z = 1$ cm by increasing beam power from 2 P_{cr} to 12 P_{cr}. The beam elipticity parameters were set as following: $q_x = 1$, $q_y = 1.1$. The lighter areas (as in all previous sections) correspond in the figure to higher intensities so that the light regions highlight the points at which the beam undergoes filamentation.

We can see, filamentation starts with appearance of the surrounding ring at an incident beam power as high as 4 P_{cr}. A definite structure consisting of a strong central filament and a pair of weaker filaments along the major axis appears after the incident beam power reaches 10 P_{cr}. An additional pair of weaker filaments appears along the minor axis when increasing beam power. At the incident input power 15 P_{cr} the picture appears to be rotated at 90 degrees with respect to that characteristic for 10 P_{cr} (in good agreement with that measured in [30]). Note that the beam evolves further (by increasing the propagation distance) then cycli-

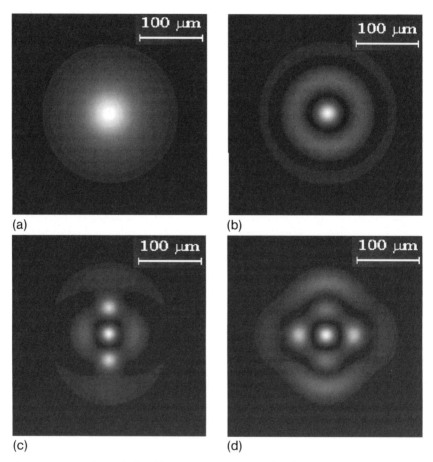

Fig. 6.11 Images of the calculated beam transverse patterns for elliptical input beams at $z = 1$ cm for different incident beam powers: 2 P_{cr} (a), 4 P_{cr} (b), 8 P_{cr} (c) and 12 P_{cr} (d), demonstrating MF scenario by nucleation on the surrounding ring.

cally reforms in the same manner (from more pronounced vertical to horizontal structures), creating additional filaments on the x and y coordinates (shown in Fig. 6.11).

On the other hand, the increasing intensity will destroy this regular structure of the multiple filament formation when inhomogeneity of the material properties or the irregularities of the incident beam are taken into account.

6.5.2
Capillary Waveguide from Femtosecond Filamentation

The influence of irreversible changes of the refractive index to the formation of multiple filaments in fused silica will be studied in depth in future work. Here we restrict ourselves to demonstrating to what extent the induced self-guiding channels influence the beam propagation and threshold of the continuum generation.

First, we have numerically computed the propagation of the axially symmetric beam at the high beam power (15 P_{cr}). In this case, a light filament, guided by the surrounding ring of diameter about 20 µm is formed and propagates over one centimeter. Beyond this distance, diffraction prevails and this marks the end of the filament. Figure 6.12 shows the field intensity distribution corresponding to this case in the bulk fused silica. This function, i.e., (normalized) distribution of the field intensity, was taken as representing permanent changes of the refractive index in the following equation:

$$\frac{\partial A}{\partial z} = \frac{i}{2k}\left(\frac{\partial^2 A}{\partial x^2} + \frac{\partial^2 A}{\partial y^2}\right) + \frac{i_0 \Delta n}{c} A - \frac{\beta^{(K)}}{2}|A|^{2K-2}A \qquad (14)$$

where $\Delta n = \delta n + n_2|A|^2$ stands for both the permanent (δn) and intensity dependent $(n_2|A|^2)$ changes in the index of refraction.

The effect of the induced permanent refractive index changes on the beam filamentation at moderate values of the incident beam power (0.9 P_{cr}, 2 P_{cr} and 4 P_{cr}) is demonstrated in Fig. 6.13. Here the calculated beam transverse intensity images for two values of the constant a are shown. The images, displayed in (a), (b), (c) in Fig. 6.13 were calculated without taking into account permanent changes of the refractive index ($a = 0$), whereas the images (d), (e), (f) displays the effects under consideration. Here $a = 0.2$, therefore, taking into account the ratio of the intensities $I_h(x, y, z)$ and $I(x, y, z)$, one could determine that the permanent refractive index becomes of the same order as the transient one (induced due to the nonlinear refractive index n_2).

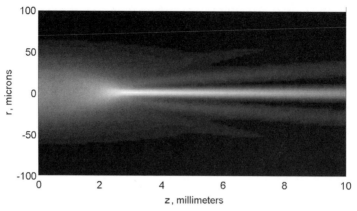

Fig. 6.12 The refractive index profile used for modeling the irreversible changes in the bulk. The profile was taken according to the beam transverse intensity for incident beam power 15 times above critical.

It is seen from Fig. 6.13(a) that diffraction prevails over self-focusing for an incident beam power $P < P_{cr}$ in the absence of the induced refractive index changes. On the other hand, at the same incident beam power (below the critical one) self-focusing appears at a distance of propagation of one centimeter, when refractive index changes are accounted for Fig. 6.13(d). In fact, the presence of permanent changes in the refractive index, result in the lowering of the critical value of the beam power for self-focusing.

When increasing the beam power above its critical value, focusing appears at the same length of propagation both for $a = 0$ (Fig. 6.13(b) and (c)) and $a = 0.2$ (Fig. 6.13(d) and (f)). Here, again, effect of the lowering of the critical beam power by induced permanent refractive index changes, should be noted. Actually, filaments guided by the surrounding rings is clearly seen both for beam power two and four times over the critical one (Fig. 6.13(d) and (f)), when refractive index changes are considered. Note that the position of the surrounding rings is not connected with those in the refractive index profile (shown in Fig. 6.12) and becomes closer to the filament core when the beam power is increased. In comparison, ignoring induced refractive index changes in the bulk, the same feature (filament supported by surrounding ring after propagating one centimeter) appears for a beam power as high as 4 P_{cr}.

It should be mentioned once again that our simulation was done for the *cw* case. Complete simulation which also includes time-related effects would qualitatively give the same results (not shown here). The main difference is that, for the time-dependent radiation the self-focusing threshold is higher than for the *cw* case. Therefore, the intensity distributions shown in the figures would be shifted on the power scale and appear at slightly higher power levels for the case of pulsed radiation.

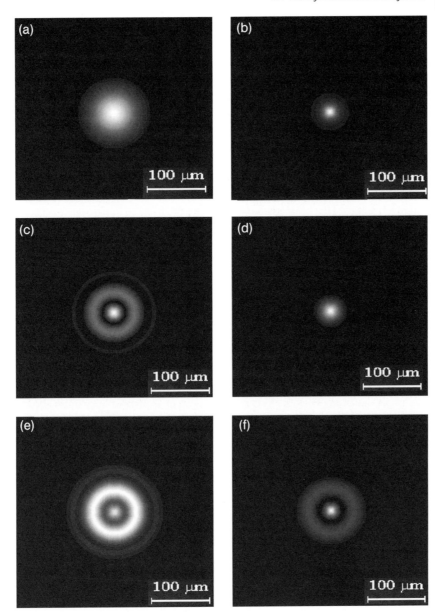

Fig. 6.13 Calculated beam transverse intensity profiles for beam power 0.9 P_{cr}, 2 P_{cr} and 4 P_{cr}: without (a–c) and with (d–f) taking into account the permanent changes in the refractive index profile shown in Fig. 6.12.

6.6
Filamentation Induced by Conical Wavepacket

Having in mind the Gauss–Bessel (GB) nature of filaments, as described above, the possibility of directly launching a powerful Bessel beam into the material and thus circumventing the internal transformation from Gaussian towards a GB beam, is intriguing. The internal transformation is governed by the nonlinear response of the material, which may limit the obtainable peak intensity of the GB beam. An external GB beam (e.g., generated using an axicon) may have sufficient power to induce and sustain extensive damage along its entire propagation path. Besides providing an additional proof in support of the "filamentation without self-channeling" model, the possibility of fabricating extended lines almost instantaneously is beneficial for laser micro-fabrication. Here we study propagation of a GB beam in transparent media. As a suitable model system we address, theoretically, a GB beam delivering powerful femtosecond laser pulses with central wavelength of 800 nm into a bulk borosilicate glass. We find that, at pulse powers well exceeding the self-focusing threshold of a Gaussian beam, a GB beam can propagate over tens of micrometers leaving a linear pattern of discrete, equidistant damage spots representing cyclical refocusing of its central part, where the absorptive losses at each focus cause the damage, but are quickly replenished due to the self-healing property of the conical wave. The theoretical analysis based on the "filamentation without self-channeling" model is presented, which yields the spot-to-spot distance of 8.6 μm, close to the value of 9 μm found experimentally.

The GB beam used as the input to the simulations is a solution of the free-space Helmholz equation [39] with electric field amplitude expressed in cylindrical coordinates as $A(r, \phi, z) \propto J_l(k_\perp r)\exp(ik_z z)\exp(il\phi)$, where J_l is the lth-order Bessel function of the first kind, $k_\perp$ and k_z are the radial and longitudinal components of the wavevector k, respectively, and l is an integer. The beam propagates along the z-axis; at $z = 0$ its central part is assumed to have a diameter $w \simeq 2$ μm and its maximum power (at the temporal peak) exceeds the critical self-focusing power by twelve times. The evolution of the complex scalar envelope of the wavepacket, $A(r, z, t)$, was deduced from the nonlinear Schrödinger equation (2) assuming cylindrical symmetry. Since the effects arising due to the conical nature of the Bessel beam were the main focus of this work, possible de-focusing due to the plasma generated by the multi-photon absorption was neglected.

The beam propagation obtained with the above assumptions is illustrated by the data in Fig. 6.14. The spatio-temporal intensity maps shown in Fig. 6.14 indicate that the central part of the beam propagates without visible loss of intensity, simultaneously exhibiting oscillations with period of 8.6 μm (twelve cycles can be seen in the 100 μm z-axis range). These oscillations can be understood as periodic focusing of the central part during propagation. This behavior is illustrated by the wavepacket structures in Fig. 6.14 corresponding to various stages of the focusing within one period at $z = 50$, and $z = 58$ μm. Similar periodicity, albeit with smaller period, was observed earlier for low intensity beams, and explained by the interference between the initial GB beam and the Gaussian constituent generated during the four-wave mixing

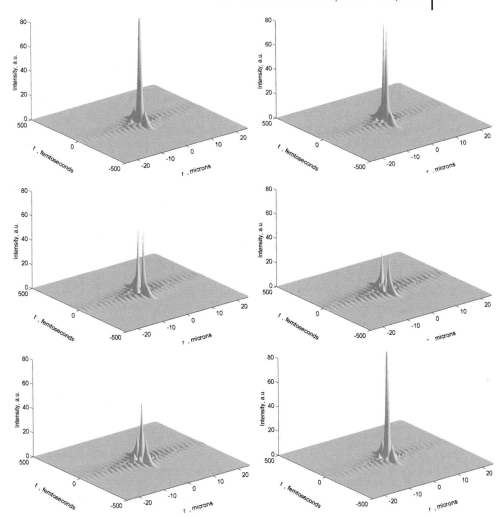

Fig. 6.14 Calculated space-time resolved energy fluence profiles of the wave-packets at distances between $z = 50$ μm and $z = 58.5$ μm, representing the refocusing cycle of the GB beam in glass.

process [40, 41]. However, we have verified that neither a Gaussian, nor a GB beam would exhibit such behavior in the present case, when the intensity strongly exceeds the self-focusing threshold of the Gaussian beam, and when the mechanisms which arrest the beam's collapse (e.g., dispersion and nonlinear losses) are active. It is helpful to note here that, at this intensity, for a Gaussian beam having a waist of 2 μm, the self-focusing length is less than 3.5 μm.

The spatial distributions of the intensity and fluency shown in Fig. 6.15 (a), (b), (c) indicate that a high power density, possibly sufficient to induce significant optical damage, might be achievable at the intensity maxima. At these positions optical losses resulting from scattering or absorption by the damaged regions should not inhibit the beam propagation owing to its self-regeneration effect. This effect can be qualitatively explained using the spectrum shown in Fig. 6.15(d). By taking into account the linear on-axis interference between the input beam $k_\perp$ and the secondary Gaussian and Bessel beams, generated during the nonlinear interaction at $k_{\perp 0} = 0$ and $k_{\perp 1} = 1.5 \times k_\perp$ respectively, oscillatory terms $\sim(k_{zj} - k_z)z$ can be obtained, where k_{zj}, $(j = 0, 1)$ are z-components of the wave vectors of secondary beams. For the distances between interference maxima the values 10.5 μm and 12.8 μm can be inferred.

The above evaluation hints that the expected distance between the damage points falls into the interval between 3.5 μm (due to the arrest of the collapse) and 10.5 μm (due to the influence of the nonlinear change in the refractive index on the interference).

6.7
Conclusion

We have described the propagation of short optical pulses in transparent media in the normal dispersion regime for intensities above the self-focusing threshold. We have shown that, above the critical *cw* self-focusing power, the onset of pulse splitting into sub-pulses, separated in time, occurs. At higher power, the details of the pulse splitting are more complex, due to multiple filamentation of the beam. Both pulse splitting and multiple filamentation can be described by the theoretical model which does not include the de-focusing effect due to the multi-photon ionization produced plasma. The beam collapse is prevent in this case by nonlinear losses, transforming the beam into the complex structure of conical waves, containing the "cold" part of the wave (reservoir), in which an enormous amount of energy may be stored. Such a conical wave generally travels along locked to a number of filaments, continually refueled by the energy exchange with the periphery. It was demonstrated that such a model, with input beam ellipticity, can lead to MF. Finally, we demonstrate that filamentation from a conical wavepacket (Bessel beam) appears as lines consisting of discrete, equidistant damage spots, extending over hundreds of micrometers. These findings are explained by self-regeneration of Gauss–Bessel beams during propagation and are potentially applicable in the laser micro-fabrication of transparent materials.

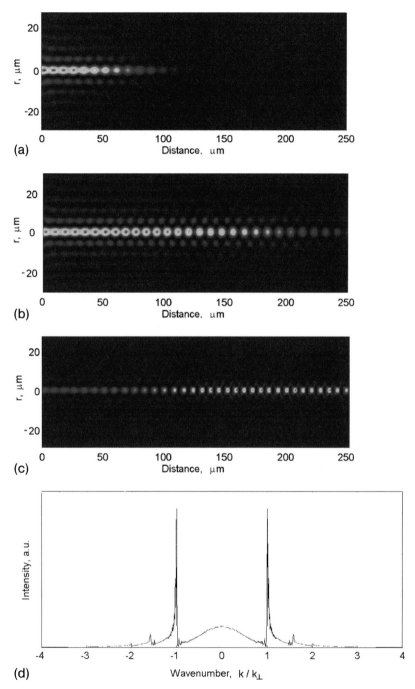

Fig. 6.15 Images of the discrete fluence treks calculated for the GB beam propagating in glass for different "Gaussian" waists of GB beam: $w_G = 10\,w_B$ (a); $w_G = 25\,w_B$ (b); and $w_G = 50\,w_B$ (c), and the angular spectrum calculated at 100 μm of the propagation distance in glass.

References

1 A. Braun, G. Korn, X. Liu, D. Du, J. Squier, and G. Mourou, Opt. Lett. **20**, 73 (1995).

2 A. Brodeur, C.Y. Chien, F.A. Ilkov, S.L. Chin, O.G. Kosareva, and V.P. Kandidov, Opt. Lett. **22**, 304 (1997).

3 H.R. Lange, G. Grillon, J.-F. Ripoche, M. Franco, B. Lamouroux, B. Prade, A. Mysyrowicz, E.T.J. Nibbering, and A. Chiron, Opt. Lett. **23**, 120 (1998).

4 L. Berge and A. Couairon, Phys. Rev. Lett. **86**, 1003 (2001).

5 M. Mlejnek, E.M. Wright, and J.V. Moloney, Opt. Lett. **23**, 382 (1998).

6 S. Tzortzakis, L. Sudrie, M. Franco, B. Prade, A. Mysyrowicz, A. Couairon, and L. Berge, Phys. Rev. Lett. **87**, 213902 (2001).

7 L. Sudrie, A. Couairon, M. Franco, B. Lamouroux, B. Prade, S. Tzortzakis, and A. Mysyrowicz, Phys. Rev. Lett. **89**, 186601 (2002).

8 V. Sirutkaitis, E. Gaižauskas, V. Kudriasov, M. Barkauskas, R. Grigonis, V. Vaicaitis, and A. Piskarskas, SPIE Proc. **4932**, 346 (2003).

9 A. Dubietis, E. Gaižauskas, G. Tamosauskas, and P. Di Trapani, Phys. Rev. Lett. **92**, 253903 (2004).

10 W. Liu, O. Kosareva, I.S. Golubtsov, A. Iwasaki, A. Becker, V.P. Kandidov, and S. L. Chin, Appl. Phys. B **76**, 215 (2003).

11 W. Liu, S. Petit, A. Becker, N. Akozbek, C.M. Bowden, and S.L. Chin, Opt. Commun. **202**, 189 (2002).

12 H. Snajalg, M. Rtsep, and P. Saari, Opt. Lett. **22**, 310 (1997).

13 P. Saari and K. Reivelt, Phys. Rev. Lett. **79**, 4135 (1997).

14 P. Saari and K. Reivelt, Phys. Rev. **69E** (2004).

15 M.A. Porras, G.Valiulis, and P. Di Trapani, Phys. Rev. **68E** (2003).

16 a. S.T.C. Conti, P. Di Trapani, G. Valiulis, A. Piskarskas, O. Jedrkiewicz, and J. Trull, Phys. Rev. Lett. **90** (2003).

17 J.K. Ranka, R. Schirmer, and A.L. Gaeta, Phys. Rev. Lett. **77**, 3783 (1996).

18 S.A. Diddams, H.K. Eaton, A.A. Zozul.ya, and T.S. Clement, Opt. Lett. **23**, 379 (1998).

19 J.K. Ranka and A.L. Gaeta, Opt. Lett. **23**, 534 (1998).

20 H. Ward and L. Berge, Phys. Rev. Lett. **90**, 053901 (2003).

21 V.E. Zakharov, N.E. Kosmatov, and V.F. Shvets, Sov. Phys. JETP Lett. **49**, 432 (1989).

22 N.E. Kosmatov, V.F. Shvets, and V.E. Zakharov, Physica D **52**, 16 (1991).

23 M. Trippenbach and Y.B. Band, Physical Review A **56**, 4242 (1997).

24 A. Dubietis, G. Tamosauskas, I. Diomin, and A. Varanavicius, Opt. Lett. **28**, 1269 (2003).

25 F. Faisal, J. Phys. B **6**, L312 (1973).

26 H.R. Reiss, Phys. Rev. A **22**, 1786 (1980).

27 S. Juodkazis, private communication (2003).

28 V. Kudriasov, E. Gaižauskas, and V. Sirutkaitis, JOSA B **22** (2005).

29 M. Kolesik and J.V. Moloney, Opt. Lett. **29**, 590 (2004).

30 A. Dubietis, G. Tamosauskas, G. Fibich, and B. Ilan, Opt. Lett. **29**, 1126 (2004).

31 S. Juodkazis, Optics Communications **162**, 261 (2003).

32 R. Butkus, R. Gadonas, J. Janusonis, A. Piskarskas, K. Regelskis, V. Smilgevicius, and A. Stabinis, Opt. Commun. **206**, 201 (2002).

33 V.I. Bespalov and V.I. Talanov, JETP. Lett. **3**, 307 (1966).

34 G. Fibich and B. Ilan, Opt. Lett. **26**, 840 (2001).

35 G. Fibich and B. Ilan, Physica D **157**, 112 (2001).

36 G. Fibich and B. Ilan, Phys. Rev. Lett. **89**, 013901 (2002).

37 G. Fibich and B. Ilan, Phys. Rev. E **67**, 036622 (2003).

38 G. Mechain, A. Couairon, M. Franco, B. Prade, and A. Mysyrovicz, Phys. Rev. Lett. **93** (2004).

39 J. Durnin, J. Miceli, and J. H. Eberly, Phys. Rev. Lett. **58**, 1499 (1987).

40 V. Jarutis, R. Paškauskas, and A. Stabinis, Opt. Commun. **184**, 105 (2000).

41 R. Gadonas, V. Jarutis, R. Paškauskas, V. Smilgevicius, A. Stabinis, and V. Vaicaitis, Opt. Commun. **196**, 309 (2000).

7
Photophysics and Photochemistry of Ultrafast Laser Materials Processing

Richard F. Haglund, Jr.

Abstract

As ultrashort-pulse lasers proliferate across the spectrum from extreme ultraviolet to mid-infrared, laser micro-fabrication is entering a new era. It is now possible to select a laser intensity, fluence, wavelength and total photon dose most appropriate to the materials and processing protocols in micro-fabrication, rather than the converse. Moreover, because the pulse durations of these lasers are shorter than typical material relaxation times, the laser–materials interaction, rather than material thermal properties, generally determines the outcome of the laser fabrication process. Thus, for example, wide-bandgap inorganic materials can be processed by multi-photon electronic excitations, while organic materials that are sensitive to photochemistry induced by electronic excitation, can be processed instead by vibrational excitation, while remaining in the electronic ground state. This chapter explores the fundamental photophysics and photochemistry of materials modification using laser pulses whose duration is short compared to relevant material relaxation times. Using illustrative examples drawn from current literature, we show how multi-photon and multi-phonon excitation, applicable in the electronic and vibrational regimes respectively, access distinctive pathways to micro- and nanostructuring and to materials modification by solid-state chemistry. Recent instrumental developments – including femtosecond solid-state lasers, free-electron lasers with pulse-repetition frequencies in the MHz range, and novel optical patterning and masking techniques–portend a greatly expanded future for ultrafast micro- and nanofabrication of materials.

7.1
Introduction and Motivation

The continuing development of high average-power femtosecond amplified near-infrared lasers [1], and of broadly tunable picosecond and sub-picosecond free-electron lasers [2] in the mid-infrared (2–20 μm) has opened up a hitherto unexplored field of nonequilibrium materials processing. This new materials process-

3D Laser Microfabrication. Principles and Applications.
Edited by H. Misawa and S. Juodkazis
Copyright © 2006 WILEY-VCH Verlag GmbH & Co. KGaA, Weinheim
ISBN: 3-527-31055-X

ing regime is made possible by three salient characteristics of these lasers: (1) the high local spatial and temporal electronic and vibrational excitation densities created by the absorption of ultrashort light pulses; (2) laser pulse durations short compared to relevant relaxation processes, so that the return to thermal and mechanical equilibrium occurs only after the deposition of laser energy; and (3) the high probability of laser-induced nonlinear processes, such as multi-photon absorption, that open new kinetic and dynamic channels to states which would not otherwise be accessible through thermal equilibrium pathways.

In this chapter, we highlight the photophysical, photochemical and photomechanical effects associated with materials modification and processing by ultrafast laser sources. Ultrafast is defined here not in terms of a fixed time interval (e.g., 100 fs), but rather in terms of a relationship to the relevant relaxation processes. Thus "ultrafast" electronic excitations means fast with respect to electron-lattice equilibration times, typically of the order of a few picoseconds; laser ablation and laser-induced melting are two examples of this process. "Ultrafast" for vibrational excitation, on the other hand, means fast with respect to anharmonic to harmonic vibrational mode coupling times, and may be somewhat longer; nonequilibrium molecular processes and charge-transfer processes during laser ablation furnish examples of this type. In all of these cases, it is necessary to take into account not only the initial interaction of the laser light with the solid, but also to understand how the properties of the material are modified at the atomic scale by the photophysical and photochemical interactions following light absorption. In some cases, especially in insulators, these interactions may involve the creation and decay of optically or chemically active defects, such as color centers; in other cases, the relaxation processes may proceed through nonthermal channels.

Classic papers on the photophysics and photochemistry of ultrafast laser-materials interactions [3, 4, 5] have already showed that this régime of laser-materials interactions would open new doors to micromachining; many of the newest developments are described in recent special issues of *Applied Physics A* [6, 7]. A few examples of progress in structuring, patterning and alteration of materials characteristics are chosen to be illustrative, rather than comprehensive. In conclusion, however, we also consider how developments in new ultrafast laser sources may affect opportunities for light-induced materials modification.

7.2
Ultrafast Laser Materials Interactions: Electronic Excitation

Materials processing with femtosecond Ti:sapphire lasers, begins with single-photon or multi-photon electronic excitations that relax by a series of complex energy-transfer processes, first among electrons in the conduction band and later with the lattice atoms. Materials processing may involve varying combinations of the following outcomes of these processes: nuclear motion (displacement, amorphization or recrystallization, ablation); changes in local electronic structure; changes in composition induced either by ejection of a component of the material or by adsorption or binding of exogenous chemistry. In this section, we consider the microscale physics of metals, semiconductors and insulators irradiated by ultrafast laser pulses in the ultraviolet to the near-infrared region of the spectrum.

A necessary (though not sufficient) condition to take advantage of the unique characteristics of ultrafast laser excitation is that the absorbed energy should be localized in the laser focal volume on a timescale short compared with to thermal diffusion times; otherwise, energy will dissipate out of the absorption zone before it is able to begin moving along the desired configuration coordinate. These conditions are straightforward, and easily related to materials properties; they are

$$\tau_p \ll \tau_{thermal} \approx L_p^2 / D_{thermal}, \quad \tau_p \lesssim \tau_s \approx L_p / C_s \tag{1}$$

where τ_p is the laser pulse duration, $\tau_{thermal}$ and τ_s are, respectively, the thermal and stress confinement times. L_p is the optical penetration depth, $D_{thermal}$ is the characteristic diffusion constant, and C_s is the speed of sound in the material. Since sound speeds are of the order 10^3 m s^{-1} in solid materials, and with $D_{thermal}$ ranging from 0.1 to 10 cm^2 s^{-1}, Eq. (1) dictates that pulse durations of 100 ps or less will be both thermally and mechanically confined even in metals. For nonmetals, on the other hand, penetration depths are much greater (of order 1–10 μm), and stress confinement may not be guaranteed even when thermal confinement is. Clearly these constraints on confinement depend critically on the strength of the electron-lattice coupling, as highlighted in Fig. 7.1.

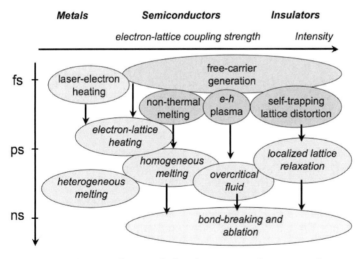

Fig. 7.1 Chart showing schematically the relevant time and intensity scales for laser interactions with metals, semiconductors and insulators, indicating the duration of initial excitations and of various relaxation processes. Adapted from Ref. [20].

A key difference between nanosecond and femtosecond laser processing, is that the former tends to be governed by fluence, while in the latter, the rates for various processes are set by the intensity I and the multi-photon cross-sections $\sigma_{(k)}$,

$$\frac{dN^*}{dt} = \eta N_o \sigma_{(k)} (I/\hbar\omega)^k \tag{2}$$

where N^* is the number density of atoms taken from the initial to the final state), η a quantum efficiency; N_o is the number of atoms or molecules in the laser-irradiated volume; $\sigma_{(k)}$ the kth-order cross section; and $\Phi \equiv (I/\hbar\omega)$ is the photon flux, the number of photons per unit time per unit area. The total integrated effect induced by the end of the laser pulse, on the other hand, is proportional to the specific energy E deposited in a volume V and hence to fluence F_L:

$$Yield \propto (E/V) \cong F_L a(\omega, I) \cong I_o \tau_L [a_0(\omega) + \beta I] \tag{3}$$

where $a_0(\omega)$ is the linear absorption coefficient, β is the nonlinear absorption coefficient, and τ_L the laser pulse duration.

Unlike nanosecond laser-materials interactions, which are generally described in terms of fluence and hence depend on equilibrium thermodynamics of the irradiated material, the use of femtosecond pulses invites us to focus on the *density of excitation*, that is, on the number of quanta deposited into the material per unit volume and per unit time. The recognition of the central role played by the spatio-temporal density of electronic excitation is largely due to Itoh [8]. The combination of the density of excitation and the strength of the electron–lattice coupling largely determine the outcome of any ultrafast laser process.

7.2.1
Metals: The Two-temperature Model

Lasers interact with metals by exciting free-free transitions of conduction-band electrons in a volume defined roughly by the product of the laser focal spot and the skin depth of the metal. Coupling with the lattice is relatively weak. For femtosecond pulses, this creates a condition quite different from that obtaining for nanosecond interactions, because the electron–phonon coupling occurs over a timescale of picoseconds, rather long compared with the pulse duration. Thus the laser-excited electrons reach a high temperature by the end of the laser pulse but are not in thermal equilibrium with the lattice until some time later. This circumstance is well described by a model that describes the time evolution of electron and ion temperatures separately, with a coupling term that connects the two differential equations:

$$C_e \frac{\partial T_e}{\partial t} = \nabla(\kappa_e \nabla T_e) - \Gamma_{e-\ell}(T_e - T_\ell) + Q(x_a, t)$$
$$C_\ell \frac{\partial T_\ell}{\partial t} = \nabla(\kappa_\ell \nabla T_\ell) + \Gamma_{e-\ell}(T_e - T_\ell) \tag{4}$$

Here the subscripts e and ℓ refer to electron and lattice parameters heat capacity (C_i) and thermal diffusivity (κ_i), respectively; the electron-lattice coupling constant $\Gamma_{e-\ell}$ represents the rate of energy transfer between the electron gas and the lattice, and Q is the laser (source) term. Because of the much larger mass of atomic nuclei and the relatively weak electron–lattice coupling in metals, the laser energy is initially converted to free-electron heating, and the extremely hot electrons are out of equilibrium with the colder lattice ions. Over a timescale of a few picoseconds, the electrons come to thermal equilibrium with the lattice by electron–phonon scattering, and reach the much lower equilibrium temperature dictated by the heat capacity of the lattice.

While the solution of these equations gives a zeroth-order picture of energy transfer from the source to the electron gas and thence to the lattice, it fails to give detailed dynamical information. The optical reflectivity of metal surfaces irradiated by femtosecond lasers has been shown by many different authors to change dramatically over a timescale of less than a picosecond; from this, a nonthermal melting mechanism was proposed and widely accepted. Recently it has become possible to image the dynamics of ultrafast melting on an atomic scale with ultrafast electron diffraction [9]. The picture that emerges from these studies confirms that the rapid heating induced by the laser pulse leads to violent oscillations of the ion cores about their equilibrium positions, corresponding, for a brief time, to temperatures far above the normal equilibrium melting temperature ("superheating"). Thereafter, the normal crystalline order disappears, replaced within a few picoseconds by the disordered structure characteristic of the liquid phase. However, the configurations sampled by the metal during this transition are all consistent with a purely thermal mechanism having an initial phase whose temperature is substantially greater than the normal melting temperature. This suggests that in metals, "nonthermal" melting should be re-christened "ultrafast melting."

If sufficient energy is deposited close to the surface of a metal by an ultrafast laser pulse, material ablation occurs. Time-resolved microscopy [10] of laser ablation in vacuum confirms that this ablation process begins with a superheated phase in which the irradiated volume is at high pressure (GPa) and superheated temperatures. This superheating leads in 20–40 ps to the formation of a bubble of low-density material much thinner than the wavelength of the incident light forming under a much higher-density interfacial layer. As the material near the surface begins an isentropic expansion into its environment, it follows the bimodal boundary (Fig. 7.2a) and develops into a two-phase mixture of vapor and liquid. In this two-phase regime, the speed of sound decreases drastically, Simultaneously, a self-similar rarefaction wave propagates forward into the vacuum and backwards toward the surface (Fig. 7.2b), while the inhomogeneous two-phase mixture expands into the vacuum. Because the energy necessary to sustain this process is deposited in the metal before thermal equilibrium sets in, it is possible to get ablation without creating a heat-affected zone, as one can see from one of the earliest papers on femtosecond laser processing [11] with femtosecond versus picosecond and nanosecond pulses (Fig. 7.3).

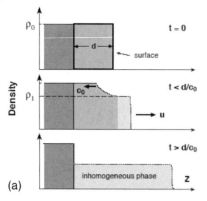

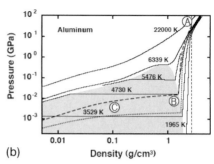

(a)

(b)

Fig. 7.2 (a) Schematic of the time evolution of ultrafast laser ablation. At $t = 0$, there are two regions: the cold target, and the hot region where the laser energy has been absorbed. For times less than d/c_0, where c_0 is the speed of sound, a spherical rarefaction wave propagates backward into the target, while a heterogeneous phase (liquid plus vapor) is expanding outward into vacuum. At times $t > d/c_0$, the rarefaction wave is spent and there remain only the cold target material and the inhomogeneous ablation plume. (b) The three phases shown here correspond roughly to the points A, B and C on the equation-of-state diagram for Al metal. From Ref. [10].

(a)

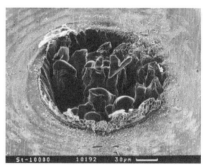

(b)

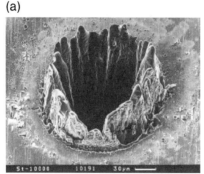

Fig. 7.3 Micromachined hole in a 100 μm thick stainless steel foil, drilled by a chirped-pulse amplified Ti:sapphire laser at a wavelength of 780 nm. (left) pulse duration 130 fs, fluence 0.5 J cm^{-2}; (center) pulse duration 5 ps, fluence 3.5 J cm^{-2}; (right) pulse duration 10 ns, fluence 4.7 J cm^{-2}. The scale bar in each case is 30 μm. Note the signs of thermal-wave propagation outward from the laser spot for the ps and ns irradiation. From Ref. [11].

Since the heating of the electron gas during laser ablation will also change the thermal and transport properties of the metal, it is necessary to go beyond the simple two-temperature model to calculate processes such as the deformation or dissolution of the lattice [12]. When excitation is weak, energy transfer from electrons to the lattice is delayed if the electrons have a nonthermal energy distribution [13]; on the other hand, for the strong excitations required to initiate melting and ablation, the electron–phonon coupling can be described by the two-temperature model and the ultrafast thermal equilibration observed in the experiments is confirmed by conventional theory, as well as by recent experiments combining femtosecond laser ablation with molecular dynamics simulations [14].

7.2.2
Semiconductors

Semiconductors differ from metals, as far as the laser-materials interaction is concerned, in having a finite energy bandgap and a somewhat larger electron–phonon coupling constant that tends to increase with ionicity. Laser interactions with semiconductors are shaped by three physics complications. (1) The finite energy gap means that photons with energies larger than the gap can introduce band-to-band transitions, even at low fluence. (2) The existence of surface states or defect states in the bandgap opens new absorption channels that do not exist in the perfect bulk material. (3) For wide-bandgap semiconductors (e.g. GaP, GaN) there are multi- or multiple-photon excitation channels that may play a role. Moreover, unlike in the case of metals, details of the crystal structure may play a role in melting and ablation; for example, computational studies predict that graphite will have two different ablation mechanisms, corresponding to breaking of intra- versus inter-plane bonds [15]. Finally, the fact that materials with finite bandgaps have differing polarizabilities means that the relative strength of the electron–lattice coupling can play a significant role in the excitation and relaxation processes that are important to laser processing.

7.2.2.1   Ultrafast Laser-induced Melting in Semiconductors
There is an enormous amount of literature on this subject, growing out of the early interest in laser annealing of semiconductor materials for the microelectronics industry. The earliest measurements relied on optical techniques – such as femtosecond time-resolved reflectivity – to infer the changes in surface reflectivity [16], collapse of the bandgap due to formation of a dense electron–hole plasma [17] and very large time-dependent variations in the dielectric function [18] occurring as the semiconductor makes the transition to thermal equilibrium after ultrafast laser irradiation.

An illustrative example of ultrafast melting in a direct-gap semiconductor is furnished by experiments on GaAs using an ultrafast time-resolved ellipsometry scheme that gives direct information on the change of the dielectric function during laser irradiation [19]. In the experiment, Cr-doped bulk GaAs (100) samples were irradiated by a 70 fs pump pulse, and probed after a variable time delay by a weak broadband pulse (1.5–3.5 eV) generated from the 800 nm pump by focusing the probe beam into a 2 mm thick CaF_2 window. By measuring the spectral reflectivity of the probe at two angles of incidence and numerically inverting the measured broadband Fresnel reflectivity, it is possible to measure simultaneously the real and imaginary parts of the complex dielectric function $\varepsilon(\omega)$ as a function of pump-probe delay time. The results are illustrated by the selection of snapshots shown in Fig. 7.4, for a fluence equal to $0.7 \cdot F_{th}$, where F_{th} is the threshold fluence at which single-shot damage can be observed with a microscope. Only 250 fs after the pump pulse is incident on the film, the measured dielectric function has already departed significantly from the room temperature dielectric function (solid and dashed curves) of the GaAs film. This evolution continues at 500 fs, until at 2 ps pump-probe delay, the dielectric function resembles much more nearly those for amorphous GaAs at room temperature; by 16 ps after irradiation, the

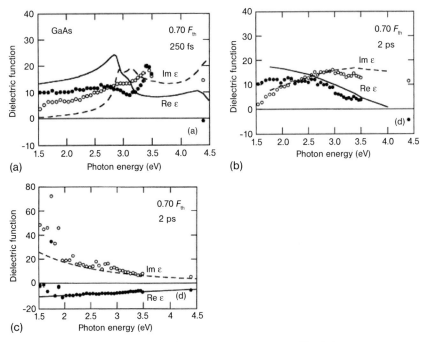

Fig. 7.4 Measured dielectric functions for GaAs film illuminated by pump pulses from a 780 nm regeneratively amplified Ti:sapphire laser, for varying fluences and time delays. Solid circles refer to the real part of $\varepsilon(\omega)$, open circles to the imaginary part of $\varepsilon(\omega)$. The probe in each case was a femtosecond supercontinuum pulse (1.5–3.5 eV). (a) Real and imaginary parts of $\varepsilon(\omega)$ at a fluence of $0.7\ F_{th}$, 250 fs after the pump pulse; the solid and dashed curves are respectively the real and imaginary parts of for GaAs at room temperature. (b) Real and imaginary parts of $\varepsilon(\omega)$ at a fluence of $0.7\ F_{th}$, 2 ps after the pump pulse; the solid and dashed curves are respectively the real and imaginary parts of $\varepsilon(\omega)$ for amorphous GaAs at room temperature. (c) Real and imaginary parts of $\varepsilon(\omega)$ at a fluence of $1.6\ F_{th}$, 2 ps after the pump pulse; the solid and dashed curves are respectively the real and imaginary parts of $\varepsilon(\omega)$ for metallic GaAs at room temperature. From Ref. [19].

dielectric functions have evolved even beyond the amorphous GaAs into something that reflects a permanently altered structure. For similar experiments carried on at $1.6 F_{th}$, the dielectric function at long pump-probe delays strongly resembles that due to a Drude metal.

7.2.2.2 Ultrafast Laser Ablation in Semiconductors

Femtosecond pump-probe microscopy has been used to form the most detailed
picture of the laser ablation process in semiconductors that we presently have. A
120 fs pulse from an amplified dye laser was used as a pump, at a wavelength of
620 nm; the fluence was of order 0.5 J cm^{-2}. At these densities, and given the
absorption in the topmost μm or so of the Si, the density of electron–hole pairs is
of order 10^{22} cm^{-3}, similar to what would be characteristic of a metal. The area
thus excited is then illuminated by a time-delayed weak probe beam, and the
reflected signal viewed from a direction normal to the surface. In the illustration
shown in Fig. 7.5, a Si wafer is irradiated at approximately 0.5 J cm^{-2}, above the
ablation threshold but below the fluence needed to form a plasma. In the first
picosecond after laser irradiation, the surface turns highly reflective, indicating
the formation of a metallic phase. As the rarefaction wave described for metals
travels from the surface into the bulk (see Fig. 7.2), alternating layers of dense and

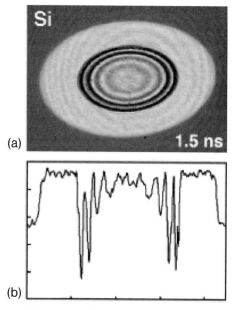

(a)

(b)

Fig. 7.5 Ultrafast ablation from Si. (a) Interferogram of the
irradiated surface, showing the Newton's rings that arise from
the generation of nearly co-planar high- and low-density regions
caused by the generation of the rarefaction wave. (b) Reflectivity
profile of the ablated region, showing the high- and low-density
ring profile from the interferogram. From Ref. [10].

rarified regions form and hydrodynamic motion begins. However, since the rare-faction wave propagates at the local speed of sound (of order 10^3 m s^{-1}), substantial motion of material out of the surface does not begin until something like 0.1–1 ns after the ultrafast pump pulse [20]. Approximately 1 ns after the pump pulse, a series of bright and dark Newton's rings appears, due to the interference of parallel interfaces between regions of high and low refractive index as the strongly heated, pressurized Si is ejected into the vacuum and simultaneously begins the slow process of equilibration with the cold region outside the absorption zone. Figure 7.5(b) shows the reflectivity profile of the interference pattern created by reflections from the high- and low-index components of the expanding laser plume.

The key to this phenomenology is the creation of a dense electron–hole plasma at near-metallic electron-density levels, due to the high density of electronic excitation, i.e., to the high density of electron–hole pairs. It is possible to reach this condition in Si, GaAs, Ge and some other low-bandgap semiconductors, where a band-to-band transition can be initiated by a single photon, thus creating the lattice-destabilizing *e-h* plasma.

7.2.2.3 Theoretical Studies of Femtosecond Laser Interactions with Semiconductors

Theoretical studies of microscopic mechanisms have shown in detail the effects of the dense, laser-excited electron–hole plasma on material structure, and the dynamical evolution of the laser-irradiated material. Stampfli and Bennemann have shown, for example [21], that the initial transition from the covalently bonded semiconducting state to the melted metallic state results from strong excitation of longitudinal optical phonon modes; this in turn leads to a softening of the acoustic modes of the semiconductor, leading to the melting of those bonds. This softening occurs very rapidly, requiring only a few picoseconds. At the same time, the fact that the absorbed energy is strongly localized provides the driving force for the hydrodynamic ejection of material observed in experiments. Indeed, the density of excitation is in some cases sufficient for homogeneous phase transitions to outrace the energetically more favored process of heterogeneous nucleation and vaporization at semiconductor surfaces.

7.2.3
Insulators

The interest in ultrafast machining and ablation of wide band-gap insulators can be dated to the publication of seminal papers by groups at the University of Michigan and the Lawrence Livermore National Laboratory. In those experiments, the former focused on bulk damage in fused silica indicated by plasma breakdown (Fig. 7.6) [22] and the latter [23] on surface damage to fused silica by ablation and thermal effects (Fig. 7.7). Using chirped-pulse amplified lasers of variable pulse durations, commercial samples of fused silica were irradiated by pulses of varying duration ranging from a few tens of femtoseconds up to nanoseconds. Both

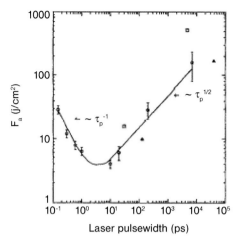

Fig. 7.6 Threshold for initiation of a plasma spark in fused silica by Ti:sapphire laser irradiation of varying pulse durations. From Ref. [22].

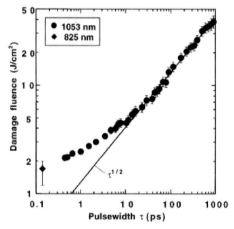

Fig. 7.7 Fluence for initiation of surface damage in fused silica as a function of pulse duration for two different wavelengths. Pulse duration (825 nm), (1053 nm). From Ref. [94].

experiments clearly indicated that, for pulse durations shorter than about 10 ps, materials modification or damage induced by multi-photon excitation had a nonthermal character, while for longer pulse durations the effects were due to strong local heating obeying the normal dependence of diffusive thermal effects on the square-root of the laser pulse duration.

Compared with semiconductors, the ultrafast laser-materials interaction with insulators adds three additional complications: (1) the larger bandgap energy means that multi-photon processes play an *essential* role [24] (i.e. one-photon valence-to-conduction-band transitions are unlikely except as noted below); (2) the lattice polarizability plays a much more important role than it does in semiconductors because of the strong Coulomb forces, through such phenomena as self-trapping of excitons; and (3) laser irradiation often leads to the formation of permanent vacancy or interstitial defects with energy levels in the bandgap, thereby altering the optical absorption of the sample, e.g., through the formation of color centers. Unlike in the case of nanosecond laser irradiation, this last effect is unimportant in single-shot materials modification, but it can be critical in multiple-shot laser-induced processing of insulators.

7.2.3.1 Ultrafast Ablation of Insulators

Ablation occurs when localized excitations lead to bond-breaking and ejection of material into the ambient. In wide-bandgap materials, such as typical optical dielectrics, single-photon excitation is generally insufficient to induce a valence-band to conduction-band electronic transition. Except for those materials, such as GaP, in which a surface state can be excited by one-photon transitions into a configuration coordinate that leads to efficient desorption and ablation, multiple- or multi-photon transitions are necessary to provide the initial injection of electrons into the conduction band. This by itself, however, does not usually provide sufficient localized energy for the few vibrational periods needed so that atoms, ions or clusters can be ejected from an insulator surface.

The detailed mechanisms of femtosecond ablation of insulators in the nonthermal regime are both complex and controversial. The conventional theory developed from studies of laser-induced surface and bulk damage has been that the femtosecond laser interaction with wide-bandgap insulators was dominated by multi-photon ionization and subsequent collisional damage done to the material by the accelerated conduction-band electrons ("electron avalanche"). Other studies, however, indicate the need for a more nuanced view of these mechanisms. For one thing, multi-photon ionization is not the only mechanism for producing free electrons in the conduction band; at very high fields, tunneling ionization dominates the picture, and it is necessary to take into account the competition between strong-field and electron-impact ionization mechanisms [25]. In fact, calculations show that for SiO_2 and with pulse durations shorter than 50 fs, there is no electron avalanche at all! There is also a significant role for electronic excitations that proceed through various defect-related channels, such as metastable or self-trapped polarons, charge carriers or excitons, that ultimately dissipate energy into the motion of atomic nuclei within the solid [26, 27]. The failure to take lattice polarization and self-trapping into account can mislead one in attributing optical breakdown to a relaxation process associated with a plasma produced by avalanche ionization [28], when in fact it is due to self-trapping effects [29]. Finally, even though most ablation products are neutral, there is evidence that ultrashort-pulse ablation produces extremely energetic ions [30, 31], and photoelectrons [32] whose effects on the ablation dynamics are still poorly understood.

Measurements of the optical breakdown threshold and ablation depths in fused silica and barium borosilicate glass, with spatially filtered pulses with durations ranging from 5 ps to 5 fs, yield a more reproducible, deterministic process of micromachining by laser ablation. In the experiments, the diffraction-limited beam from a 1 mJ, 1 kHz Ti:sapphire laser system was spatially filtered, and the ablation depth measured microscopically after fifty shots at a fluence of 5 J cm^{-2}. There were significant shifts in ablation thresholds compared to the earlier work, possibly through the elimination of hot spots in the beam; this gave much better quality damage spots (Fig. 7.8a). Whereas ablation depth in fused silica was almost independent of pulse duration (Fig. 7.8b), the ablation depth curves for borosilicate glass show significant variation with pulse duration, although the differences are much less pronounced with increasing pulse duration (Fig. 7.8c). Ablation thresholds were also found to be strongly influenced by incubation effects [33], with a nearly four-fold decrease in threshold when the fused silica sample was irradiated by fifty laser shots on a single site. This indicates that the local electronic structure of the fused silica is gradually altered by continuing irradiation at a single site.

Modeling and theoretical understanding of these results remain incomplete at present. The early data from Du et al. (Fig. 7.6) were explained by assuming that

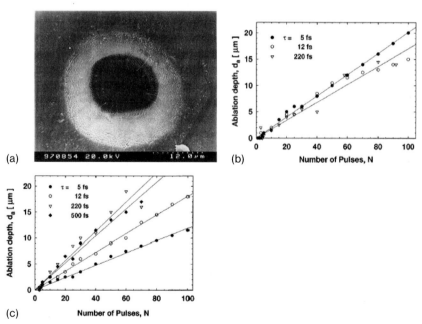

Fig. 7.8 Surface damage experiments carried out with pulse durations of 5–500 fs in fused silica (FS) and barium-borosilicate glass (BBS). (a) Scanning electron micrograph of front-surface damage on FS. Pulse duration 5 fs, fluence 6 F_{th}. From Ref. [95]. (b) Ablation depth in FS as a function of number of pulses and pulse duration, fluence 5 J cm^{-2}. Lines are linear fits to the data. (c) Ablation depth in BBS as a function of number of pulses and pulse duration, fluence 5 J cm^{-2}. Lines are linear fits to the data. From Ref. [33].

the electron avalanche scaled with the square-root of the laser intensity, with only the seed electrons produced by MPI. The data of Stuart et al. (Fig. 7.7) were modeled by assuming that the density of avalanche electrons scales linearly with laser intensity; given the Keldysh model for the MPI rate, it appeared that MPI might be the dominant mechanism for producing the avalanche electrons. This scaling was verified in these later experiments covering the range of pulse durations appropriate to the nonthermal effects, with the added benefit of much higher beam quality [34]. However, in contrast to the Stuart prediction of an ablation threshold of 0.1 J cm^{-2} for fused silica, these experiments found a threshold of 1.5 J cm^{-2} for pulse durations less than 10 fs.

This ablation threshold, in turn, implies MPI rates that are orders of magnitude lower than those predicted by the Keldysh theory, leaving another puzzle to be addressed. One possibility has been suggested in a recent paper by Rethfeld, which proposes a multiple rate-equation model to take into account the fact that, once an electron is in the conduction band, it takes only a single photon to add energy and increase the probability for more avalanche electrons by collisional processes [35]. It may also be that the MPI rate is reduced, compared to the Keldysh prediction, by the self-trapping mechanisms that are known to operate in many wide band-gap dielectrics; this hypothesis has the additional attraction of including the variations in lattice polarizability through the well-known electron–phonon coupling constants. The changes in sample absorption due to the formation of self-trapped defects have been shown to occur on nanosecond timescales [36], and are therefore potentially important especially for materials modifications wrought by mode-locked femtosecond lasers. Ablation thresholds in insulators (and also in semiconductors) are also affected by adatoms, steps and terraces on atomically well-defined surfaces [37].

7.2.3.2 Self-focusing of Ultrashort Pulses for Three-dimensional Structures

Self-focusing, a nonlinear optical effect that arises from the third-order susceptibility of a material, is easily observed in femtosecond laser-materials interaction due to the high intensities generated in the focal plane. Sub-surface structuring of transparent materials was reported almost a decade ago [38]. The nonlinear index term βI in Eq. (3) becomes significant as one approaches a focal point, leading to self-focusing, dielectric breakdown and irreversible materials modification. Among the more spectacular demonstrations of this concept is the use of femtosecond pulses for three-dimensional structuring of transparent materials to create structures for three-dimensional data storage, waveguides, and even optical amplifying media. An early demonstration of three-dimensional optical data storage in poly(methyl methacrylate) (PMMA) was soon followed by a demonstration that similar three-dimensionally structured "bits" could be formed in fused silica [39]. Later papers have demonstrated various useful morphologies in transparent polymers and read-out schemes for such optical memory structures [40]. Most recently, variations on the nonlinear refraction processing scheme have appeared, such as making use of the glass transition temperature in poly(methacrylate) to increase the spatial density of bits [41].

All of these demonstrations were based on the idea that, above a threshold intensity or fluence, catastrophic self-focusing due to the third-order susceptibility of the transparent material (e.g., fused silica) would lead to irreversible modifications of the material, ranging from structural changes to local alterations in composition. A simple theory for self-focusing indicates that the distance of a lens at which catastrophic self-focusing occurs is given by [96]

$$ z_F^{-1} = \frac{1}{K} \left[\sqrt{\frac{P}{P_{cr}}} - 0.852 \right], \quad P_{cr} = \frac{(1.22)^2 \pi \lambda^2}{32 n_0 n_2}, \quad K = 0.367 R_d \tag{5} $$

where P is the laser power, n_0 and n_2 are, respectively, the linear and nonlinear indices of refraction, and R_d is the Rayleigh length, which in turn depends on both the wavelength and the laser spot diameter. This equation is known from many experiments to be reasonably accurate when $P > 1.5 P_{cr}$. In fact, the materials modification occurs experimentally at fluences below the critical power for self-focusing. However, the power dependence of the modification depth z_M follows the square-root of the laser power P as expected (Fig. 7.9). Experiments on fused silica show that the depth at which modifications occur depends on pulse duration, as shown in Fig. 7.10. At femtosecond pulse durations, the morphology

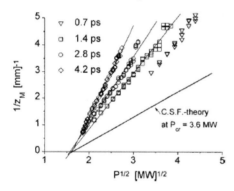

Fig. 7.9 Scaling of materials-modification depth with pulse duration. From Ref. [96].

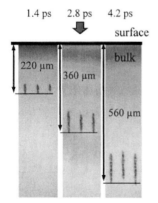

Fig. 7.10 Micrographs showing sub-surface laser-induced modifications in sapphire for varying pusle durations, following the scaling laws shown in the previous figure. From Ref. [96].

of the structures is strongly influenced by the details of the laser pulse and focusing properties, as shown in experiments showing single-pulse modification to borosilicate glass (Corning 0211) [42]. In these experiments, it appeared that the changes in refractive index accompanying the structural modifications might be due either to strong localized melting followed by nonuniform re-solidification, or explosive vaporization near the focal spot, with a subsequent ejection of hot ions and electrons into the surrounding material. It is also possible that pulse duration itself can play a role. Experiments with laser irradiation of fused silica and other dielectrics and with pulse durations ranging from 5 fs to 5 ps [43] seem to indicate that pulses below 50 fs duration can structure dielectrics almost deterministically; that is, the threshold for ultrafast laser ablation is not described by a statistical process, but can be predicted from physical criteria.

Two other phenomena associated with the laser ablation process in insulators should be mentioned here. One is the fact that nonlinear effects in surface micromachining are also affected by nonlinear absorption effects when the process takes place in air. These nonlinear processes can produce significant effects on ablation rates and the shape of machined structures [44]. The other effect of interest is the generation of charge separation in the focal volume of ultrafast lasers. This can produce a huge local imbalance of Coulomb forces in the laser-irradiated material, leading to the explosive ejection of ions from the surface [30]. The energetic ion species liberated in these processes can have kinetic energies ranging up to 1 keV. This means that ultrafast laser ablation of insulators can be a source of energetic ion species that can be used for other purposes [31]. But it also means that ultrafast pulsed laser deposition, as noted below, may be strongly affected precisely by these extremely energetic ions.

7.2.3.3 Color-center Formation by Femtosecond Laser Irradiation

A distinctive feature of wide-bandgap insulators is the lattice polarization and relaxation that accompanies laser interactions with materials. Of these, the most famous is probably the self-trapped exciton found in alkali halides and fused silica [45, 46]; but other metastable vacancy and interstitial defects are also created by laser irradiation. These defects change the absorption characteristics of materials, and therefore may change laser processing parameters. Femtosecond lasers, with their high intensities and high probabilities of multi-photon excitations, can also produce defects that make transparent insulators photosensitive and therefore more amenable to micromachining, patterning and structuring. This makes possible three-dimensional microstructuring even in materials that are not photosensitive polymers, photorefractive crystals or photochromic glasses [47].

The first experiments to show coloration of optically transparent glasses by femtosecond radiation were carried out by Efimov et al. [48]. At intensities of order 10^{12} W cm^{-2}, two orders of magnitude below the intensity threshold for bulk damage, a slight coloration was observed in high-purity borosilicate and alkali silicate glasses; no such coloration was observed in fused silica, however. The spectra of the colorations were remarkably similar to those obtained under irradiation by

γ rays from a ^{60}Co source. Other curious phenomena were observed, including dark tracks along the laser-beam direction. In all cases, the coloration was reversed by gentle heating (150 °C) for a few minutes, suggesting that the coloration was due to the formation of shallow (i.e., low activation-barrier) traps for charged defects.

In a pair of recent experiments [49, 50], the kinetics of coloration in soda-lime glass (SLG, an aluminosilicate glass) and sodium chloride crystals were measured following irradiation by repeated exposure to 120 fs pulses from an amplified Ti:sapphire laser (800 nm). Changes in absorption as a function of irradiation parameters (fluence, number of laser shots) and time following irradiation were measured by ordinary transmission spectroscopy. As shown in Fig. 7.11, both soda-lime glass (SLG) and NaCl single crystals are efficiently colored by femtosecond irradiation. In fact, the efficiency can be even larger than that of X-ray sources. The defects in SLG are trapped hole centers, linking nonbonding oxygen holes and neutral Na atoms, that absorb at 460 and 620 nm; the 620 nm center was probed by a 633 nm He–Ne laser, the 460 nm center by a blue diode laser. Lifetime studies show that there is an initial rapid partial recovery after 1000 laser shots, followed by a much longer-lived recovery that leaves approximately 30% of the laser-initiated defects in a stable condition.

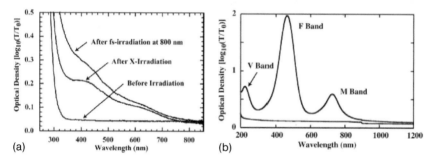

Fig. 7.11 (a) UV-visible spectra of soda-lime glass before irradiation, after X-ray irradiation and after fs-laser irradiation at a wavelength of 800 nm. (b) UV-visible absorption spectra of NaCl before and after coloration by fs-laser pulses at a wavelength of 400 nm. From Ref. [50].

The three peaks in the darkened spectrum of NaCl correspond to the *F*-center (alkali vacancy associated with a trapped electron), the *M*-center (two *F* centers on adjacent sites) and the *V*-center (a collective name for several hole traps with miscellaneous properties, probably associated with impurities). In NaCl as in SLG, the rate of defect formation depends strongly on the laser pulse energy, or, more accurately on the intensity.

7.3
Ultrafast Laser-materials Interaction: Vibrational Excitation

Although picosecond IR laser sources are available at fixed frequencies, the development of tunable, picosecond and femtosecond free-electron lasers (FELs) in the mid-infrared (MIR) makes possible selective excitation by tuning to resonant vibrational modes of irradiated materials. These anharmonic MIR modes are usually substantially more energetic than the harmonic modes that constitute the phonon bath of the material; a typical Debye temperature for the highest occupied phonon modes might correspond to an energy of 200 cm^{-1}, whereas typical mid-infrared vibrational modes have five or ten times this energy. Because of the differing spatial characteristics of the wave functions for the harmonic and anharmonic oscillator modes, the coupling between the resonantly excited anharmonic modes and the phonon bath is not instantaneous, but occurs on a timescale of a few picoseconds; this means that nuclear motion and bond-breaking can begin, if the density of excitation is sufficiently high, before the energy leaks out of the excited anharmonic mode. This makes possible new kinds of resonant multi-photon materials processing. Moreover, vibrational excitation does not generate electronic excitations, and thus avoids relaxation pathways that lead to photo-fragmentation or structural alterations.

At the intensities characteristic of the FEL micropulses, the discussion of mechanisms should be in terms of absorbed intensity I rather than fluence. Integrating Eq. (3) over the laser pulse for a process with a characteristic relaxation frequency γ, one finds that

$$Yield \propto E_v \cong \int_0^{\tau_L} \{I[a_o(\omega) + \beta I] - \gamma E_v\}dt = (1 - e^{-\gamma \tau_L}) \frac{I}{\gamma}[a_o(\omega) + \beta I]$$

$$\equiv \left(\frac{1 - e^{-\gamma \tau_L}}{\gamma \tau_L}\right) F_L \tag{6}$$

The density of energy in the vibrational mode E_v has units of J m^{-3}, γ is the decay constant of the mode, a_0 is the linear optical absorption coefficient, β is the nonlinear absorption coefficient, I is the laser intensity and τ_L the laser pulse duration. The first factor in the last expression has a maximum value of 1 for small values of $\gamma \tau_L$, corresponding to pulse durations shorter than $1/\gamma$; it falls roughly as $1/\gamma \tau_L$ as τ_L becomes much larger than $1/\gamma$. Because the FEL micropulse duration is short compared with the relaxation time of the initial anharmonic vibrational excitation, we are justified in considering the 1 ps FEL micropulses as "ultrafast" in the same sense as this terminology is usually applied to femtosecond laser processing by electronic excitation.

An estimate of the probability of a k-photon excitation can be made [51] by calculating the probability P_k for k photons to be simultaneously in the volume L^3 occupied by a unit cell or a typical polymer (for example, a random-walk model for polystyrene of average mass 10^4 Da predicts a volume of order 10^3 nm^3) and its nearest neighbors, when the average number of photons per unit volume is m:

$$P_k = \frac{m^k}{k!}, \quad m = \frac{I}{hc/\lambda}\frac{L}{c/n}L^2 = \frac{n\lambda IL^3}{hc^2} \tag{7}$$

where n is the index of refraction and h is the Planck constant. For conditions typical of an FEL in recent experiments on poly(tetrafluorethylene) [52], $P_2 \sim 0.25$, $P_3 \sim 0.06$ and $P_4 \sim 0.01$ for a polymer of moderate size. Hence there is a non-negligible probability that a given unit cell will experience a multi-quantum vibrational transition during a micropulse, generating strong localized nuclear motion and intermolecular bond-breaking *without* electronic excitation. These probabilities are enhanced by the bandwidth (ca. 10–20 cm^{-1}) of the FEL micropulses, which is large enough to overcome the anharmonicity that otherwise tends to inhibit vibrational ladder-climbing.

7.3.1
Ablation of Inorganic Materials by Resonant Vibrational Excitation

The inherent tunability of the free-electron laser provides an opportunity to understand ultrafast laser ablation that proceeds via vibrational excitation. The absorption coefficient of fused silica rises by five orders of magnitude between 4 μm and 9.4 μm, the absorption maximum in the Si-O stretching mode. Changing the fluence of the FEL macropulse delivered to a fused silica sample, thus determines the local density of vibrational excitation.

In experiments carried out at Vanderbilt, an FEL with a nominal 4 μs macropulse was used, containing some 20 000 micropulses of 1 ps duration and 1–2 μJ energy each. For wavelengths around 4 μm, and irradiation using the full macropulse, the radiation is not thermally confined, because the absorption of the silica is very weak. Hence, one has deep penetration and thermomechanical ablation in large chunks. At a wavelength of 8 μm, the thermal diffusion length and the optical penetration depth are nearly equal, so that one can expect to see normal boiling and evaporation. Close to the maximum absorption of the fused silica, the absorption length is much smaller than the thermal diffusion length, and one sees signs of explosive vaporization, with thin layers of silica expelled from the surface. SEM micrographs, in fact, show a melted zone that is an order of magnitude shallower than the linear optical absorption would predict, indicating substantial nonlinear absorption (Fig. 7.12). As the micropulse intensity in this case is approaching 10^{10} Wcm^{-2}, this is hardly surprising. Similar results are seen when exciting the resonant modes of $CaCO_3$ and $NaNO_3$ near and away from the resonant 7 μm wavelength due to the $\nu_2-\nu_4$ stretch vibration of the carbonate and nitrate groups [53]. The combination of resonant multi-photon excitation and the attendant nonlinear absorption leads to shallow, smooth ablation on resonance, and chunk ablation due to fracture at greater depths off resonance.

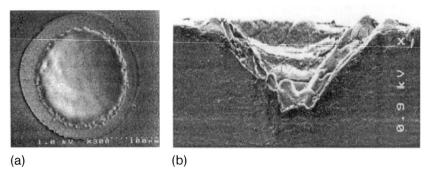

(a)   (b)

Fig. 7.12 (a) Single-shot laser damage at 9.4 μm wavelength and 80 J cm^{-2} (4 μs FEL macropulse) exhibiting efficient material removal from the near-surface in the high-intensity center of the pulse and a spreading thermal wave as some absorbed laser energy diffuses outward. (b) Cross-sectional scanning electron micrograph of a multiple-shot ablation crater in fused silica at a FEL wavelength of 9.4 μm. The dark region underneath the crater indicates roughly the depth of the melted region. The laser focal spot is approximately 200 μm in diameter. From Ref. [53].

7.3.2
Ablation of Organic Materials by Resonant Vibrational Excitation

High-quality, carefully controlled films of organic and polymeric materials are needed for many applications in electronics, photonics, sensing and protective coatings. Traditional methods of organic thin-film deposition – aerosols, spin and dip coating, vacuum thermal evaporation – all have one or another deficiencies that drive a continuing search for efficient, conformal vapor-phase coating techniques for organic molecules and polymers.

In recent experiments at Vanderbilt using the same FEL parameters described above, it has been shown that several polymers – including poly(ethylene glycol) [54], poly(styrene) [55], poly(tetrafluoroethylene) [52] and poly(glycol-lactides) [56] – can be efficiently ablated and transferred intact into the gas phase. The most detailed mechanistic studies have been carried out on poly(ethylene glycol) (PEG), and show that, not only is resonant infrared-pulsed laser deposition (RIR-PLD) more efficient in doing this than are ultraviolet lasers, but also that deposition of energy into the resonant IR mode is more effective than IR radiation detuned from the resonance. In the experiment which produced the data illustrated in Fig. 7.13, a drop-cast target of PEG was irradiated at the end-group O-H stretch mode (2.94 μm) and off resonance at 3.34 μm. The ablated material was collected on an NaCl flat for FTIR analysis and on a Si wafer for subsequent analysis by size-selective chromatography and matrix assisted laser desorption-ionization mass spectrometry (MALDI-MS). The FTIR spectrum gives information on integrity of the local bonding arrangements in the ablated polymers. A more severe test of structural integrity is imposed by the MALDI mass spectra showing how the mass distributions (and hence the polydispersities) before and after ablation com-

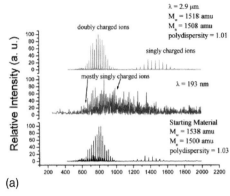
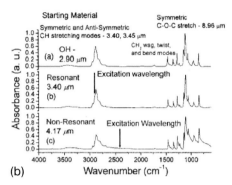

(a) (b)

Fig. 7.13 Characteristics of poly(ethylene glycol) ablated by infrared FEL and UV irradiation. (a) MALDI mass spectra of singly and doubly-charged PEG ions (bottom to top) observed from the starting ablation target material, ablated from the target by 193 nm nanosecond laser and by resonant infrared radiation at 2.9 μm. From Ref. [54]. (b) Fourier-transform infrared spectra for PEG (top to bottom) from the PEG starting material, resonant ablation of the C–H stretching mode at 3.4 μm, and nonresonant ablation at 4.17 μm. From Ref. [57].

pare. By both measures, the RIR-ablated material is essentially the same as the starting material, while nonresonantly ablated material exhibits fundamental differences. In the mass spectra, the damage incurred in the ablation process is seen in the fragmentation which is so evident in the mass distribution; in the FTIR spectra, the nonresonantly ablated material shows peak broadening and shifting, indicating bond-breaking and cross-linking.

Gel-permeation chromatography (GPC), sometimes called size-selective chromatography, measures the hydrodynamic volume of the eluents, and is thus sensitive both to mass and shape distribution of the ablated polymers. The elution profiles of the bulk starting material and the ablated polymers have been studied in PEG as a function of wavelength [57]. The results show that the most effective and least destructive ablation occurs at the frequency of the O–H stretch mode of the polymer endgroup; irradiation at the C–H stretch of the side chains produces minor fragmentation, while irradiation at the C=O backbone-mode frequency causes rather significant bond-breaking and fragmentation. Hence weakly resonant excitation appears to be optimal.

7.4
Photochemistry in Femtosecond Laser-materials Interactions

As observed by Cavanagh et al. [58], photochemistry at surfaces can be initiated by lasers: (1) through surface thermal chemistry; (2) through reactions that are initiated by excitations localized via an adsorbate; (3) through carrier-mediated chemical reactions; and (4) through reactions mediated by surface states. It can be involved in ultrafast laser-materials interactions: first, through femtosecond gas-surface chemistry for materials immersed in an ambient, and second, through femtosecond solid-state chemistry in the solid substrate. At present the data are too sparse to be certain that femtosecond photon-assisted chemistry is really essential, as many useful processes – such as selective chemical etching – require only photons of a distinct wavelength. Below are presented two studies that have at least compared the differences between photochemical processes induced by ns- versus fs-laser irradiation.

7.4.1
Sulfidation of Silicon Nanostructures by Femtosecond Irradiation

Pulsed laser deposition experiments early on showed that the ablation target was structured after repeated exposure to nanosecond pulse, ultraviolet lasers. This discovery led very quickly to the demonstration that silicon surfaces with remarkably regular, conical structures could be prepared by irradiating silicon in various gases, notably SF_6. (The formation of the structures requires a reactive gas; without it, or in vacuum, there is only ablation.) In typical experiments, a clean Si surface is irradiated in a vacuum chamber filled with approximately 1 Bar SF_6; large areas, up to 1 cm², can be prepared by rastering the beam across the target. The structure, composition and optical properties of laser-microstructured Si prepared using nanosecond (KrF laser, 248 nm) versus femtosecond (amplified Ti:sapphire laser, 800 nm) pulses have been compared [59]. The microscale surface morphology of the samples prepared in the two regimes is quite different, with the former smooth but with half-micron-sized protrusions, while the latter is covered with nanoparticles 10–50 nm in diameter. The photoconductivities of the two kinds of material are comparable, as is the gross morphology.

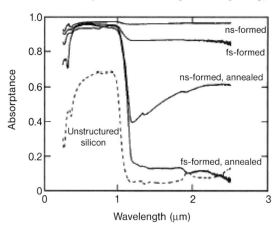

Fig. 7.14 Optical spectra of silica nanocones produced by ns- and fs-laser irradiation in low-pressure SF_6, before and after annealing. From Ref. [59].

The most intriguing result of this work is that, although the fractional incorporation of sulfur in the top 100–200 nm of the conical structures is very similar (~0.5%) for nanosecond versus femtosecond illumination, the near-infrared absorbance is substantially different, as shown in Fig. 7.14. Whereas the ns-annealed samples show an absorbance around 0.5 with some wavelength dependence, the fs-annealed samples are essentially transparent from 1.0 µm out to 2.5 µm wavelength. This may result from the fact that the sulfur-enriched surface layer in the fs-annealed samples is amorphous, while it is significantly single-crystalline in the ns-annealed material, with the sulfur largely incorporated into substitutional sites.

7.4.2
Nitridation of Metal Surfaces Using Picosecond MIR Radiation

Nitridation is an extremely important way of improving the tribological properties of metal surface. The process turns Ti, for example, from a rather undesirable, somewhat soft material, into a material with reasonable wear resistance and all of its other desirable properties, such as bio-inertness. Titanium substrates (impurity concentration less than 100 ppm) were irradiated by 0.5–0.6 ps micropulses from the Jefferson National Accelerator Facility's infrared free-electron laser (FEL) [60]. The wavelength chosen was 3.1 µm, and the duration of each macropulse was 50–1000 µs; the average power during a macropulse was 750 W. The surface nitrogen concentration was subsequently measured by resonant nuclear reaction analysis, in the reaction $^{15}N(p,a\gamma)^{12}C$ (429.6 keV); crystallographic analysis and surface morphology were measured by standard X-ray diffraction and electron microscope methods. Interestingly, direct comparisons of FEL, Ti:sapphire ultrafast laser and excimer nitridation, show a much more mixed picture [61]. The highest-efficiency nitriding of iron and steel was achieved with the ns excimer laser; the FEL was not effective. On the other hand, all lasers showed good efficiency for nitriding titanium, a result attributed to a combination of low optical reflectivity, thermal conductivity and entropy of TiN formation. Nitriding of Al is not effective with ultrashort-pulse lasers because of the larger reflectivity compared to Ti at NIR wavelengths. How radiation in the MIR spectrum might alter this picture is as yet unknown.

7.5
Photomechanical Effects at Femtosecond Timescales

Because a 100 fs laser pulse has a duration comparable to, or even shorter than, lattice relaxation times, it is possible to use femtosecond lasers to modify materials photomechanically by the generation of coherent phonons [62] and shock waves [63]. This opens the way to generating phase transitions, producing new states of matter under extreme pressures, or quenching incipient phase transitions.

Shock waves produced by femtosecond laser pulses can be shaped to a planar profile suitable for studying the shock response of materials by using a combination of ultrafast laser ablation and Kerr-lens focusing to spatially flatten the pulse [64]. The shock-wave profiles in Fig. 7.15 were produced by focusing femtosecond pulses from a CPA Ti:sapphire laser onto thin metal films (0.05–2.0 μm thick) vapor-deposited onto borosilicate glass cover slips. As the fluence of the femtosecond pulse increases, the quasi-Gaussian spatial profile is gradually flattened. Moore et al. suggest that this flattening is caused by dielectric breakdown and incipient ablation of the glass substrate, which blocks the central high-intensity Gaussian peak due to nonlinear absorption. This inference is based on the observation that the fluence being used in the experiments is close to the ablation threshold.

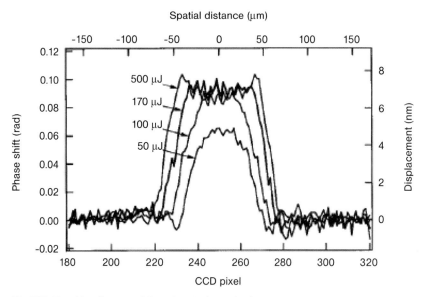

Fig. 7.15 Transition from quasi-Gaussian to planar shock profile as measured by interferometric phase shift in an Al film illuminated by an ultrafast laser. From Ref. [64].

7.5.1
Shock Waves, Phase Transitions and Tribology

A good example of fs shock waves influencing a phase transition comes from an experiment on laser quenching of the ε phase of iron [65]. The normal α-bcc iron structure converts to a γ fcc phase at higher temperatures, and to the ε-hcp phase at still higher pressures. However, shock-wave experiments show that 180 ns are required to change α-bcc iron into ε-hcp iron by a shock-induced transition. The Debye temperature of the ε-hcp phase is about 700 K; this corresponds to a vibrational period of 7×10^{-14} s. The 120 fs laser pulse corresponds to only two phonon vibrations, and thus cannot cause the phase transition; but the femtosecond shock wave loosed by this laser pulse can quench the ε-hcp phase of iron.

One important area of mechanically induced materials modification, is that observed in tribology, when a lubricant is subject to a rapid mechanical deformation. A recent experiment [66] simulated this effect by driving a shock wave into a layer of an alkane self-assembled monolayer (SAM) using the geometry shown in Fig. 7.16. As seen in the figure, the shock is generated by absorption of the femtosecond pump pulse in a thin Au film; the planar shock wave accelerates the Au surface to a velocity of 0.5 nm ps^{-1}. The response of the alkane groups was probed simultaneously by sum-frequency generation combining probe pulses from broadband and narrow-band signals derived from the Ti:sapphire laser. The shock wave produces a purely elastic compression in pentadecane thiol; that is, the recovery of the structure of the SAM is complete. However, in octadecane thiol (ODT), the SFG signal returns to only a fraction of its initial value; analysis indicates that, up to a normalized volume change $DV{\sim}0.07$, the ODT response is elastic. Above that value, however, there is a mechanical failure of the ODT structure resulting in a *trans*-to-*gauche* isomerization of the ODT that produces *gauche* defects in the ODT.

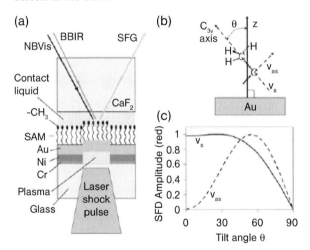

Fig. 7.16 (a) Geometry of an experiment using a femtosecond laser generated shockwave together with narrow-band (NB) and broad-band (BB) IR laser light to generate sum-frequency radiation (SFG) that is indicative of (b) the angle of the molecular orientation in the self-assembled monolayer (SAM). (c) shows how the SFG amplitude varies with tilt angle to enable quantitative analysis of the SAM orientation. From Ref. [66].

7.5.2
Coherent Phonon Excitations in Metals

Initiation of phase transitions by ultrafast laser excitation in aluminum and bismuth have been studied using coherent-phonon generation; the atomic-scale structural changes during the transition from one state to another have been confirmed by femtosecond X-ray or electron diffraction. While these techniques are still so difficult that they have few practitioners, they are providing valuable tests of our understanding of femtosecond-laser induced photomechanical modifications in model materials, such as Bi and Al.

In Bi, a semi-metal, the stable structure is a weak rhombohedral distortion of a cubic unit cell. A 120 fs pump pulse corresponding to a fluence of 6 mJ cm^{-2} incident on the (111) surface of a thin Bi film was used to launch an optical phonon mode into the Bi film [67]. An incoherent X-ray probe pulse, created by focusing a portion of the Ti:sapphire laser pulse onto a Ti wire, was used to probe the diffracted signal from the (111) and (222) lattice planes as a function of the time delay between the Ti:sapphire pump and the X-ray probe pulses. The results, shown in Fig. 7.17(a), indicate a coherent oscillation of the crystal lattice planes at an oscillation frequency ν_{1g} = 2.12 THz, evidently significantly downshifted from the normal A$_{1g}$ phonon frequency of 2.92 THz. Excitation at 20 mJ cm^{-2}, on the other hand, produced an X-ray signal without any sign of oscillations, evidently due to damage and disordering of Bi that is close to its melting point (Fig. 7.17(b)). The oscillations shown in Fig. 7.17(a) are indicative of a substantial change in the nearest-neighbor distances compared to their equilibrium values, of order 5–8%. Given that the Lindemann criterion for melting suggests a change of 10–20% in the nearest-neighbor distance, this interpretation seems quite reasonable.

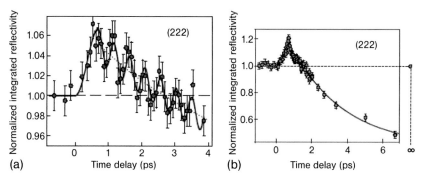

Fig. 7.17 (a) X-ray signals from the (222) plane of Bi film undergoing laser irradiation at 6 mJ cm^{-2} showing oscillations due to coherent phonon generation. (b) X-ray signals from the (222) plane of Bi film undergoing laser irradiation at 20 mJ cm^{-2} showing that the Bi has been driven into a disordered phase at higher laser fluence. From Ref. [67].

A different kind of detailed information about phase transitions driven by ultrafast coherent phonons in solids has been made possible by the recent development of a femtosecond electron-diffraction source [68]. In this case, the fs pump is used to drive the transition as well as to provide the femtosecond photoelectrons from commercial photocathodes. The electron-diffraction approach has two important advantages for time-resolved spectroscopy of condensed phases: the high cross-section of electron–electron interactions compared to X-rays; and the short mean-free path in condensed phases, corresponding to high time resolution. The duration of the electron pulse can be measured by cross-correlation techniques [69]. Figure 7.18 shows the radial electron-density function for a melting transition initiated by launching coherent phonons into an aluminum film. The darkest curve, taken before the laser pulse strikes the sample, shows distinct oscillations approximately every Å out to 13 Å, indicating the existence of the long-range correlations that are to be expected in the fcc aluminum lattice. At 6 ps after the coherent phonons are launched, and thereafter, the longer-range correlations have largely disappeared, although it also appears that the liquid phase is not fully equilibrated even by 50 ps. A similarly detailed picture of the ultrafast melting process is conveyed by the time-dependent pair-correlation function (not shown), permitting one to extract a time-dependent picture of nearest-neighbor distances and vibrational frequencies.

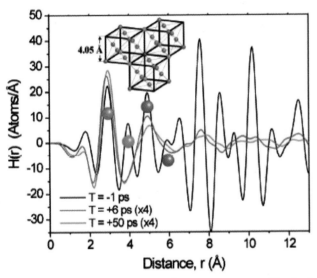

Fig. 7.18 Electron-density distribution function H® measured by ultrafast electron diffraction as a function of time following the launching of a coherent phonon excitation by a femtosecond laser. From Ref. [9].

7.5.3
Ultrafast Laser-induced Forward Transfer (LIFT)

The fact that "cold" laser ablation – that is, without the diffusive thermal effects seen in nanosecond laser ablation – can be driven by picosecond and femtosecond laser pulses opens up interesting new opportunities in material transfer applications, such as laser-induced forward transfer (LIFT). These can be characterized generically as "impulse-driven" materials transfer processes. In the standard geometry, a thin film of the material to be deposited by LIFT is made on a quartz or other transparent wafer; a UV laser is incident from the glass side, and the receiving surface is located a fraction of a mm from the film material to be transferred.

Deposition of metal and metal-oxide structures with sub-micron spatial resolution has been carried out recently with a 248 nm, 0.5–0.6 ps laser [70]. Thin films of In_2O_3 (50–450 nm thick) and Cr (40, 80 and 200 nm) were prepared by reactive pulsed-laser deposition and sputtering or e-beam evaporation, respectively. The most important result was the demonstration of sub-micron spatial resolution, in contrast to LIFT with excimer lasers having pulse durations of 10s of nanoseconds; there feature sizes of 20–100 µm have been published. The most important advantage of the sub-ps pulses has been the reduction in the LIFT threshold, which means that the material ejected from the surface is relatively cold and therefore undergoes little spreading upon reaching the receiving surface, despite the fact that the LIFTed material travels at Mach 0.75. An electron micrograph of a computer-generated holographic pattern by fs Cr microdeposition, exhibiting sub-micron resolution, is shown in Fig. 7.19.

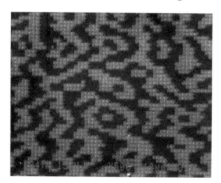

Fig. 7.19 Holographic pattern created by laser-induced forward transfer of Cr by a 500 fs UV laser. The pixel size is 3 µm; the LIFT target was a 400 Å Cr film, illuminated from the rear. From Ref. [70].

7.6
Pulsed Laser Deposition

Pulsed laser deposition (PLD) with ultrafast lasers has a much more recent history than ultraviolet (UV) PLD, and the results that have so far been achieved are still being weighed. Nevertheless, several characteristics have appeared that bear on the ability to use ultrafast PLD to synthesize thin films. The salient characteristics of the ablation source seem to be the following. For fs-NIR ablation, the ablation plume is much more forward-directed than ns-NIR or ns-UV ablation; it has a higher ion fraction, and the ions are much more energetic than those produced by ns-NIR or UV ablation. This creates both new opportunities and new challenges. For ps-IR-PLD, the ablation plume is much colder (less electronic excitation), and the ion content even lower than that observed in fs-NIR ablation. Here we present the results of some representative experiments in this field in which one capitalizes on the unique properties of femtosecond NIR lasers and tunable MIR free-electron lasers.

7.6.1
Near-infrared Pulsed Laser Deposition

Femtosecond lasers were used almost as soon as the CPA laser was developed for NIR-PLD experiments. The first results showed clearly that ultrafast lasers were superior to nanosecond lasers in PLD of many materials systems, for many of the reasons already identified in our catalogue of fundamental properties of laser-materials interactions at the picosecond and sub-picosecond timescales. Specifically, in the case of thin-film growth by laser-assisted deposition, the nonthermal character of ultrafast laser ablation led to reduced ablation thresholds, dramatically reduced particulate formation [3], and an auxiliary source of energy for film growth in the form of energetic neutral and ionic species formed by collisional processes in the ablation plume. In some cases, this processing, regime enables unusually high-quality thin-film growth, as in the recent demonstration of epitaxial SnO_2 film growth on sapphire (Fig. 7.20) [71]. In other cases, the unusual energetics of fs PLD have made it possible to observe growth mechanisms – such as

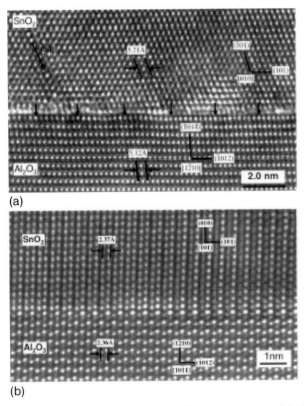

(a)

(b)

Fig. 7.20 High-resolution transmission electron micrographs of the interface between the substrate (Al_2O_3) and the fs-PLD deposited SnO_2 viewed in the SnO_2 (a) [010] direction and (b) [$\bar{1}01$] direction. From Ref. [71].

step-flow growth – that are not observable in conventional MBE or ns-PLD growth processes [72].

Comparative studies using both ns and fs or sub-ps lasers for PLD are relatively scarce, but those that exist clearly benefit from combining the generic characteristics of PLD (e.g., stoichiometric mass transport) with the ultrafast mechanisms that privilege fs over ns laser ablation. In PLD of AlN, for example [73], it was observed that the ablation target did not metallize when illuminated by a 0.5 ps, 248 nm laser pulse. Apparently, because of this, the AlN targets deposited by the sub-ps laser do not show the excess aluminum content that is characteristic of ns PLD of AlN [74]. A time-of-flight and optical emission study of the ns versus fs ablation plumes showed a characteristic difference: whereas the ns-ablation plume showed a preponderance of Al^+ ions with thermal velocities, the fs-ablation plume was dominated by AlN^+ ions, with a bimodal ion distribution having both thermal (~1 eV) and hyperthermal (~10 eV) components. This would seem to indicate a quite different mechanism of plasma formation in the two cases. The sub-ps PLD films displayed accurate stoichiometry and reasonable morphologies.

However, comparisons can be deceptive unless one studies the characteristics both of the ablation process and the deposited films across a range of materials. Another comparative study [75] of ns versus fs PLD of ZnO on sapphire substrates, concentrating this time on the film characteristics, found distinctive differences and not all favorable to fs PLD. In both cases, smooth, dense and stoichiometric ZnO films were grown, with the correct hexagonal structure and without measurable microparticulate content. Whereas the ns-PLD ZnO films showed crystallites roughly 50 nm in size, the crystallites in the fs-PLD ZnO film were only one-third as large, suggesting a distinctive growth pattern, perhaps the stacking of nanocrystallites of ZnO. The fs-PLD films, however, showed less film stress, but more defects – possibly caused by the higher-energy ions in the ZnO plume. Plume images showed the characteristic jet-like forward-directed plume for fs ablation, and a more nearly hemispherical expanding plume for ns-PLD.

7.6.2
Infrared Pulsed Laser Deposition of Organic Materials on Micro- and Nanostructures

Poly(tetrafluoroethylene), or PTFE, is an addition polymer prized for its biological inertness, high dielectric strength and excellent tribological properties – all leading to applications for films of PTFE. However, because PTFE is insoluble, solution-based coating processes in use for other polymers do not work; moreover, UV-PLD [76, 77, 78] and plasma deposition both "unzip" the polymer [79], so that the PTFE vapor consists primarily of monomers. Hence, PTFE films can be deposited by UV-PLD only if the substrate is heated to 600 °C or so, to re-polymerize the deposited material. These temperatures render such vapor-phase processes unsuitable for many applications, such as coating micro-electro-mechanical system (MEMS) components. Femtosecond PLD with Ti:sapphire produces only PTFE-*like* films [80].

In experiments with a FEL, crystalline PTFE was deposited successfully on crystalline substrates by resonant IR-PLD, choosing a wavelength of 8.26 μm (C–F$_2$ stretch) [52]. The deposition was highly efficient, with an ablation threshold around 0.26 J cm^{-2}; that the RIR-PLD was nonthermal was shown by the very shallow penetration depths, and the calculated modest temperature rise, well below the melting point for the FEL wavelength. Comparison with ablation behavior at 4.14 μm (the first overtone of the C–H stretch) showed both a higher ablation threshold (3.5 J cm^{-2}) and a higher temperature reached at threshold; the ablation products seemed to be the mix of liquid and vapor which is typical of explosive vaporization. At 8.26 μm, the FEL-PLD films were largely free of the particulates seen in UV-PLD [81].

X-ray photoelectron spectroscopy and Fourier-transform infrared spectroscopy of the deposited films showed that their electronic structure reproduced that of the starting material very well, whether that starting material was pressed pellets or solid Teflon® rod stock. Crystallinity and surface morphology (as seen by atomic-force microscopy) of the deposited films were somewhat improved by heating to temperatures well below the melting point of the PTFE. However, it should be stressed that even samples deposited at room temperatures appeared to be reasonably crystalline. To ascertain the quality of films that could be deposited on microstructures, PTFE was ablated and deposited on transmission electron-microscope grids, as shown in Fig. 7.21. Surface roughness of the films on Si substrates, it was of order ±10 nm. Of particularly interest in this deposition experiment is the fact that the deposited film follows the contours of the grid without overrunning the edges; this quite uniform recession from the edges is not understood, but is reproducible.

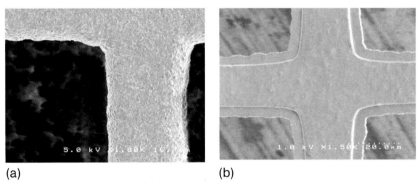

(a)   (b)

Fig. 7.21 Scanning electron micrograph of nickel transmission electron-microscope grid (a) as delivered and (b) after coating with a 135 nm film of PTFE (Teflon). Note the fidelity to the grid shape, and the sharp edge of the PTFE film that does not extend over the edge of the film. The scale bars are respectively 16.7 μm (a) and 20 μm (b). From Ref. [52].

7.7
Future Trends in Ultrafast Laser Micromachining

Although their commercial application is still limited, the promise of femtosecond laser-materials interactions to provide smaller heat-affected zones, higher intensities and machining capability, for a great variety of materials, is pushing the field forward. The development of laser systems with high pulse-repetition frequencies, higher pulse energies, sub-femtosecond pulse durations and ever broader tunability, hints at interesting developments to come in ultrafast micromaching and other materials processing applications.

7.7.1
Ultrashort-pulse Materials Modification at High Pulse-repetition Frequency

Recent proposals for laser processing of materials using ultrashort pulses at high pulse-repetition frequency (PRF) represent a watershed in thinking about laser materials processing [82]. The use of ultrashort pulses minimizes collateral thermal damage and particulate formation. Because reaction *rates* are proportional to intensity and cross-section, rather than to fluence, at some point the combination of high PRF and high intensity will win out over high fluence and lower PRF in overall yield or throughput. In addition, at the higher intensities typical of femtosecond lasers, nonlinear effects may produce an additional yield of desirable products. Moreover, the low-fluence, high-PRF processing régime results in an overall lowering of process temperatures.

There have been few systematic tests of this novel materials-processing paradigm. Mode-locked Nd:YAG lasers used in early high-PRF experiments had 100 ps pulses; here the intensities are too low and the material rapidly reaches thermal equilibrium, although films of amorphous carbon made in this way [83] exhibit intriguing magnetic properties [84]. Moreover, given the fixed frequencies of most high-PRF lasers, it is not possible to optimize the spatio-temporal density of excitation during ablation by tuning to material resonances.

However, a recent experiment on bulk heating of transparent materials shows that sub-surface structuring with low pulse energies at high-PRF is feasible. In these experiments, bulk Corning 0211 Zn-doped silicate glass was irradiated by 100 fs 800 nm pulses; the pulse energies ranged up to 300 nJ, and a 25 MHz pulse-repetition frequency was made possible by a long cavity [85]. Using high numerical aperture optics (N.A. 0.5–1.4), subsurface waveguides were written in Corning 0211 with pulse energies as small as 5 nJ; Fig. 7.22 shows the profile of 633 nm laser light exiting one of those waveguides. One of the questions that has to be answered is the extent to which the damage in the glass, the increase in refractive index, is the result of melting or some other process. A thermal diffusion model, assuming an absorption coefficient of 0.30, fits the size of the features made by the fs irradiation up to about 4 μm radius as shown in Fig. 7.21; for the larger features seen microscopically, it appears that the difference between model and experiment is explained by the lower heat conduction of the halo surrounding the focal spot.

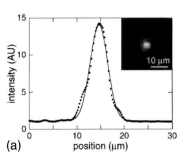

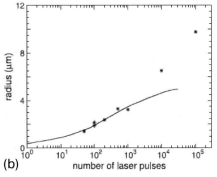

(a)

(b)

Fig. 7.22 (a) Intensity profile of a 633 nm laser beam transmitted through the 5 µm end aperture of a waveguide created in bulk Corning 0211 by a 25 MHz Ti:sapphire oscillator focused to create an explosive expansion of the electron–hole plasma initiated by the laser pulse. From Ref. [85]. (b) Radius of the waveguide structure created by a train of 30 fs, 5 nJ, 800 nm wavelength pulses from a 25 MHz Ti:sapphire oscillator, as a function of number of laser pulses. The curve is a one-dimensional calculation of the melt diameter; at very large shot numbers, the model breaks down, as explained in the text. From Ref. [98].

7.7.2
Pulsed Laser Deposition at High Pulse-repetition Frequency

In a recent paper [86], the Luther-Davies group have proposed a design for a table-top high repetition-rate, ultrashort-pulse laser processing that could be particularly attractive for pulsed laser deposition. Ideally, laser-ablation thin-film deposition is accomplished by a vaporization mechanism that employs relatively modest pulse energies to ablate a small amount of material; relatively high intensity to enhance cross-section; and high pulse repetition frequency (PRF) to optimize throughput. At high PRF, this makes the PLD process almost continuous, since the accommodation time of the vapor arriving on the substrate is typically many tens of nanoseconds or longer. Since each pulse of vapor carries relatively little material, accommodating the arriving atoms or clusters is also more efficient.

7.7.2.1 Deposition of Inorganic Thin Films
The principal difficulty with fs-PLD using Ti:sapphire laser technology is that the laser systems, with Hz to kHz pulse repetition frequency and low average power (~1 W), are not able to obtain high production rates. The Jefferson National Accelerator Facility's free-electron laser (FEL) offers a unique combination of laser parameters for PLD: high micropulse intensity (10^{10}–10^{11} W cm^{-2}) but relatively low micropulse energy (a few µJ), broad tunability (2–10 µm), and high average power (kW!) achieved through high micropulse repetition frequency (up to 75 MHz). This means that the ablation process mimics to perfection the regime envisioned by Gamaly et al. [87]. Using this source over a wavelength range 2.9–3.1 µm, with

micropulse durations of 650 fs, micropulse energies of 5–20 μJ, films of $Ni_{80}Fe_{20}$ (permalloy) were deposited; for comparison purposes, an amplified Ti:sapphire system was also used to deposit the same films, but with pulse durations of 150 fs, pulse energies of 0.5–0.7 mJ at pulse repetition rates of up to 1 kHz.

This direct comparison of low and high pulse-repetition rate PLD shows clear contrasts generally consistent with expectations. The FEL-PLD film is much smoother (see Fig. 7.23) than the fs-PLD film; the ablation plume in the case of FEL-PLD has a predominantly black-body character with temperatures on the order of 1700–2400 K, while the fs-PLD film shows a preponderance of line emission. The deposition rate for the fs Ti:sapphire system was 10^{-3} Å per fs pulse, while that for the FEL was 5×10^{-7} Å per micropulse; nevertheless, the much higher pulse repetition frequency of the FEL (75 MHz versus 1 kHz) yielded an overall growth rate of 17 Å s^{-1} as compared to 1 Å s^{-1} for the amplified Ti:sapphire laser. An interesting test of the quality of the film is a measurement of the magnetization of the Permalloy film as a function of applied magnetic field. The fs-PLD film showed a low magnetization with substantial coercivity (Fig. 7.23), while the FEL-PLD film showed high magnetization and virtually no coercivity. These differing behaviors reflect a better organized grain structure in the FEL-PLD film, as one would infer from the AFM comparisons of the film surfaces, which indicate a much smoother film in this case.

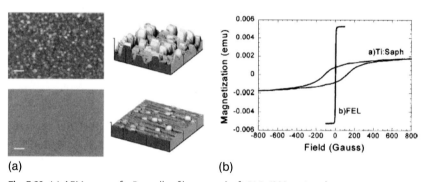

(a)   (b)

Fig. 7.23 (a) AFM scans of a Permalloy film grown by fs-PLD (800 nm) at the top, and by FEL-PLD (3 μm) at the bottom. (b) Comparison of the magnetization curves for the Permalloy films grown by these same two methods. From Ref. [99].

Here the wavelength of the FEL is almost certainly not an issue, as the penetration depths in the ablation target are very small even for mid-IR wavelengths. However, the difference in processing outcomes due to the high intensity, low pulse energy and quasi-continuous deposition of small quantities of material, suggest that similar protocols are likely to be effective with other metals and perhaps even with other inorganic films.

7.7.2.2 Deposition of Organic Thin Films

Laser-assisted deposition of organic materials for sensors is a particularly intriguing potential application for the future [88, 89]. In such applications, precise thickness control, while maintaining the electronic structure, chemical functionality and thermo-mechanical properties of the polymer, is paramount. This is because most sensing schemes rely on sensitive monitoring of thermal, chemical, electronic or physical responses of the film to adsorbates, all of which may be thickness dependent. For example, in a surface acoustic-wave (SAW) sensor, the resonant response (Q-factor) of the sensor is extremely sensitive to nonuniformities in film thickness, which degrade device performance. One such molecule is fluoropolyol; along the molecular backbone, a functional group is attached which has a strong affinity for the nerve agent Sarin (GB). The local structure, chemical functionality and polydispersity of fluoropolyol films deposited either on NaCl plates for later optical analysis, or on Si substrates from which the polymer was dissolved for mass spectral analysis, compared favorably with the bulk starting material [90].

Another example is the fluorinated polysiloxane SXFA, a branched polymer also suited for sensor applications. For SXFA, as for several other polymers, the target material was frozen in liquid nitrogen before mounting in the vacuum chamber on an unheated, rotating stage; the growth substrate, maintained at room temperature, was 4–6 cm from the surface of the target. Typical irradiation times were a few minutes per film. Films of various polymers were deposited on both Si (111) and NaCl substrates for subsequent analysis by gravimetry, atomic-force and optical microscopy, Fourier transform infrared (FTIR) spectrophotometry and gel-permeation chromatography (GPC). Films were also deposited on cantilevers to test the uniformity and sensitivity of the films under conditions approximating those for sensors, as shown in Fig. 7.24. Related RIR-PLD tech-

(a) (b)

Fig. 7.24 (a) Fluorinated polysiloxane (SXFA) coated cantilever sensor structure, with the SXFA deposited by a free-electron laser tuned to 2.94 μm. (b) Molecular structure of SXFA, showing the functional side group designed to detect explosive vapors.

niques have been applied successfully to the transfer of biomolecules (DNA and proteins) for biomedical sensors.

7.7.3
Picosecond Processing of Carbon Nanotubes

Single-walled carbon nanotubes (SWNTs) have been produced by arc discharge, chemical vapor deposition and pulsed laser vaporization, invariably as a fraction of the total number of carbon nanotubes produced. However, because the rates of production are typically too low to make their use in many applications practical, a major challenge is to increase the yield of SWNTs without sacrificing the integrity of the nanotube wall. Preliminary experiments using the Jefferson National Accelerator Facility's free-electron laser (FEL) gave a production rate of 1.5 g h^{-1}, without optimization of wavelength or other parameters [100]. The advantage of the FEL in this case was the combination of sub-ps pulses with 200 W average power on target. However, in this case the micropulses, each with an energy of a few µJ, arrive at a rate of 75 MHz, one approximately every 13 ns. The ablation protocol effectively mimics the conditions envisioned by Gamaly et al. [87]: an ultrashort pulse ablates the target at a rate much faster than typical diffusive relaxation times, but with much less energy per pulse. Hence FEL-PLD is much more like a continuous ablation process than ablation with conventional ns or fs lasers at much lower repetition rates, where a pulse with relatively large energy and carrying away a substantial amount of target mass arrives only infrequently, with a consequently large target overpressure and often with the undesirable particulates that are the bane of PLD experimenters.

7.7.4
Sub-micron Parallel-process Patterning of Materials with Ultraviolet Lasers

Much of the laser structuring that has been done up to now with Ti:sapphire laser systems is limited, by intensity considerations, to serial processing. However, industrial scale micromachining and materials processing will generally require much higher throughputs. This leads to requirements for higher pulse energies and higher photon energies to permit the deployment of mask processes (parallel processing) and the use of optical interference techniques to allow the structuring of large areas. By using well-known optical techniques to smooth the pulse via two-photon absorption, and sequence the pulses in space and time over the mask, impressive results in both throughput and material quality are achieved.

An important development in this area is the use of a multi-pass KrF amplifier to dramatically amplify the output of the third harmonic from a Ti:sapphire oscillator [91, 92]. The amplified output of a master oscillator-power amplifier (MOPA) system at 248 nm – with 300 fs pulse duration, 30 mJ per pulse and 350 Hz prf – offers the possibility of structuring a great many more materials (e.g., polymers, insulators and of course semiconductors and metals) and doing so in a parallel-processing protocol. Figure 7.25 shows simulated and actual patterns with sub-

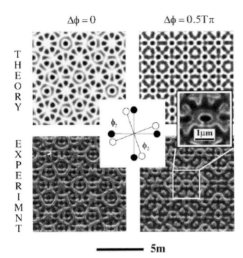

$\Delta\phi = 0$ $\qquad\qquad \Delta\phi = 0.5T\pi$

T
H
E
O
R
Y

E
X
P
E
R
I
M
E
N
T

1μm

————— 5m

Fig. 7.25 Complex structures written in polycarbonate by 300 fs pulses (248 nm) using a phase grating structure with two different phases as shown above the drawing. The horizontal scale bar is 5 μm long, not 5 m. The "Theory" structures are those calculated by a computer simulation of the laser interaction modulated by the phase grating. From Ref. [91].

micron resolution prepared in polycarbonate using the MOPA and advanced optical techniques, such as four-beam interference using phase and/or amplitude gratings. Given the advantages of parallel processing and the marriage of ultrafast laser pulses with advanced optical processing schemes, this kind of development must be regarded as auspicious. Marry these technologies to more sophisticated laser ablation protocols involving active control of the ablation plume [93], and it appears that there are many new possibilities for microfabrication.

7.8
Summary and Conclusions

Ultrafast laser micromachining is enabled by unique characteristics of femtosecond lasers and of the laser–materials interaction. For NIR lasers, these include a high density of electronic excitation due to a relatively small focal spot, short pulse duration compared to electron–phonon relaxation times, variable pulse duration and a pulse repetition frequency as high as 100 MHz, and very high peak intensities. These make possible the critical capabilities needed for machining metals (avoidance of a large heat-affected zone); semiconductors (dense electron–hole plasma to destabilize the lattice and drive the material into ablation or new states); and insulators (high multi-photon excitation probability and strong nonlinear effects). MIR free-electron lasers offer higher pulse energies at high prf (μJ instead of nJ) and the much higher average power that can be obtained using accelerator technology; the penalty, of course, is the substantially greater expense.

Among the most exciting opportunities are: (1) the attosecond frontier, made possible by novel techniques for generating subfemtosecond pulses in the soft X-ray region of the spectrum; (2) a variety of possibilities for compact, less expensive ultrafast MIR lasers based on optical parametric amplifier technology; and (3) the marriage of femtosecond Ti:sapphire oscillators to amplifiers with other gain media and to advanced optical techniques. Up to the present time, the pulse energies of attosecond laser sources are too low to make attosecond materials processing a realistic possibility. Nevertheless, the fact that such lasers could produce band-to-band transitions in insulators with single-photon transitions, suggests that we may soon see some applications other than materials spectroscopy. This looks like a promising long-term development. Although there have been a few experiments using free-electron laser technology to amplify femtosecond pulses from solid-state sources, this work is still in its infancy and is likely to be a serious development only in the medium term. The last set of developments – the marriage of hybrid laser technologies and advanced optical masking techniques – is here already, and looks like the link to near-term commercial developments in femtosecond laser micromachining, using all the many advantages that the femtosecond photophysics, photochemistry and photomechanics provide.

Acknowledgment

The author thanks the Alexander von Humboldt Foundation for financial support through a Senior Scientist Award, and the faculty of the Fachbereich Physik, Universität Konstanz for their hospitality during the writing of this article.

References

1 P. Maine, D. Strickland, P. Bado et al., Ieee Journal of Quantum Electronics **24** (2), 398 (1988); M. D. Perry and G. Mourou, Science **264** (5161), 917 (1994).

2 G.R. Neil, C.L. Bohn, S.V. Benson et al., Physical Review Letters **84** (22), 5238 (2000).

3 P.P. Pronko, S.K. Dutta, D. Du et al., Journal of Applied Physics **78** (10), 6233 (1995).

4 D. Ashkenasi, G. Muller, A. Rosenfeld et al., Applied Physics A – Materials Science & Processing **77** (2), 223 (2003).

5 D. Baeuerle, *Laser Processing and Chemistry*, 3rd ed. (Springer-Verlag, Berlin, 2000).

6 Applied Physics A – Materials Science & Processing **79** (7), *Special Issue on "Femtosecond and Attosecond Phenomena"* (2004).

7 Applied Physics A – Materials Science & Processing **79** (4–6), *Special Issue on "Laser Ablation"* (2004).

8 N. Itoh and A.M. Stoneham, *Materials Processing by Electronic Excitation.* (Oxford University Press, Oxford, 2001).

9 B.J. Siwick, J.R. Dwyer, R.E. Jordan et al., Science **302** (5649), 1382 (2003).

10 K. Sokolowski-Tinten, J. Bialkowski, A. Cavalleri et al., Physical Review Letters **81** (1), 224 (1998).

11 B.N. Chichkov, C. Momma, S. Nolte et al., Applied Physics A – Materials Science & Processing **63** (2), 109 (1996).

12 L.A. Falkovsky and E.G. Mishchenko, Journal of Experimental and Theoretical Physics **88** (1), 84 (1999).

13 B. Rethfeld, A. Kaiser, M. Vicanek et al., Physical Review B **65** (21), 214303 (2002).

14 Y. Hirayama and M. Obara, Journal of Applied Physics **97** (6) (2005); Y. Hirayama and M. Obara, Applied Surface Science **197**, 741 (2002).

15 H. O. Jeschke, M. E. Garcia, and K. H. Bennemann, Physical Review Letters **87** (1), 015003 (2001).

16 C.V. Shank, R. Yen, and C. Hirlimann, Physical Review Letters **51** (10), 900 (1983).

17 Y. Siegal, E.N. Glezer, L. Huang et al., Annual Review of Materials Science **25**, 223 (1995).

18 L. Huang, J.P. Callan, E.N. Glezer et al., Physical Review Letters **80** (1), 185 (1998).

19 J.P. Callan, A.M.T. Kim, L. Huang et al., Chemical Physics **251** (1–3), 167 (2000).

20 B. Rethfeld, K. Sokolowski-Tinten, D. von der Linde et al., Applied Physics A – Materials Science & Processing **79** (4–6), 767 (2004).

21 P. Stampfli and K.H. Bennemann, Applied Physics A – Materials Science & Processing **60** (2), 191 (1995); P. Stampfli and K. H. Bennemann, Physical Review B **49** (11), 7299 (1994).

22 D. Du, X. Liu, G. Korn et al., Appl. Phys. Lett. **64** (23), 3071 (1994).

23 B.C. Stuart, M.D. Feit, A.M. Rubenchik et al., Physical Review Letters **74** (12), 2248 (1995).

24 P. Daguzan, S. Guizard, K. Krastev et al., Physical Review Letters **73** (17), 2352 (1994).

25 A. Kaiser, B. Rethfeld, M. Vicanek et al., Physical Review B **61** (17), 11437 (2000).

26 S. Guizard, P. Martin, G. Petite et al., Journal of Physics-Condensed Matter **8** (9), 1281 (1996); F. Queré, S. Guizard, P. Martin et al., Applied Physics B – Lasers and Optics **68** (3), 459 (1999).

27 F. Quere, S. Guizard, and P. Martin, Europhysics Letters **56** (1), 138 (2001).

28 M. Li, S. Menon, J.P. Nibarger et al., Physical Review Letters **82** (11), 2394 (1999).

29 G. Petite, S. Guizard, P. Martin et al., Physical Review Letters **83** (24), 5182 (1999).

30 R. Stoian, D. Ashkenasi, A. Rosenfeld et al., Physical Review B **62** (19), 13167 (2000).

31 R. Stoian, A. Rosenfeld, D. Ashkenasi et al., Physical Review Letters **88** (9), 097603 (2002).

32 A.N. Belsky, H. Bachau, J. Gaudin et al., Applied Physics B – Lasers And Optics **78** (7–8), 989 (2004).

33 M. Lenzner, J. Kruger, W. Kautek et al., Applied Physics A – Materials Science & Processing **69** (4), 465 (1999).

34 M. Lenzner, J. Kruger, S. Sartania et al., Physical Review Letters **80** (18), 4076 (1998).

35 B. Rethfeld, Physical Review Letters **92** (18), 187401 (2004).

36 S. Iwai, A. Nakamura, K. Tanimura et al., Solid State Communications **96** (10), 803 (1995); S. Iwai, T. Tokizaki, A. Nakamura et al., Physical Review Letters **76** (10), 1691 (1996).

37 N. Itoh, J. Kanasaki, A. Okano et al., Annual Review of Materials Science **25**, 97 (1995).

38 D. Ashkenasi, H. Varel, A. Rosenfeld et al., Applied Physics Letters **72** (12), 1442 (1998).

39 E.N. Glezer, M. Milosavljevic, L. Huang et al., Optics Letters **21** (24), 2023 (1996).

40 K. Yamasaki, S. Juodkazis, M. Watanabe et al., Applied Physics Letters **76** (8), 1000 (2000).

41 S. Juodkazis, K. Yamasaki, V. Mizeikis et al., Applied Physics A – Materials Science & Processing **79** (4–6), 1549 (2004).

42 C.B. Schaffer, A.O. Jamison, and E. Mazur, Applied Physics Letters **84** (9), 1441 (2004).

43 M. Lenzner, J. Kruger, W. Kautek et al., Applied Physics A – Materials Science & Processing **68** (3), 369 (1999).

44 L. Shah, J. Tawney, M. Richardson et al., Ieee Journal of Quantum Electronics **40** (1), 57 (2004).

45 N. Itoh, Pure and Applied Chemistry **67** (3), 419 (1995).

46 N. Itoh, T. Shimizuiwayama, and T. Fujita, Journal of Non-crystalline Solids **179**, 194 (1994).

47 M. Richardson, L. Shah, J. Tawney et al., Glass Science and Technology **75**, 121 (2002).

48 O.M. Efimov, K. Gabel, S.V. Garnov et al., Journal of the Optical Society of America B – Optical Physics **15** (1), 193 (1998).

49 J.B. Lonzaga, S.M. Avanesyan, S.C. Langford et al., Journal of Applied Physics **94** (7), 4332 (2003).

50 J.T. Dickinson, S. Orlando, S.M. Avanesyan et al., Applied Physics A – Materials Science & Processing **79** (4–6), 859 (2004).

51 D.L. Andrews, American Journal of Physics **53** (10), 1001 (1985).

52 M.R. Papantonakis and R.F. Haglund, Applied Physics A – Materials Science & Processing **79** (7), 1687 (2004).

53 D.R. Ermer, M.R. Papantonakis, M. Baltz-Knorr et al., Applied Physics A – Materials Science & Processing **70** (6), 633 (2000).

54 D.M. Bubb, J.S. Horwitz, J.H. Callahan et al., Journal of Vacuum Science & Technology A – An International Journal Devoted to Vacuum Surfaces and Films **19** (5), 2698 (2001).

55 D.M. Bubb, M.R. Papantonakis, J.S. Horwitz et al., Chemical Physics Letters **352** (3–4), 135 (2002).

56 D.M. Bubb, B. Toftmann, R.F. Haglund et al., Applied Physics A – Materials Science & Processing **74** (1), 123 (2002).

57 D.M. Bubb, M.R. Papantonakis, B. Toftmann et al., Journal of Applied Physics **91** (12), 9809 (2002).

58 R.R. Cavanagh, D.S. King, J.C. Stephenson et al., Journal Of Physical Chemistry **97** (4), 786 (1993).

59 C.H. Crouch, J.E. Carey, J.M. Warrender et al., Applied Physics Letters **84** (11), 1850 (2004).

60 E. Carpene, M. Shinn, and P. Schaaf, Applied Physics A – Materials Science & Processing **80** (8), 1707 (2005).

61 E. Carpene, P. Schaaf, M. Han et al., Applied Surface Science **186** (1–4), 195 (2002).

62 A.M. Weiner, D.E. Leaird, G.P. Wiederrecht et al., Science **247** (4948), 1317 (1990).

63 K.T. Gahagan, D.S. Moore, D.J. Funk et al., Physical Review Letters **85** (15), 3205 (2000).

64 D.S. Moore, K.T. Gahagan, J.H. Reho et al., Appl. Phys. Lett. **78** (1), 40 (2001).

65 T. Sano, H. Mori, E. Ohmura et al., Applied Physics Letters **83** (17), 3498 (2003).

66 J.E. Patterson, A. Lagutchev, W. Huang et al., Physical Review Letters **94** (1), 015501 (2005).

67 K. Sokolowski-Tinten, C. Blome, J. Blums et al., Nature **422** (6929), 287 (2003).

68 B.J. Siwick, J.R. Dwyer, R.E. Jordan et al., Chemical Physics **299** (2–3), 285 (2004).

69 B.J. Siwick, A.A. Green, C.T. Hebeisen et al., Optics Letters **30** (9), 1057 (2005).

70 I. Zergioti, S. Mailis, N.A. Vainos et al., Applied Surface Science **129**, 601 (1998).

71 P.P. Pronko, P.A. VanRompay, C. Horvath et al., Physical Review B **58** (5), 2387 (1998).

72 M.S. Hegazy and H.E. Elsayed-Ali, Journal of Vacuum Science & Technology A – Vacuum Surfaces and Films **20** (6), 2068 (2002).

73 C. Ristoscu, I.N. Mihailescu, M. Velegrakis et al., Journal of Applied Physics **93** (4), 2244 (2003).

74 E. Gyorgy, C. Ristoscu, I.N. Mihailescu et al., Journal of Applied Physics **90** (1), 456 (2001).

75 J. Perriere, E. Millon, W. Seiler et al., Journal of Applied Physics **91** (2), 690 (2002).

76 Graciela B. Blanchet and S. Ismat Shah, Appl. Phys. Lett. **62** (9), 1026 (1993).

77 N. Huber, J. Heitz, and D. Bauerle, European Physical Journal – Applied Physics **29** (3), 231 (2005).

78 S.T. Li, E. Arenholz, J. Heitz et al., Applied Surface Science **125** (1), 17 (1998).

79 Graciela B. Blanchet, Journal of Applied Physics **80** (7), 4082 (1996).

80 Melissa Womack, Monica Vendan, and Pal Molian, Applied Surface Science **221** (1–4), 99 (2004).

81 J. Heitz, E. Arenholz, and J.T. Dickinson, Applied Physics A: Materials Science & Processing **69** (Suppl.), S467 (1999).

82 E.G. Gamaly, A.V. Rode, and B. Luther-Davies, Journal of Applied Physics **85** (8), 4213 (1999).

83 A.V. Rode, B. Luther-Davies, and E.G. Gamaly, Journal of Applied Physics **85** (8), 4222 (1999).

84 A.V. Rode, R.G. Elliman, E.G. Gamaly et al., Applied Surface Science **197**, 644 (2002); A.V. Rode, E.G. Gamaly, A.G. Christy et al., Physical Review B **70** (5), 054407 (2004).

85 C.B. Schaffer, A. Brodeur, J.F. Garcia et al., Optics Letters **26** (2), 93 (2001).

86 B. Luther-Davies, V.Z. Kolev, M.J. Lederer et al., Applied Physics A – Materials Science & Processing **79** (4–6), 1051 (2004).

87 E.G. Gamaly, A.V. Rode, and B. Luther-Davies, J. of Appl. Phys. **85** (8), 4213 (1999).

88 Robert W. Catrall, *Chemical Sensors*. (Oxford University Press, Oxford, England, 1997).

89 Ursula E. Spichiger-Keller, *Chemical Sensors and Biosensors for Medical and Biological Applications*. (Wiley-VCH, Weinheim, Germany, 1998).

90 D.M. Bubb, J.S. Horwitz, R.A. McGill et al., Applied Physics Letters **79** (17), 2847 (2001).

91 J.H. Klein-Wiele, J. Bekesi, and P. Simon, Applied Physics A – Materials Science & Processing **79** (4–6), 775 (2004).

92 J.H. Klein-Wiele, G. Marowsky, and P. Simon, Applied Physics A – Materials Science & Processing **69**, S187 (1999); J. H. Klein-Wiele and P. Simon, Applied Physics Letters **83** (23), 4707 (2003).

93 E.G. Gamaly, A.V. Rode, O. Uteza et al., Journal of Applied Physics **95** (5), 2250 (2004).

94 B.C. Stuart, M.D. Feit, S. Herman et al., Physical Review B **53** (4), 1749 (1996).

95 M. Lenzner, F. Krausz, J. Kruger et al., Applied Surface Science **154**, 11 (2000).

96 D. Ashkenasi, H. Varel, A. Rosenfeld et al., Applied Physics A – Materials Science & Processing **63** (2), 103 (1996).

97 I. Zergioti, S. Mailis, N.A. Vainos et al., Applied Physics A – Materials Science & Processing **66** (5), 579 (1998).

98 C.B. Schaffer, J.F. Garcia, and E. Mazur, Applied Physics A – Materials Science & Processing **76** (3), 351 (2003).

99 A. Reilly, C. Allmond, S. Watson et al., Journal of Applied Physics **93** (5), 3098 (2003).

100 P.C. Eklund, B.K. Pradhan, U.J. Kim et al., Nano Letters **2** (6), 561 (2002).

8
Formation of Sub-wavelength Periodic Structures Inside Transparent Materials

Peter G. Kazansky

Abstract

Progress in high-power ultrashort pulse lasers has opened new frontiers in the physics and technology of light–matter interactions and laser micro-machining. Surface ripples with a period equal to the wavelength of incident laser radiation have been observed in many experiments involving laser deposition and laser ablation. Such gratings are generated as a result of interference between the light field and the surface plasmon-polariton wave launched because of initial random surface inhomogenities. However, until now there has been no observation of periodic structures being generated within the bulk of a material just by a single writing laser beam, and the mechanism of its appearance has not been fully understood. Recently, new phenomena of light scattering peaking in the plane of polarization during direct writing with femtosecond light pulses in glass have been reported. The phenomena were interpreted in terms of the angular distribution of photoelectrons and sub-wavelength ripple-like index inhomogenities. Another experiment demonstrated uniaxial birefringence of structures in fused silica written by femtosecond light pulses. The index change for light polarized along the direction of polarization of the writing beam was much stronger than for the orthogonal polarization. A further anisotropic property in optical materials, after being irradiated by a femtosecond laser, is strong reflection from the modified region occurring only along the direction of polarization of the writing laser. This can arise from a self-organized periodic sub-wavelength refractive index modulation. The femtosecond-laser-induced birefringence is therefore likely to be caused by these laterally-oriented small-period grating structures. Birefringence of this nature is well known as "form" birefringence. The mechanism of self-organized nano-gratings formation in transparent materials and recent observation of the smallest embedded structures ever created by light, are discussed.

3D Laser Microfabrication. Principles and Applications.
Edited by H. Misawa and S. Juodkazis
Copyright © 2006 WILEY-VCH Verlag GmbH & Co. KGaA, Weinheim
ISBN: 3-527-31055-X

8.1
Introduction

The use of a femtosecond laser source to directly write structures deep within transparent media has recently attracted much attention due to its capability of writing in three-dimensions [1–3]. Tight focusing of the laser into the bulk of material causes nonlinear absorption only within the focal volume, depositing energy that induces a permanent material modification [4, 5]. A variety of photonic devices have already been created by translating a sample through the focus of a femtosecond laser [6, 7]. Although molecular defects caused by such intense irradiation have been identified in fluorescence, ESR and other studies [8], the mechanism of induced modifications in glass is still not fully understood. New phenomena of light scattering [9, 10] and Cherenkov third-harmonic [11] generation, peaking in the plane of polarization during direct writing with ultrashort light pulses in glass, have been reported. These observations were unexpected because the scattering of polarized light in the plane of light polarization in an isotropic medium such as glass is always weaker compared with the orthogonal plane, since a dipole does not radiate in the direction of its axis. The phenomena were interpreted in terms of angular distribution of photoelectrons and sub-wavelength anisotropic index inhomogenities. Another experiment demonstrated uniaxial birefringence of structures in fused silica written by femtosecond light pulses [12]. The index change for light polarized along the direction of polarization of the writing beam was much stronger than for orthogonal polarization. A further anisotropic property, observed in silica after being irradiated by a femtosecond laser, is strong reflection from the modified region occurring only along the direction of polarization of the writing laser [13]. This can arise from a self-organized periodic sub-wavelength refractive index modulation. Surface ripples, with a period equal to the wavelength of incident laser radiation and that are likewise aligned in a direction orthogonal to the electric field, have been observed in experiments involving laser deposition [14] and laser ablation [15]. In this chapter the evidence of self-organized periodic nanostructures (much smaller than the wavelength of incident light) being generated within the bulk of a material, is presented. The femtosecond-laser-induced birefringence is likely to be caused by these laterally-oriented small-period grating structures. Birefringence of this nature is well known as "form" birefringence [16]. Direct observation of the smallest embedded structures ever created by light; and the mechanism of self-organized, nano-gratings formation in transparent materials are discussed [17]. The analysis suggests that self-organized sub-wavelength periodic structures are also the primary cause of all anisotropic phenomena reported in the experiments on direct writing with ultrashort pulses in glass, including self-organized form birefringence [18]. The anisotropic phenomena and related nanogratings should be useful in many monolithic photonic and micro-optic devices and can be harnessed for information storage where nanoscale periodic structuring is required.

8.2
Anomalous Anisotropic Light-scattering in Glass

Anomalous anisotropic light scattering was observed during experiments on femtosecond direct writing in silica glass. A regeneratively-amplified mode-locked Ti:Sapphire laser (120 fs pulse duration, 200 kHz repetition rate) operating at $\lambda = 800$ nm was used in these experiments. The collimated laser beam passed through a variable neutral density filter and half-wave plate before a dichroic mirror reflecting only in the 400–700 nm region. The infrared laser light traveled through the mirror and was focused through a 20x objective into the bulk of the sample, down to a beam waist diameter estimated to be ~ 4.6 µm. The silica sample was mounted on a computer-controlled linear-motor 3D translation stage. To simultaneously observe the writing process, a CCD camera with suitable filters and white light source was used.

During the experiments on Ge-doped (GeO_2 ~ 8 mol %) silica glass strong blue luminescence (with a centre wavelength at 410 nm) of defect states (Ge–Si wrong bonds with a concentration of 10^{19} cm^{-3}) was observed. This luminescence (triplet luminescence) can be excited via the singlet–singlet transition by absorption of three pump photons or one UV photon of the third harmonic of the pump followed by quick nonradiative decay (with a decay time of 1 ns) to the long-lived triplet level. When the pump (50 mW average power, 2 MW peak power, 1.2×10^{13} $\frac{W}{cm^2}$ intensity in the focus of a beam) was focused slightly (~ 50 µm) above the surface of the sample, the shape of the spot of the blue luminescence, imaged via the microscope and CCD camera, was circular. Unexpectedly, it has been discovered that when the pump was focused inside the sample the spatial isotropy of the blue luminescence can be broken (Fig. 8.1): the luminescence scattering increases along the direction of the pump polarization, while the circular shape of the pump beam remains unchanged. The elongated pattern of the blue luminescence followed the direction of the pump polarization, rotated by using a half-wave plate. It should be noted that the blue luminescence was not polarized and self-focusing was not observed at peak powers used in the experiments. The

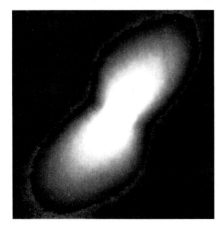

Fig. 8.1 Anisotropic blue luminescence in Ge-doped glass observed via a microscope.

phenomenon was called the "propeller effect" due to the propeller-like shape of the luminescence spot in the focus of the pump beam. This effect represents the first evidence of anisotropic light scattering which peaks in the plane of light polarization in isotropic media.

The phenomenon can be explained as follows. Firstly, let us estimate the size of a light spot which is produced by the isotropically emitted luminescence in the focal plane of the microscope objective. Assuming that the luminescence is excited by the three-photon absorption of the pump at wavelength $\lambda = 800$ nm in a Gaussian beam with radius $r_0 = 2.3$ µm or by the one-photon absorption of UV (267 nm) third harmonic of the pump, and that it is emitted isotropically all along the length of a beam waist, the size of the light spot a can be estimated as: $a = \pi r_0^2 \, n \, / \, \lambda = 30$ µm, where $n = 1.45$ is the refractive index of silica glass. This estimate is in good agreement with the transverse size of the blue propeller, which could be justified by ordinary (isotropic) luminescence. However, the longitudinal size of the blue propeller (~100 µm) is about 4.5 times larger than its transverse size. The fact that the blue luminescence is elongated along the pump polarization indicates that some additional momentum is acquired by the photons along this direction. Such a transformation of the momentum can be caused by the photoelectrons moving along the direction of pump polarization. Microscopic (much less than a wavelength of light) displacements of the photoelectrons along the direction of light polarization can lead to *anisotropic* fluctuations of the dielectric constant. Such fluctuations are obviously stronger along the direction of light polarization (in the direction of electron movement) compared to the perpendicular direction. The fluctuations of dielectric constant along the direction of light polarization induce index inhomogeneities which are elongated in the direction perpendicular to the pump polarization and which have $\mathbf{k_w}$ vectors of spacial harmonics parallel to the direction of polarization. The anisotropic inhomogeneities scatter photons (e.g., the ultraviolet photons of the third harmonic of the pump) in the plane of light polarization. Considering the angle of scattering $\varphi = 80°$ ($\tan \varphi = 3b/2z_0$, where $b = 100$ µm is the longitudinal size of the "blue propeller", $2z_0 = 2\pi r_0^2 \, n \, / \, \lambda = 60$ µm is the waist length of a pump beam), the size of these inhomogenities can be estimated as: $d = \lambda_{uv} \, / \, (2 \, n \, \sin \varphi) = 90$ nm, where $\lambda_{uv} = 267$ nm.

It should be pointed out that the scattering phenomenon described above must have strong wavelength dependence (λ^{-4}), which is similar to the wavelength dependence of Rayleigh scattering of light. Rayleigh scattering is normally caused by isotropic density inhomogenities and the anisotropy in the scattering (the scattering is stronger in the direction perpendicular to the light polarization) is explained by the fact that a dipole does not radiate along its axis. In contrast to Rayleigh scattering the anisotropy in the observed scattering is caused by the anisotropy of the inhomogenities itself. The strong dependence of the scattering on the wavelength can explain the absence of noticeable changes in the shape of the infrared pump.

8.3
Anisotropic Cherenkov Light-generation in Glass

In another experiment nondoped silica glass with weak absorption in the UV, and a regeneratively amplified mode-locked Ti:Sapphire laser operating at 1 kHz repetition rate, were used [11]. The laser radiation in the Gaussian mode was focused via a lens with a focal distance of 6 cm into a fused silica (SiO_2) glass sample of 3 mm thickness. The pump spot size in the focus of the beam was about 14 μm. After passing through the sample, the radiation was imaged on a screen of white paper.

When the pump (4 μJ energy, 4 mW average power, 33 MW peak power, 2.1×10^{13} W cm^{-2} intensity in the focus of a beam) was focused near the input surface, or in the middle of the sample, generation of a white light continuum was observed. When the radiation was focused closer to the output surface of the sample, the generation of the white light continuum terminated. At that focus position a spectacular blue light pattern appeared on the screen and intensified (Fig. 8.2).

Fig. 8.2 The pattern of ultraviolet light generated in a fused silica sample by an intense, linearly polarized, infrared pump. The ultraviolet light is visualized on the screen via luminescence of the paper. Notice that the crescent-like lobes are located on both sides of the pump (the central bright spot of the pattern), along the polarization of the pump. The silhouette of a sample and a spot in the focus of the beam are on the lower right.

The pattern consisted of two crescent-like lobes on both sides of the pump, along the direction of light polarization. The intensity of light in the pattern increased over time and saturated after about 10 seconds of irradiation of the sample. A filter transmitting visible light, placed between the sample and the screen, completely blocked the blue pattern on the screen. This indicated that ultraviolet radiation generated in the sample is responsible for the observed pattern, which is visualized on the screen via luminescence of the paper in the blue spectral range. Analyzing the spectrum of radiation from the output of the sample we confirmed the presence of 267 nm ultraviolet light, which is the third harmonic of the pump at 800 nm and which is generated via the third-order optical nonlinearity of glass. The angle between the direction of propagation of the crescent-like ultraviolet light and the pump was measured to be about 21.6°.

The phenomenon could be explained as follows. It should be remembered that commonly, in nonlinear optical harmonic generation, e.g., third-harmonic generation, collimated nonlinear polarization with wave vector $\mathbf{k'}_{3\omega}$ ($k'_{3\omega} = 3\omega/c' = 3\omega n_\omega/c$) is generated and emits coherent radiation with wave vector $\mathbf{k}_{3\omega}$ ($k_{3\omega} = 3\omega/c = 3\omega n_{3\omega}/c$) in the direction of $\mathbf{k'}_{3\omega}$ in agreement with Huygens' principle and the conservation of transverse momentum. The radiation efficiency is at a maximum when the condition for phase matching and conservation of total momentum is fulfiled $\mathbf{k}_{3\omega} = \mathbf{k'}_{3\omega}$. If, however, there is a transverse surface discontinuity (abrupt boundary) in a medium or in a light beam (uniform beam with sharp cut-off), the surface or Cherenkov radiation mechanism, conserving only longitudinal momentum in which $\mathbf{k}_{3\omega}$ is not collinear with $\mathbf{k'}_{3\omega}$ is possible [19]. Cherenkov radiation is allowed when the polarization phase velocity c' exceeds the velocity c of free radiation in the medium, and is emitted on the "Cherenkov cone" (for the cylindrical boundary) of the half-angle inside the medium a_i given by $\cos a_i = k'/k = c/c' = n_\omega/n_{3\omega}$. The angle between the direction of propagation of Cherenkov third-harmonic light outside the sample and the direction of a pump propagation is given by $\sin a = n_{3\omega}\{1 - (n_\omega / n_{3\omega})^2\}^{1/2}$, where n_ω is the refractive index at the fundamental frequency and $n_{3\omega}$ is the refractive index at the frequency of the third harmonic. The Cherenkov angle of the third-harmonic propagation is estimated to be 20.9° using $n_{267\ nm} = 1.499$ and $n_{800\ nm} = 1.453$ for fused silica. This estimate is very close to the measured angle of propagation of the ultraviolet light, which confirms that the crescent-like light is generated via a Cherenkov mechanism of third-harmonic generation. Two conclusions can be made on the basis of experimental observation. Firstly, the Cherenkov mechanism of the third harmonic generation gives clear indication that some kind of transverse surface discontinuities appear in the medium under intense irradiation. Secondly, the anisotropic distribution of the Cherenkov light with a maximum in the plane of pump polarization, in contrast with the isotropic "Cherenkov cone", indicates that the discontinuities appear only along the direction of polarization of intense light.

The observed phenomena of anisotropic light-scattering and third harmonic generation, represent unique optical effects in which information on light polarization is revealed macroscopically via enhanced light-scattering and generation. Both phenomena give evidence of anisotropic index inhomogenities appearing in glass during the interaction with intense ultrashort pulses.

8.4
Anisotropic Reflection from Femtosecond-laser Self-organized Nanostructures in Glass

A further anisotropic property in silica after being irradiated by a femtosecond laser, is strong reflection from the modified region occurring only along the direction of polarization of the writing laser [13].

The laser radiation in the Gaussian mode, produced by a regeneratively amplified mode-locked Ti:Sapphire laser (150 fs pulse duration, 250 kHz) was focused via a 50 × (NA = 0.55) objective into the sample. The pump spot size in the focus of the beam was 1.5 μm. The silica samples were mounted on a computer-controlled linear-motor 3D translation stage of 20 nm resolution and an electronic shutter was used to control the duration of exposure.

A range of embedded gratings with overall dimensions 700 μm × 700 μm, each consisting of 100 rulings with a 7 μm period, were directly written towards the edges of the plate at a depth of 0.5 mm below the front surface. In every case, the speed of writing was 200 μm s^{-1} and each grating ruling had only one pass of the laser. Pairs of embedded gratings were created with orthogonal writing polarizations directed parallel and perpendicular to the grating rulings respectively, with average fluence ranging from 270 mW (~1.1 μJ per pulse) down to 26 mW (~0.1 μJ per pulse). Additionally, pairs of embedded single lines of length 1mm were written by the same method. Finally, a regular array of 40 × 40 "dots" with a pitch of 10 μm was directly written into the corner of the plate. Each "dot" was produced by holding the sample translation stage stationary at each writing point, and irradiating for 3 ms (~750 pulses) using the electronic shutter.

After writing, the samples were viewed through the silica plate's polished edges using a 200× microscope incorporating a color CCD camera. During inspection, the structures were illuminated with a randomly-polarized white light source in a direction along the viewing axis, either from below the structure (opposite side to microscope objective), or above the structure (through the microscope objective). The embedded structures were examined through the edge nearest to them, and the array of dots was examined in two orthogonal directions. A striking reflection was observed in the blue spectral region, from a number of the structures, when illuminated via the viewing objective. Closer analysis revealed that the reflection *only* occurred when the viewing axis was both parallel to the electric field vector of the writing beam and the structure was written with a pulse energy greater than ~0.5 μJ. This indicates that the observed reflectivity is both fluence dependent and highly anisotropic. Fluorescence cannot account for the observation due to the directional dependence. Figure 8.3 shows a schematic of the reflection phenomenon. As can be seen, the macroscopic shape of the photonic structures does not determine the direction of the anisotropic reflection.

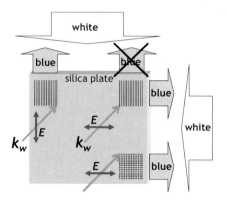

Fig. 8.3 Schematic showing the anisotropic reflection from embedded photonic structures. Reflection only occurs for incident light parallel to the electric-field vector of the incident writing laser. The magnified region (bottom) illustrates the laser-induced self-organized periodic nanostructuring responsible.

Figure 8.4 shows microscope images of the reflection from several directly-written embedded structures. The illuminating light in all cases was incident above the samples through the viewing objective and set to a level that ensured that the weak-contrast microstructure itself was not imaged.

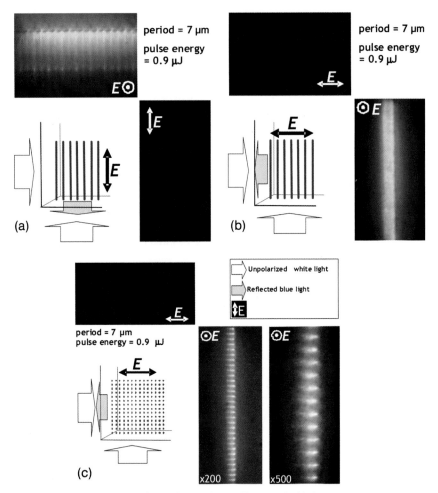

Fig. 8.4 CCD camera images of the reflection from different embedded structures directly-written with the laser. (a) Reflection from 100 lines one behind another into the page. (b) Reflection from a single line. (c) Array of "dots" which show reflectivity dependent only on the writing polarization orientation.

The spatial position of the embedded objects relative to the focus of the microscope objective was checked beforehand by illuminating from below, when the embedded structure in all cases could be clearly observed. The displayed images were chosen from regions of modification created with a pulse energy of ~0.9 μJ. In each example the orientation of the directwrite laser's electric field is indicated, while the k_w vector marks the incident direction of the writing laser beam. Figure 8.4(a) displays the reflection from a single line which has dimensions ~1.5 μm into the page due to the focal width of the beam, and ~30 μm down the page due to the beam's confocal parameter, enhanced by self-focusing effects. Figure 8.4(b) shows the reflection from the side of a 100-line grating, with its rulings going into the depth of the page. Not all of the 100 lines of the grating contribute to the recorded reflection because of the ~2 μm focal depth of the imaging objective. Nevertheless, the reflection is considerably enhanced compared with the single line in Fig. 8.4(b). Figures 8.4(a, b) also show the result of imaging identical structures, but written with orthogonal polarization. From this orientation, no reflecting structure at all can be observed. Figure 8.4(c) shows a section of the 40 × 40 array of "dots" described above, once again producing strong reflection along the writing beam polarization axis. These particular structures are interesting because they are approximately circular with a diameter of ~1.5 μm when viewed from the direction of the writing laser, and therefore have a uniform cross-section. However, when viewed from a direction orthogonal to the axis of the writing beam's polarization, there is no reflecting component as Fig. 8.4(c) clearly demonstrates. The reflected light observed from the structure shown in Figure 8.4(b) is analyzed, yielding the spectrum displayed in Fig. 8.5. This data shows a strong peak at 460 nm, which accounts for the blue color when observed under a microscope.

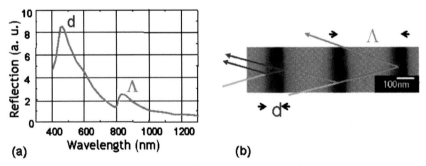

(a)

(b)

Fig. 8.5 (a) Spectrum of the reflected light from the nanostructure shown in Fig. 8.4; (b) SEM image of the reflecting nanostucure.

It was suggested that the anisotropic reflectivity can only be explained as a consequence of Bragg reflection from a self-organized periodic structure. Indeed, a modulation in the refractive index of period Λ~150 nm, produced only along the direction of the incident laser's electric field, can account for the observed anisotropic reflection at λ~460 nm ($\Lambda = \lambda/2n$). Alternatively the maximum at 460 nm can be explained by the reflection from a single layer structure of thickness $d = \lambda/4n$ ~ 85 nm. Such a structure does not reflect when viewed edge-on. The orientation and the size of the nanostructure (85 nm) are almost identical to the nanostructure (90 nm) implicated in the phenomenon of anisotropic light scattering [9]. Closer inspection of Fig. 8.5 shows an additional smaller peak at 835 nm. This suggests that an extra grating component may be formed, which has double

the periodicity of the laser-induced structures ($\Lambda = 300$ nm). Recent scanning electron microscope imaging of the induced structures in the sample producing blue
reflection has revealed stipe-like regions of about 80 nm thicknesses and the period of about 340 nm (Fig. 8.5b). Surface ripples with a period equal to the wavelength of incident laser radiation and which are likewise aligned in a direction orthogonal to the electric field, have been observed in experiments involving laser
deposition [15]. Nevertheless, these data are the first reported evidence of self-organized periodic nanostructures (much smaller than the wavelength of incident light)
being generated within the bulk of a material.

8.5
Direct Observation of Self-organized Nanostructures in Glass

Experiments were carried out to directly observe self-organized nanostructures in
glass [17]. The laser radiation in the Gaussian mode, produced by regenerative
amplified mode-locked Ti:Sapphire laser (150 fs pulse duration, 200 kHz repetition rate) operating at a wavelength of 800 nm, was focused via a 100× (NA = 0.95)
microscope objective into the silica glass samples placed on the XYZ piezo-trans-
lation stage. The beam was focused at ~ 100 μm below the surface and the beam
waist diameter was estimated to be ~ 1 μm.

After laser irradiation, the sample was polished to the depth of the beam waist
location. The surface of the polished sample was analyzed by scanning electron
microscope (SEM, JEOL, model JSM-6700F) and Auger electron spectroscopy
(AES, PHI, model SAM-680). Secondary electron (SE) images and backscattering
electron (BE) images of the same surface were compared (Fig. 8.6).

It is well known that the SE image reveals the surface morphology of a sample,
while the BE image is sensitive to the atomic weight of the elements, or the density of material constituting the observation surface. The SE images of the polished silica sample indicate that the morphology of an irradiated sample in the
examined cross-section does not change, namely, a void does not exist. On the
other hand, the BE images reveal a periodic structure of stripe-like dark regions
with low density of material and of ~ 20 nm width which are aligned perpendicular to the writing laser polarization direction. It was speculated, based on the fact
that the elements constituting the sample are silicon and oxygen (average molecular weight of SiO_2 glass ~ 60.1), that the oxygen defects were formed in the regions
corresponding to dark domains of the BE image, which reduce the average molecular weight in these regions (SiO_{2-x} ~ 60.1–16x). To test this suggestion, Auger
spectra mapping of silicon and oxygen on the same surface by Auger electron
spectroscopy, were carried out. The Auger spectra signal of the oxygen in the
regions corresponding to dark domains in the BE image is lower compared with
other regions, indicating low oxygen concentration in these domains. Furthermore, there is some indication that the intensity of the oxygen signal is stronger
in the regions between the dark domains of the BE image. On the other hand, the
intensity of the silicon signal is the same in the whole imaged region. These

results indicate that the oxygen defects (SiO_{2-x}) are periodically distributed in the focal spot of the irradiated region. The Auger signal intensity is proportional to the concentration of the element constituting the surface, which gives an estimate of the value $x \sim 0.4$.

Fig. 8.6 (a) Secondary electron images of silica glass surface polished close to the depth of focal spot. (b) Light "fingerprints": backscattering electron images of the same surface. The magnification of the upper and lower images is ×10 000 and ×30 000 respectively.

A decrease in the grating period, with an increase in the exposure time, was also observed. The grating periods were about 240 nm, 180 nm and 140 nm for the number of light pulses of 5×10^4, 20×10^4 and 80×10^4, respectively, and for a pulse energy of 1 µJ. This indicated a logarithmic dependence of the grating period Λ on the number of light pulses N_{pulse}. The dependence of the observed periodic nanostructures on pulse energy, for a fixed exposure time, was also investigated, and an increase in the period with the pulse energy was observed. Grating periods of 180 nm, 240 nm and 320 nm were measured at pulse energies of 1 µJ, 2 µJ and 2.8 µJ, respectively, and for the number of light pulses of 20×10^4.

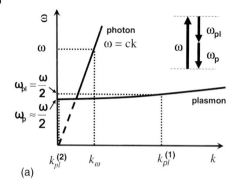

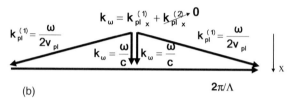

Fig. 8.7 Two-plasmon mechanism of the nanostructure formation: (a) dispersion dependences and energy conservation for light and bulk plasmon at half of the light frequency; (b) momentum conservation and plasmon interference producing periodic nanostructure.

8.6
Mechanism of Formation of Self-organized Nanostructures in Glass

The following explanation of the observed phenomenon is proposed [17]. The light intensity in the focus of the beam is high enough for multiphoton ionization of glass matrix. Once a high free electron density is produced by multi-photon ionization, the material has the properties of plasma and will absorb the laser energy via the one-photon absorption mechanism of inverse Bremsstrahlung (Joule) heating. The light absorption in the electron plasma will excite bulk electron plasma density waves. These are longitudinal waves with the electric field component parallel to the direction of propagation. Such an electron plasma wave can couple with the incident light wave only if it propagates in the plane of light polarization. Initial coupling is produced by inhomogeneities induced by electrons moving in the plane of light polarization [9]. The coupling is increased by a periodic structure, created via a pattern of interference between the incident light field and the electric field of the bulk electron plasma wave, resulting in the periodic modulation of the electron plasma concentration and the structural changes in glass. A positive gain coefficient for the plasma wave will lead to an exponential growth of the periodic structures, oriented perpendicular to the light polarization, which become frozen within the material. Such behavior is common for self-organized structures in light–matter interactions [14]. The electron plasma wave is efficiently generated only by wave vector k_{pl} ($k_{pl} = \omega_{pl}/v_{pl}$, where v_{pl} is the speed and ω_{pl} is the angular frequency of the plasma wave) in the plane of light polarization and only in the direction defined by conservation of the longitudinal component of the momentum. The latter condition is similar to the condition in Cherenkov's mechanism of nonlinear wave generation [19]. The period of the grating is defined by this momentum conservation condition: $k_{gr} = 2\pi/\Lambda = \sqrt{k_{pl}^2 - k_{ph}^2}$, where $k_{ph} = \omega n/c$ is the wave vector, ω is the angular frequency, n is the refractive index and c is the speed of light. This relation can also be used to estimate the value of the electron plasma wave vector $k_{pl} = \sqrt{k_{ph}^2 + (2\pi/\Lambda)^2}$ which gives $k_{pl} = 43.5 \times 10^4$ cm^{-1}, assuming $\omega = 2.36 \times 10^{15}$ s^{-1} and $\Lambda = 150$ nm. The dispersion relation for the electron plasma density waves or Langmuir waves is as follows:

$$\omega_{pl}^2 = \omega_p^2 + \frac{3}{2} v_e^2 k_{pl}^2,$$

where $\omega_p = \sqrt{N_e e^2/\varepsilon_0 m_e}$ is the plasma frequency, N_e is the electron density, m_e is the electron mass, e is the electron charge, $v_e = \sqrt{2k_B T_e/m_e}$ is the thermal speed of the electrons, T_e is the electron temperature and k_B is the Boltzmann constant. Taking into account the energy conservation condition $\omega_{pl} = \omega$, the momentum conservation relation and the above dispersion relation it is possible to obtain an analytical expression for the grating period versus the electron temperature and density:

$$\Lambda = \frac{2\pi}{\sqrt{\frac{1}{T_e}\left(\frac{m_e \omega^2}{3k_B} - \frac{e^2 N_e}{3\varepsilon_0 k_B}\right) - k_{ph}^2}} \tag{1}$$

This dependence shows that the grating period increases with an increase in the electron temperature and electron concentration. This dependence becomes strong when the electron concentration approaches

$$N_e^{cr} - 3\varepsilon_0 k_B k_{ph}^2 T_e e^{-2} \approx N_e^{cr}$$

where N_e^{cr} is the critical plasma density ($\omega_p(N_e^{cr}) = \omega$, $N_e^{cr} = 1.75 \times 10^{21}$ cm^{-3}, $\lambda_\omega \cong 800$ nm). It also follows from above that for a given grating period the electron temperature linearly decreases with an increase in the electron concentration:

$$T_e = A(\Lambda)) - B(\Lambda)) N_e,$$

where

$$A(\Lambda) = \frac{m_e \omega^2}{3k_B k_{pl}}, \quad B(\Lambda) = \frac{e^2}{3\varepsilon_0 k_B k_{pl}}, \quad A(\Lambda)/B(\Lambda) = N_e^{cr}.$$

This gives a wide range of realistic [20, 21] electron temperatures (e.g., up to $T_e = \frac{m_e \omega^2}{3 k_B k_{pl}} = 5 \times 10^7$ K for $\Lambda = 150$ nm) and densities (up to 1.75×10^{21} cm^{-3}) which could explain a certain grating period, including the periods observed in our experiments.

It should be worth mentioning an alternative explanation of the nanostructures formation involving two-plasmon decay [22]. The major difference to the previous explanation (involving only one plasmon) is the excitation of two bulk plasmons of about half of the photon energy: $\omega = \omega_{pl1} + \omega_{pl2}$. This process is possible for $N_e = N_{cr}/4 \sim 4 \times 10^{20}$ cm^{-3}, which is close to the electron concentrations observed in the experiments. An interesting case is when only one plasmon is propagating ($\omega_{pl1} = \omega_{pl}$) and the second plasmon is the plasma oscillation ($\omega_{pl2} = \omega_p$) (Fig 8.7a). Interference between two plasmons of the same frequency propagating in the opposite directions and satisfying Cherenkov mechanism of momentum conservation can produce the periodic nanostructure (Fig 8.7b). The fact that plasmon frequency in this case could be very close to the plasma frequency ω_p reduces significantly the temperature of electrons (down to less than 10^4 K), which can explain the formation of the nanostructures.

Based on the above mechanisms of grating formation, it is possible to give the following explanation of the observed formation of stripe-like regions with low oxygen concentration. The interference between the light wave and the electron plasma wave (one plasmon mechanism) or between two plasmon waves (two plasmon mechanism) will lead to modulation of the electron-plasma concentration. The plasma electrons are created in the process of breaking of Si–O–Si bonds via multi-photon absorption of light which is accompanied by the generation of a Si–Si bonds, nonbridging oxygen–hole centers (NBOHC, $\equiv$Si–O$^-$) and interstitial oxygen atoms (O$_i$). Such oxygen atoms are mobile and can diffuse from the regions of high concentration. Negatively charged oxygen ions can also be repelled from the regions of high electron concentration. We measured photoluminescence and ESR (Electron Spin Resonance) spectra, which confirmed the presence of nonbridging oxygen defects and E centers in the irradiated samples. The small thickness of these regions, compared with the period of the grating, could be explained by a highly nonlinear dependence of the structural changes on the electron concentration. Major changes in composition take place after the attainment of thermal equilibrium. Hot electrons are almost instantaneously excited by ultrashort laser pulses and, subsequently, decay into the lattice. Realistic lattice temperatures are about two orders of magnitude lower than electron temperatures due to the difference in heat capacity of electrons and ions. Our view is that electrons locally undergo interference with the light first, and then the resultant periodic structure will remain throughout the subsequent interactions. Structural changes involve formation and decay of defect states, such as oxygen vacancies, which is a prominent feature in silica. The detailed mechanism of the structural changes responsible for nanograting formation is still under investigation.

It should be pointed out that the irradiated regions in glass, reflected light only in the direction parallel to the polarization of the writing laser and the periods of the gratings in these experiments are very close to 300 nm, which is the period of the gratings responsible for the reflection phenomenon.

Apart from the fundamental importance of the observed phenomenon, as the first direct evidence of interference between light and electron density waves, the observed light "fingerprints" are the smallest embedded structures ever created by light, which could be useful for optical recording and photonic crystal fabrication.

8.7
Self-organized Form Birefringence

Experiments were carried out to clarify the relationship between the phenomena of anisotropic reflection, self-organized nanograting formation and the birefringence induced by femtosecond irradiation [18]. Negative index changes as high as -5×10^{-3} and -2×10^{-3} were measured for light polarized along the direction of polarization of the writing laser and in the perpendicular direction, respectively. Irradiated glass samples were positioned between cross polarizers and it was observed that the onset of birefringence occurred at a writing-fluence level of ~ 0.5 µJ per pulse, equal to that found in the case of anisotropic reflection (Fig. 8.8).

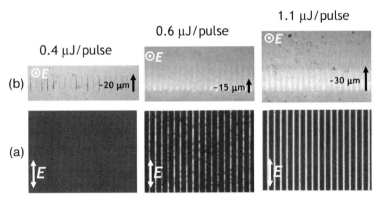

Fig. 8.8 Images of embedded diffraction gratings in cross-polarizers (a) and side-views of the gratings, demonstrating strong anisotropic reflection (b).

This strongly suggests that the mechanism responsible for inducing reflection along the writing beam polarization axis, is the same mechanism that causes birefringence of directly written structures. The femtosecond-laser-induced birefringence is therefore likely to be caused by the laterally-oriented small-period grating structures. Birefringence of this nature is well known as "form" birefringence [16]. Form birefringence is described by the following equations:

$$n_{xy(//)} = n_0 - \Delta n_{xy} = \sqrt{n_1^2 q + n_0^2 (1 - q)}$$

$$n_{xx(\perp)} = n_0 - \Delta n_{xx} = \cfrac{1}{\sqrt{\cfrac{1}{n_1^2} q + \cfrac{1}{n_0^2} (1 - q)}}$$

where n_{xy} is the index of refraction for light polarized parallel to the layers, and n_{xx} is the index of refraction for light polarized perpendicular to the layers, n_1 is the index of refraction of the layer of thickness t_1 and n_0 is the index of refraction of the neighboring layer, Λ is the period of the structure and $q = t_1/\Lambda$ is the filling factor. Index change in oxygen deficient layers ($n_1 - n_0$) of the nanostructure versus filling factor q is shown in Fig. 8.9.

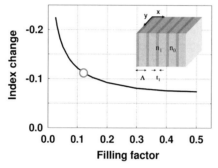

Fig. 8.9 Index change in oxygen-deficient layer versus filling factor t_1/Λ. Average index changes of birefringent structure are $n_{xx} = 0.005$ and $n_{xy} = -0.002$. Experimental conditions are indicated by circle.

Taking into account experimental parameters of the periodic structure $t_1 = 20$ nm, $\Lambda = 150$ nm, it is possible to estimate the index change in the oxygen-deficient layers as high as −0.13. This indicates that this is the strongest embedded periodic structure ever created by light.

Experiments were carried out in different glass samples and sapphire listed in Table 8.1. The red shift of the reflection spectrum in sol-gel silica and Ge-doped silica indicates the smaller period of self-induced grating structures.

Tab. 8.1 Experimented results for birefringence and reflection for different glass samples and sapphire.

Samples	Birefringence	Reflection
Soda-lime glass	No	No
Nanocrystal glass	No	No
BK7	No	No
Sapphire	Yes	Yes, weak
Ge-doped silica	Yes	Yes, redish/blue
Fused silica	Yes	Yes, blue
Sol–gel silica	Yes	Yes, red

Birefringent Fresnel zone plates were fabricated in silica glass and tested for evidence of nanostructure formation. Figure 8.10 shows the SEM image in the back-scattering configuration of a portion of a Fresnel zone plate. The processed zone was written, translating the sample along concentric circular paths of 1.5 μm width, defined by the size of the focused beam. The picture clearly shows that the nanogratings written during two adjacent scans are spatially coherent, which gives unambiguous indication that self-organization is responsible for the phenomenon of nanograting formation.

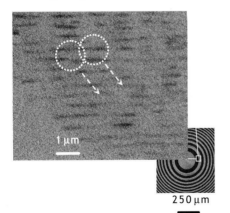

1 μm

250 μm

Fig. 8.10 Image from scanning electron microscope in back-scattering configuration of a portion of Fresnel zone plates (experimental details in [8]). The processed zone was written translating the sample along concentric circular paths of 1.5 μm width. The picture clearly shows that the nano-gratings, written during two adjacent scans, are spatially coherent.

8.8
Conclusion

The evidence of self-organized periodic nanostructures (much smaller than the wavelength of incident light) being generated within the bulk of a material, has been presented. Self-organized structures created by femtosecond laser irradiation are the smallest and strongest embedded structures ever created by light. The analysis of the experimental results suggests that self-organized sub-wavelength periodic structures are also the primary cause of all anisotropic phenomena reported in the experiments on direct writing with ultrashort pulses in transparent materials, including anisotropic scattering, reflection and self-organized form birefringence. The anisotropic phenomena and related nanogratings should be useful in many monolithic photonic and micro-optic devices and can be harnessed for information storage where nanoscale periodic structuring is required.

References

1 K.M. Davis, K. Miura, N. Sugimoto, and K. Hirao, Opt. Lett. **21**, 1729 (1996).

2 K. Miura, J. Qiu, H. Inouye, T. Mitsuyu, and K. Hirao, Appl. Phys. Lett. **71**, 3329 (1997).

3 E.N. Glezer and E. Mazur, Appl. Phys. Lett. **71**, 882 (1997).

4 M.D. Perry, B.C. Stuart, P.S. Banks, D. Feit, V. Yanovsky, and A.M. Rubenchick, J. Appl. Phys. **85**, 6803 (1999).

5 K. Yamada, T. Toma, W. Watanabe, K. Itoh, and J. Nishii, Opt. Lett. **26**, 19 (2001).

6 Y. Kondo, K. Nouchi, T. Mitsuyu, M. Watanabe, P.G. Kazansky, and K. Hirao, Opt. Lett. **24**, 646 (1999).

7 D. Homoelle, S. Wielandy, A.L. Gaeta, N.F. Borrelli, and C. Smith, Opt. Lett. **24**, 1311 (1999).

8 H. Sun, S. Juodkazis, M. Watanabe, S. Matsuo, H. Misawa, J. Nishii, J. Phys. Chem. B **104**, 3450 (2000).

9 P.G. Kazansky, H. Inouye, T. Mitsuyu, K. Miura, J. Qiu, K. Hirao, and F. Starrost, Phys. Rev. Lett. **82**, 2199 (1999).

10 J. Qiu, P. G. Kazansky, J. Si, K. Miura, T. Mitsuyu, K. Hirao, and A. Gaeta, Appl. Phys. Lett. **77**, 1940 (2000).

11 P.G. Kazansky, H. Inouye, T. Mitsui, J. Qiu, K. Hirao, F. Starrost, "Anisotropic Cherenkov light generation by intense ultrashort light pulses in glass,"

in Quantum Electronics and Laser Science, (Optical Society of America, Washington, D.C., 2000), paper QFA3.

12 L. Sudrie, M. Franko, B. Prade, A. Mysyrowicz, Opt. Commun. **171**, 279 (1999).

13 J. Mills, P.G. Kazansky, E. Bricchi, J. Baumberg, Appl. Phys. Lett. **81**, 196 (2002).

14 S.R.J. Brueck and D.J. Ehrlich, Phys. Rev. Lett. **48**, 1678 (1982).

15 D. Ashkenasi, H. Varel, A. Rosenfeld, S. Henz, J. Hermann, and E.E.B. Cambell, Appl. Phys. Lett. **72**, 1442 (1998).

16 M. Born and E. Wolf, *Principles of Optics* (Cambridge University Press, UK, 1999), p. 837.

17 G. Kazansky, Y. Shimotsuma, J. Qiu and K. Hirao, Phys. Rev. Lett. **91**, 247405 (2003).

18 E. Bricchi, B. Klappauf and P.G. Kazansky, Opt. Lett. **29**, 119 (2004).

19 D.H. Auston, K.P. Cheung, J.A. Valdmanis, and D.A. Kleinman, Phys. Rev. Lett. **53**, 1555 (1984).

20 B.C. Stuart, M.D. Feit, A.M. Rubenchik, B.W. Shore, and M.D. Perry, Phys. Rev. Lett. **74**, 2248 (1995).

21 C.H. Fan, J. Sun, and J.P. Longtin, J. Heat Transf. **124**, 275 (2002).

22 N.A. Ebrahim, H.A. Baldis, C. Joshi, and R. Benesch, Phys. Rev. Lett. **45**, 1179 (1980).

9

X-ray Generation from Optical Transparent Materials by Focusing Ultrashort Laser Pulses

Koji Hatanaka and Hiroshi Fukumura

Abstract

In this chapter, studies on interactions of intense laser fields with transparent materials such as glasses, polymer targets, and solutions are reviewed from the viewpoint of the high-energy photon emission of EUV, soft X-rays, and hard X-rays. Fundamentals of the mechanisms leading to high-energy photon emission are summarized on the basis of experimental results, like X-ray emission spectra. Experimental setups for laser-induced hard X-ray emission from solutions and solids are provided and the effects of air plasma formation and sample self-absorption of X-rays on X-ray generation are described. Photo-ionization processes of transparent materials and effects of the addition of electrolytes are also considered. The effects of multi-shot and double-pulse irradiation of laser pulses on X-ray emission from solid materials and solutions are explained. Finally, some possible applications using X-rays from transparent materials are suggested.

9.1
Introduction

Recent developments and progresses of laser technology, femtosecond lasers in particular, have enabled us to perform experiments with extreme intense laser lights with the power at 10^{11}–10^{17} W cm^{-2} in laboratories. Under such extreme conditions, the resulting phenomena are various from laser ablation and plasma formation to emission of extreme ultraviolet light (EUV) and X-ray. Fundamental studies on the mechanisms of interaction between intense laser pulses and metals leading to the emission of EUV and X-rays have been well summarized in [1–6]. From an application viewpoint, the generation of EUV has been extensively studied due to its great potential for nanomaterial fabrication [7]. Femtosecond laser-based X-ray pulses have also potential for a pulse source for time-resolved measurements of X-ray diffraction [8] and X-ray absorption fine spectroscopy [9], which would contribute to progress in basic science. There is also hope for the application of such high-energy light sources in medicine [10].

3D Laser Microfabrication. Principles and Applications.
Edited by H. Misawa and S. Juodkazis
Copyright © 2006 WILEY-VCH Verlag GmbH & Co. KGaA, Weinheim
ISBN: 3-527-31055-X

For the case of transparent materials such as solutions, there have been a lot of studies on laser-induced breakdown and optical damage, or multi-photon ionization since the early reports on the bubble formation of water when irradiated by 10^3 W cm^{-2} laser pulses [11], on stimulated Brillouin scattering [12] and spark emission [13] from water when irradiated by 10^7–10^8 W cm^{-2} laser pulses, and on laser-induced shock wave formation in water when irradiated by 5×10^6 W cm^{-2} laser pulses [14]. Progresses in studies on laser-induced breakdown of water were summarized by Kennedy et al. [15]. Recently, femtosecond laser-induced breakdown was studied by Fan et al. [16], Noack and Vogel [17], and Schaffer et al. [18]. Ejected electron decay and solvated electron dynamics under intense (10^{12}–10^{13} W cm^{-2}) laser irradiation conditions were also studied [19]. Recently, the group of Sawada et al. reported that excess electron ejection was induced by intense laser irradiation (532 nm, 40 ps, 2×10^{14} W cm^{-2}) into water and found that stimulated Raman scattering intensity was enhanced by excess electrons [20].

As for glasses and polymers, Wood summarized studies on laser-induced damages to these materials [21]. Preuss et al. reported femtosecond laser ablation of lithium niobate, poly(tetrafluoroethylene) and poly(methylmethacrylate) by using femtosecond laser pulses (500 fs, 248 nm, 10^{13} W cm^{-2}) [22]. Schaffer et al. studied laser-induced breakdown damage in bulk transparent materials induced by tightly focused femtosecond laser pulses [23].

Although there are a lot of studies on interactions between intense laser pulses and transparent materials, studies on laser-based photon emission of EUV and X-rays from transparent materials are much less frequent, compared with the studies on metals. In this chapter, we have reviewed studies on interactions of intense laser fields with transparent materials such as glasses, polymer targets, and solutions from the viewpoint of high-energy photon emission of EUV, soft X-rays, and hard X-rays. In this section, studies of transparent materials under an intense laser field have been briefly summarized. In Section 9.2.1, studies on EUV and soft X-ray generation using various transparent materials such as organic solvents, water, and glass materials are reviewed. In Section 9.2.2, fundamentals of mechanisms leading to high-energy photon emission are summarized on the basis of experimental results. In Section 9.2.3, X-ray emission spectra of transparent materials are presented and it is indicated that the laser-induced characteristic X-ray intensity depends on the atomic numbers of elements involved in the materials. In Section 9.3.1, experimental setups for laser-induced hard X-ray emission from solutions and solids are provided. In Section 9.3.2, the effects of air plasma formation and sample self-absorption of X-ray are considered. In Section 9.3.3, photo-ionization processes of transparent materials and effects of the addition of electrolytes are taken into account. In Section 9.3.4, the effects of multi-shot laser pulses on X-ray emission from solid materials, are explained. In Section 9.3.5, the increase in X-ray intensity under double-pulse irradiation conditions when aqueous solutions are used as samples is introduced. In Section 9.4, some possible applications using X-rays from transparent materials, are suggested.

9.2
Laser-induced High-energy Photon Emission from Transparent Materials

9.2.1
Emission of Extreme Ultraviolet Light and Soft X-ray

In this section, we review reports on the laser-induced emission of EUV and soft X-rays. The group of Richardson et al. reported EUV emission from a micrometer-sized frozen water droplet when irradiated by Nd^{3+}:YAG laser pulses (20 ns, 1064 nm, 400 mJ, 10 Hz) [24]. In these experiments, the peak laser power at the lens focus was 1.26×10^{12} W cm^{-2}. Simulation results indicated that the electron temperature and the electron density can increase to 50 eV and 10^{24} cm^{-3}, respectively, at most. Under experimental conditions, mainly strong emission lines due to 3d → 2p (17.3 nm, 72 eV), 3p → 2s (15.0 nm, 83 eV), 4d → 2p (13.0 nm, 95 eV), and 4d → 2s (11.6 nm, 107 eV) transitions of Li-like oxygen (O^{5+}) were clearly observed with a broad background of bremsstrahlung. The energy conversion efficiency from the laser pulse to EUV was estimated to be 0.63%/4π sr.

Rajyaguru et al. reported systematic optimization of several parameters, such as laser energy, spot size, and water-jet size, to maximize the conversion efficiency from laser light to EUV by using a large continuous water jet [25]. The laser used was a Nd^{3+}:YAG laser (1064 nm, 10 ns). When the laser power was 8×10^{11} W cm^{-2}, the maximum conversion efficiency obtained was 0.82% at 13.0 nm (95 eV) in 2.5% bandwidth and 4π sr for a jet with a diameter of 100 μm. They also reported EUV emission at 13.5 nm (Li^{2+}, 92 eV) from a lithium-based liquid jet when irradiated by nanosecond Nd^{3+}:YAG laser pulses (532 nm, 10 ns, 6×10^{11} W cm^{-2}) [26].

Vogt et al. reported EUV (11–15 nm, 83–113 eV) emission from a laminar water jet when irradiated by Nd^{3+}:YLF laser pulses with different laser pulse widths (30 ps, 300 ps, or 3 ns, 1047 nm, 250 kHz) [27]. The laser power at the focus was varied between 10^{11} and 10^{15} W cm^{-2}. The EUV intensity fluctuation was estimated to be about 20% by using an EUV photodiode while the laser intensity fluctuation was 2%. In the case of 30 ps laser pulse irradiation, 3d → 2p transition of O^{6+} was clearly observed at 12.85 nm (96 eV). The maximum energy conversion efficiency was about 0.12% in 2.2% bandwidth and 4π sr at 13 nm (95 eV) when the laser pulse width was 3 ns with laser power of 1.1×10^{12} W cm^{-2}.

Dusterer et al. reported EUV emission from 20 μm-diameter water droplets and from 2 mm thick glass (SiO_2) plates when irradiated by near-IR laser pulses with different laser pulse widths (200 fs to 6 ns, 795 nm, 10 Hz) [28]. The energy conversion efficiency obtained for droplets was about 0.23% with 6 ns, 8×10^{12} W cm^{-2}, while the conversion efficiency for glass targets increased logarithmically over, at most, five orders of magnitude as a function of laser pulse width, between 200 fs and 6 ns and reached 2.5% at the maximum.

The group of Hertz et al. extensively studied laser-induced EUV emission from ethanol, ammonium hydroxide, fluorocarbon, liquid nitrogen, and cryogenic liquid-jet targets of nitrogen and xenon [29]. They used picosecond laser pulses of frequency-doubled Nd^{3+}:YAG laser (532 nm, 10 Hz, 120–140 ps, $\sim 4 \times 10^{14}$ W cm^{-2}) as an excitation source. They observed C^{4+} [3.50 nm (35 eV) and 4.03 nm (31 eV)], C^{5+} [3.37 nm (2p $\rightarrow$ 1s), 37 eV], O^{6+}, and O^{7+} in the range of 1–4 nm (0.31–1.24 keV) by using ethanol droplets as a target under 4×10^{14} W cm^{-2} irradiation conditions. The absolute emission at 3.4 nm was estimated to be 0.5 $\times 10^{12}$ photons per sr line pulse [30]. Later, by introducing a nitrogen gas flow in front of the sensitive optical components to be protected, debris emission was reduced by approximately 30 times [31]. Ammonium hydroxide droplets (32% NH_3 in water by volume) or urea [$CO(NH_2)_2$] were also used as targets [32]. The laser power at the focus was $\sim 5 \times 10^{14}$ W cm^{-2}. The absolute emission at 2.88 nm (N^{5+}, 431 eV) was estimated to be about 1×10^{12} photons per sr pulse. In the case of ethanol, the main debris substance was carbon-related, while in the case of ammonium hydroxide, the debris was due to nitrogen-related gaseous components. As a result, the amount of debris was expected to decrease considerably and this was verified experimentally where the debris emission was < 0.01 pg per sr pulse which was two orders of magnitude less than in the case of ethanol droplets. As one of the debris-free compounds, liquid nitrogen was also tried as a target [33]. The laser power at the focus was 4×10^{14} W cm^{-2}. For higher EUV intensity, a pre-pulse was applied. Line emission from N^{6+}[1s–2p (2.48 nm, 0.50 keV) and 1s–3p] and N^{5+}[1s^2–1s2p (2.88 nm, 0.43 keV)] was observed. These lines were applicable for EUV imaging because the lines are in the water-window region at 2.4–4.4 nm. Absolute emission at 2.88 nm (N^{5+}) was estimated to be about 4.5×10^{11} photons per sr pulse. As one of compounds containing fluorine, which gives a line emission in the region of 1.2–1.7 nm (0.73–1.03 keV) for lithography, fluorocarbon was also used as a target compound [30, 33, 34]. The laser power at the focus was about 8×10^{14} W cm^{-2}. In order to increase the X-ray emission, the laser was operated to yield a pre-pulse having an energy of approximately 10%, 7 ns before the main pulse. The line emission assigned to F^{8+} (1.495 nm, 0.83 keV) and F^{7+} [1.681 nm (0.74 keV) and 1.446 nm (0.86 keV)], was observed. The absolute emission intensity at 1.681 nm (F^{7+}, 0.74 keV) was estimated to be 2×10^{12} photons per sr line pulse, which corresponds to an X-ray conversion efficiency of $\sim 5\%$.

The group of Hertz et al. also reported a study of pre-pulse irradiation effects on EUV emission [35]. A pre-pulse (< 3 mJ, 355 nm, 120 ps) was irradiated to liquid ethanol droplets at a time 2–7.5 ns before the main pulse (65 mJ, 532 nm, 120 ps) irradiation. The photon flux of EUV was enhanced, when the double-pulse irradiation was applied, more than eight times if compared with single main-pulse irradiation conditions.

As for glass targets, Dunne et al. reported pre-pulse irradiation effects on soft X-ray emission (7–17 nm, 73–177 eV) from cerium-doped borosilicate glasses when irradiated by a main pulse with a pre-pulse of Nd^{3+}:YAG laser (170 ps, 1064 nm, 9×10^{11} W cm^{-2}) [36]. The maximum soft X-ray intensity was obtained with a delay time of 5.1 ns and the conversion efficiency was 4.8% in a 3% bandwidth at 8.8 nm.

Nakano et al. also reported soft X-ray (5–20 nm, 60–250 eV) emission from a gold-doped (10^{-4} wt%) glass target (UV-cut filter, HOYA, L-1B) when irradiated by femtosecond laser pulses (130 fs, 800 nm, 10 Hz) with a laser power of 5×10^{15} W cm^{-2} at the focus [37]. The soft X-ray intensity was 10^8 photons per sr at 14 nm (89 eV). The soft X-ray emission intensity using a gold-doped glass target was about 40% of that from a solid gold target, while the density of gold in the doped glass was less than 0.001 vol% and the target was transparent at the wavelength of laser light (800 nm). The pulse duration of the soft X-ray was measured with an X-ray streak camera to be 7–9 ps. They also reported that, in the case of a neodymium-doped glass target, the soft X-ray emission near 8 nm (N-shell transition of a neodymium, 155 eV) increased without broadening the soft X-ray pulse width [38]. Also, by introducing a pre-pulse (6.4×10^{14} W cm^{-2}) at 50 ns before the main pulse, the soft X-ray intensity was enhanced more than 150 times and 1% energy conversion efficiency from the laser pulse (3.2×10^{16} W cm^{-2}) into the soft X-ray at 8 nm, was achieved.

Since emission of EUV and soft X-rays is related to various transitions between electron orbits, while hard X-ray emission is mostly related to a inner-shell transitions, the emission spectra of EUV and soft X-ray are very informative of how much ionization is induced if compared with hard X-ray emission spectra. Figure 9.1 shows the peak energies of EUV and soft X-rays from ions as a function of the peak power of the incident laser pulses, which is plotted from the references described above. Obviously, the valence numbers of the ions increased as the laser power increased, for instance, from O^{5+} at a laser power of 10^{12} W cm^{-2} to O^{7+} at a power of 10^{15} W cm^{-2}. Furthermore, the photon energy also increased from 100 eV

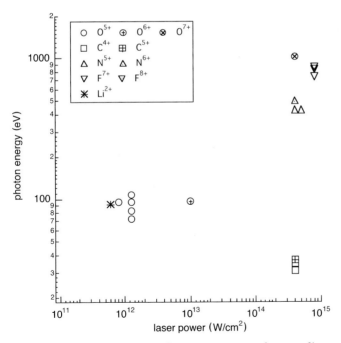

Fig. 9.1 Energy of photon emission from various ions as a function of laser power.

from O^{5+}, to 1 keV from O^{7+}. These observations strongly indicate that conductive electrons with equivalent energy are ejected by the irradiation of laser pulses at a power of 10^{11}–10^{15} W cm^{-2} and collide with the surrounding atoms and ions, inducing further ionization.

9.2.2
Fundamental Mechanisms Leading to High-energy Photon Emission

Reports on the solution of hard X-ray emission from transparent materials, in particular, when irradiated by femtosecond laser pulses, are quite limited so far and only a few studies have been reported. Tompkins et al. reported 1 kHz repetition rate X-ray generation in the 5–20 keV spectral region, induced by the interaction of femtosecond laser and copper nitrate solution or ethylene glycol liquid-jet targets in a vacuum chamber [39]. The peak laser intensity on the sample solution surface, was of the order of 10^{16} W cm^{-2}. They observed characteristic Kα X-ray [$^2P_{1/2}$ (Kα_2) or $^2P_{3/2}$ (Kα_1) to $^2S_{1/2}$] of copper (8.05 keV) in copper nitrate solution and reported that the X-ray photon flux was of the order of 10^6 photons per s sr in the spectral region. Donnelly et al. reported X-ray emission from ~1 μm water droplets when irradiated by 35 fs laser pulses at a laser intensity of up to 7×10^{17} W cm^{-2} [40]. They observed X-ray emission above 100 keV and reported that hot electron temperatures observed in the case of micron-sized droplets were significantly higher than those observed in the case of solid planar plastic targets.

Hatanaka et al. extensively studied X-ray emission from aqueous solutions of electrolytes such as cesium chloride in a laser intensity range of the order of 10^{15} W cm^{-2} when the laser intensity was 500 μJ per pulse [41–46]. Figure 9.2 shows the X-ray intensity as a function of laser intensity with different laser polarization [45]. The sample solution was a cesium chloride aqueous solution (4 mol dm^{-3}) and the laser incident angle was 60°. Apparently, the X-ray intensity was much higher in p-polarized laser irradiation than in the s-polarized one. The X-rays started to be detected at 20 μJ per pulse in the case of p-polarization, while in the case of s-polarization the X-rays were detected at 300 μJ per pulse. X-ray emission spectra were also measured by changing the laser intensities as shown in Fig. 9.3(a) [43]. Here the spectra were corrected by the absorption effects of air and a beryllium window attached to a detector [47] and plotted on a logarithmic scale, and the laser was p-polarized. Some peaks observed were assigned to cesium characteristic X-ray lines [Lα_2 = 4.27 keV ($^2D_{3/2}$ to $^2P_{3/2}$), Lα_1 = 4.29 keV ($^2D_{5/2}$ to $^2P_{3/2}$), Lβ_1 = 4.62 keV ($^2D_{3/2}$ to $^2P_{1/2}$), Lβ_2 = 4.94 keV ($^2D_{5/2}$ to $^2P_{3/2}$), Kα (30.9 keV), and Kβ [34.9 keV ($^2D_{3/2}$ to $^2S_{1/2}$)] [48]. As the laser intensity increased, the X-ray intensity also increased and the slope of the broad component became gentle. As in the usual manner [49], electron temperatures were calculated from the slope on the assumption that the electron temperature reached equilibrium;

$$X(E, T_e) = \exp(E \,/\, k_B T_e) \times const. \tag{1}$$

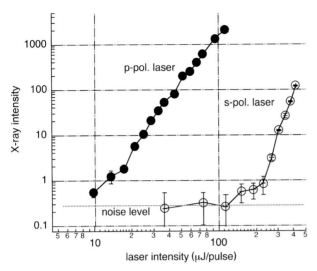

Fig. 9.2 X-ray intensity as a function of laser intensity with s- and p-polarization to the solution surface. The sample was a CsCl aqueous solution (4.0 mol dm^{-3}).

Here $X(E, T_e)$, E, T_e, and k_B represent a broad component of the X-ray emission spectrum, the photon energy, an electron temperature, and the Boltzmann constant, respectively. The calculation results are shown in Fig. 9.3(b) [43]. Two different components of the electron temperatures were obtained; the lower component stayed low at around 2 keV, even though the laser intensity increased; while the higher component increased almost linearly from 3 to 10 keV as the laser intensity

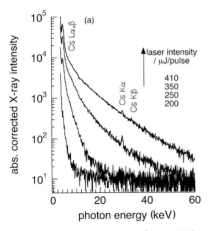

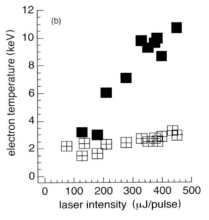

Fig. 9.3 X-ray emission spectra from a CsCl aqueous solution (6.5 mol dm^{-3}) when irradiated by different laser (p-pol.) intensities (a); and calculated electron temperatures as a function of laser intensity (b). The solid-state detector used here was a highly pure Ge detector. Spectra shown here were corrected by the absorption effects of air and the beryllium window at the detector's input.

increased from 120 to 450 µJ per pulse. Similar dependences of two different electron temperatures on the atomic numbers and solute concentrations were observed [43]. These two different electron temperatures indicate that there are two different mechanisms for the acceleration of electrons during a laser pulse, spatially or temporally.

Figure 9.4 shows the X-ray intensity as a function of the laser incident angle with different laser polarization, where the sample was a cesium chloride aqueous solution (4 mol dm^{-3}) and the laser intensity was fixed at 300 µJ per pulse [45]. In the case of s-polarized laser irradiation, the X-ray intensity decreased monotonically as the incident angle increased. This is reasonable, because the laser intensity per unit area (laser fluence) decreased as the laser incident angle increased. On the other hand, in the case of p-polarized laser irradiation, the X-ray intensity had a peak at the incident angle of 60°. Refractive indices of distilled water and a cesium chloride aqueous solution (4 mol dm^{-3}) are 1.33 and 1.38, respectively, and the Brewster angles for these refractive indices at an air–solution interface are calculated to be 53.06° and 54.07°, respectively. This indicates that the X-ray intensity peak observed at 60° does not relate directly to the reflection or transmission of laser pulses and that there is a different mechanism of interaction between laser pulses and solutions when the laser is p-polarized.

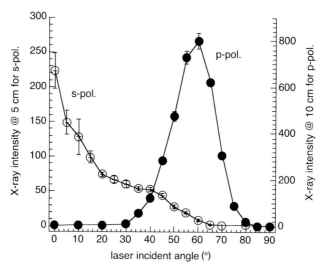

Fig. 9.4 X-ray intensity as a function of laser incident angle with s- and p-polarization. The sample was a CsCl aqueous solution (4.0 mol cm^{-3}) and the laser intensity was 300 µJ per pulse.

The mechanisms of intense laser interactions with metal surfaces leading to X-ray generation are well summarized in references [1–5]. Once free electrons or conductive electrons are ejected at the leading edge of a laser pulse, the initial process of interaction between such conductive electrons and laser pulses is inverse bremsstrahlung or, classically speaking, collisional absorption (Fig. 9.5). Due to

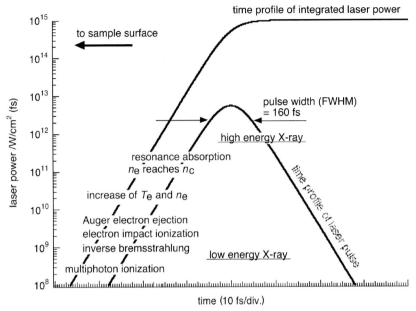

Fig. 9.5 A conceptual diagram of various processes leading to X-ray emission. The pulse shape assumed here is $(\mathrm{sech})^2$.

the laser electric field oscillation, electrons are forced to oscillate and collide with surrounding atoms and ions frequently. Then the electron kinetic energy increases and electron collisions induce inner-shell electron ejection and, as a result, Auger electrons are also ejected. During such processes, it is not only the electron temperature but also the electron density which increases nonlinearly, although electron impact ionization becomes ineffective when the electron energy is high. For instance, the electron impact ionization cross-section of oxygen for 100 eV electrons is 1.338×10^{-16} cm^{-2}, and that for 1 keV electrons is smaller at 4.377×10^{-17} cm^{-2} [50]. Needless to say, that relatively low energy X-rays are emitted during these processes as bremsstrahlung (literally, the braking emission) or as a result of transitions between internal orbits. Once the electron density reaches the critical density, in the case of insulators like aqueous solutions, the sample surface becomes metal-like. The critical density, n_c, is calculated as follows.

$$n_c = \varepsilon_0 \, m_e \, \omega_L^2 \, / \, e^2 = 4\pi^2 \, \varepsilon_0 \, m_e \, c^2 \, / \, e^2 \, \lambda_L^2 \tag{2}$$

where ε_0, m_e, ω_L, e, c, and λ_L represent the dielectric constant, the electron mass, the laser frequency, the elementary electric charge, the speed of light, and the laser wavelength, respectively. In the case of the laser wavelength at 780 nm, the value n_c is calculated to be 1.9×10^{21} cm^{-3}. Laser pulses incident on a sample surface, with the critical density, are reflected. In the case when the incident laser is s-polarized, little interaction between the incident laser and the critical surface is expected. On the other hand, in the case of a p-polarized laser pulse, during reflection the laser electric field component that is parallel to the electron density gradation can effectively excite plasma oscillation at the critical surface. This is another process of electron acceleration called resonance absorption that is a linear process by which p-polarized light is partially absorbed by conversion into an electrostatic wave at the critical surface. One experimental proof for this process is the second harmonic generation that is induced as a result of interaction between incident laser pulses and a metal-like surface. Indeed, second harmonic generation was observed when p-polarized laser pulses were used in X-ray emission from aqueous solution [45]. Due to this effective absorption process, electrons can be accelerated more and the electron temperature increases. As a result, higher energy X-ray emission can be observed. This may be the reason why two different components of electron temperatures were observed, as in Figure 9.3(b).

9.2.3
Characteristic X-ray Intensity as a Function of Atomic Number

Hard X-ray emission from various transparent solid materials when irradiated by focused femtosecond laser pulses in air, was also reported. Figure 9.6 shows X-ray emission spectra from glass plates such as color glass filters (Toshiba, B46 and A75S) and a conventional slide glass plate (Matsunami, Micro Slide Glass, S7225, soda-lime glass) [51]. Spectra shown here were not corrected by absorption effects of the air and beryllium window attached to the detector's input. Therefore X-ray intensity in the lower photon energy region is lower, since absorption coefficients of atoms in the X-ray region are higher in the lower photon energy region [47, 48]. The slide glass plate contains silicon oxide (72%) and calcium oxide (8%) mainly, and characteristic $K\alpha$ (3.69 keV) and $K\beta$ (4.01 keV) X-ray peaks of calcium [48] were clearly observed, in addition to a broad component. The characteristic $K\alpha$ X-ray of silicon which is the main component of the glass plate was also observed at 1.74 keV [48] when the detection distance was shorter. In cases of color glass filters, B46 and A75S, manganese ($K\alpha$; 5.89 keV), iron ($K\alpha$; 6.40 keV), copper ($K\alpha$; 8.04 keV, $K\beta$; 8.91 keV), and zinc ($K\alpha$; 8.63 keV, $K\beta$; 9.57 keV) [48] were also detected. Trace elements such as cobalt and arsenic were also detected faintly in the spectrum of color glass filter B46 at 6.93 keV (Co $K\alpha$) and 10.5 keV (As $K\alpha$) [48].

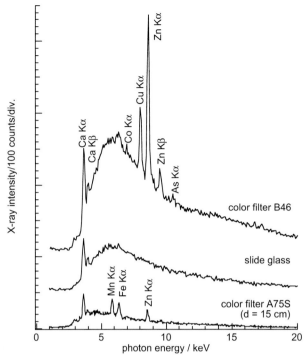

Fig. 9.6 X-ray emission spectra from a slide glass and color filters (Toshiba, B46 and A75S). The detection distance was 25 cm except in the case of the color filter A57S. The solid-state detector used here was a Si(Li) detector.

Figure 9.7 shows X-ray emission spectra from poly(vinyl alcohol) and poly(vinyl chloride) films [51]. The X-ray emission spectrum of poly(vinyl alcohol) was broad and had no line peaks. Characteristic X-rays of the polymer components such as oxygen (characteristic K X-ray = 524.9 eV, 2.36 nm) [48], were not observed because the characteristic X-rays are out of the detectable energy region. Other polymer films of poly(methyl methacrylate), poly(ethyl methacrylate), poly(vinyl carbazole), and polystyrene showed similar broad X-ray emission spectra. Nitrogen in poly(vinyl carbazole) cannot be observed because the characteristic K X-ray of nitrogen is at 392.4 eV [48]. On the other hand, in the case of poly(vinyl chloride), one line peak was clearly observed at around 2.6 keV (0.48 nm) [51]. Evidently the peak can be assigned to the characteristic $K\alpha$ X-ray of chlorine (2.62 keV) [48]. It appeared that the peak of the broad X-ray emission component shifted towards the higher energy region when compared with the spectrum of poly(vinyl alcohol). This is simply due to a difference in the detection distance which leads to a difference in the absorption effect by air. If such an X-ray emission spectrum was observed with much shorter detection distance, 5 cm for instance, as shown in the figure, the intensity of the characteristic $K\alpha$ X-ray of chlorine increased and the sum peaks at about 5.2, 7.8, and 10.4 keV, due to high intensity, were also observed.

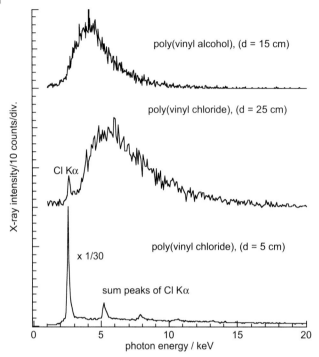

Fig. 9.7 X-ray emission spectra from films of poly(vinyl alcohol) and Poly(vinyl chloride). The undermost spectrum intensity is reduced by a factor of 30. The solid-state detector used here was a Si(Li) detector.

Figure 9.8 shows X-ray emission spectra from water, a cesium chloride aqueous solution, and a potassium bromide aqueous solution [41–43]. As in the case of polymer films of poly(vinyl alcohol) shown in Fig. 9.7, in the case of water, a broad component was observed without the characteristic X-ray. In the case of potassium bromide aqueous solution, on the other hand, characteristic $K\alpha$ X-ray peaks of potassium (3.31 keV) and bromine (11.9 keV) [48] were clearly observed. These spectra indicate that highly energetic electrons randomly collide with surrounding atoms and ions irrespective of polar characters like cations and anions. This results in hole formations in inner-shells and instantaneous characteristic X-ray emission because the excited state lifetime is very short, of the order of femtoseconds [52]. The energy conversion efficiency of the laser pulse to the X-ray pulse in the range 3–60 keV, in the case of a cesium chloride aqueous solution (6.5 mol dm^{-3}), was calculated to be ~ 10^{-8} under the assumption that X-ray radiation was spherically homogeneous [41].

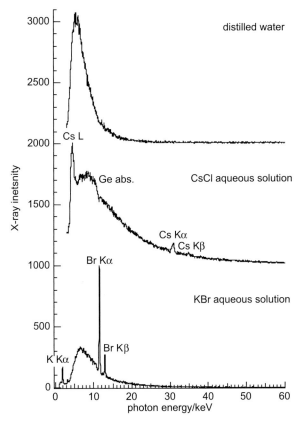

Fig. 9.8 X-ray emission spectra from distilled water and aqueous solutions of CsCl (6.5 mol cm^{-3}) and KBr (4.0 mol dm^{-3}). The solid state detector used here was a highly-pure Ge detector. A dip at 11 keV observed in the spectrum of CsCl aqueous solution is due to the germanium absorption edge.

The high intensity of the chlorine characteristic X-ray in the case of poly(vinyl chloride) film may not be due only to the air absorption effect or a high concentration of chlorine in the film. Fundamentals of intense laser–matter interaction mechanisms leading to X-ray emission have been discussed in the previous section. In the final stage of processes leading to X-ray emission, high-energy electrons collide with K-shell electrons, which results in characteristic K X-ray emission. On the basis of the above mechanism, characteristic X-ray intensity is considered to be a function of the electron energy distribution, K-shell ionization cross-section, and characteristic K X-ray emission yield. Here we try to calculate the relative intensity of the characteristic K X-ray as a function of the atomic numbers. The electron temperature ($k_B T_e$) can come to equilibrium within a laser pulse; here the value was assumed to be 3, 5, and 10 keV. As for the K-shell ionization cross-section, values are reported in a reference paper [53]. In case of chlorine, the cross-section for electrons with energy near the chlorine K absorption edge (2.9 keV) is of the order of 1.49×10^{-22} cm². X-ray emission and Auger electron ejection are competing processes and the characteristic K X-ray emission quantum yield, ϕ_K, can be calculated by using an empirical formula [54],

$$\phi_K /(1 - \phi_K) = [-0.03795 + 0.03426 \times Z - 0.11634 \times 10^{-6} \times Z^3]^4 \tag{3}$$

Here, Z represents an atomic number. In the case of chlorine, ϕ_K is calculated to be about 0.08. Based on these variables, the relative intensities of the characteristic K X-ray with different electron temperatures are plotted as a function of atomic numbers, as shown in Fig. 9.9. The characteristic K X-ray emission yield ϕ_K is also plotted in the same figure. The intensity decrease in lower and higher atomic numbers is due to a lower characteristic X-ray emission yield (higher Auger electron ejection yield) and lower population of electrons, respectively. This estimation indicates generally that the characteristic K X-ray intensity of a lighter atom is higher than that of a heavier atom. Although the relative intensity peak can shift towards higher energy due to the air absorption effect, in real experiments, the observed high intensity of the chlorine characteristic X-ray is a logical outcome.

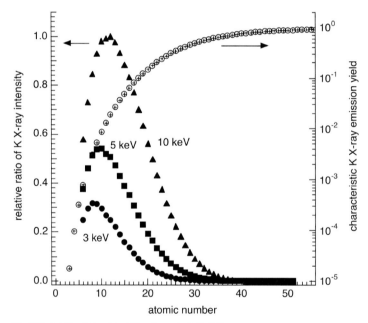

Fig. 9.9 Relative intensity of the characteristic K X-ray emission yield as a function of atomic number with different electron temperatures (the left axis) and characteristic K X-ray emission yield (the right axis).

9.3
Characteristics of Hard X-ray Emission from Transparent Materials

9.3.1
Experimental Setups for Laser-induced Hard X-ray Emission

In cases of hard X-ray emission, the air absorption effect is relatively small if com-pared with cases of EUV and soft X-ray emission. Thus, experiments can be per-formed in atmospheric pressure and the setups for laser-induced hard X-ray emis-sion spectroscopy and intensity measurements are quite simple. As one example, Fig. 9.10 shows the experimental setup for laser-induced hard X-ray emission from solutions [41–46]. A sample solution film was prepared by using a glass or a titanium nozzle with a rectangle outlet with a dimension of 0.1×5 mm^2 and a circulation pump. The nozzle body was attached to a three-dimensional and hori-zontally-rotational stage, so that the position and angle of the solution film, with respect to the laser incidence, can be controlled precisely. The solution flow rate was about 120 ml min^{-1} and a laser pulse irradiates a fresh solution surface every time. The thickness of the solution irradiated by laser pulses was estimated to be about 20 μm. Femtosecond laser pulses (160 fs, 780 nm, 1 kHz) were used as excitation pulses and focused tightly onto solid samples vertically by using an objective lens (NA = 0.28) in air. Different types of solid-state detector were used for X-ray emission spectroscopy, for instance, highly-pure Ge, Si(Li), and CdZnTe detectors. The photon energy resolution of these detectors are 150–200 eV at ~ 10 keV. A Geiger-Mueller counter was used for X-ray intensity measurements.

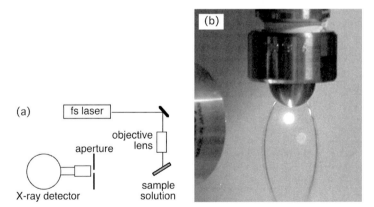

Fig. 9.10 The experimental setup for femtosecond laser-induced X-ray emission spectroscopy and intensity measure-ments of aqueous solutions. The objective lens used here was Mitutoyo, M Plan Apo 10× (NA = 0.28).

In the case of solid materials, for instance, glass plates and polymer films, the experimental setup for laser-induced hard X-ray emission is shown in Fig. 9.11 [51, 55, 56]. Femtosecond laser pulses (260 fs, 780 nm, 1 kHz) were used as excitation pulses and focused tightly onto solid samples, vertically, by using an objective lens (NA = 0.28) in air. Sample solids were mounted on a two-dimensional (vertical to the laser incident beam, x-y plane) motorized stage. In one controlling mode, the stage was moved in a zigzag direction horizontally (Fig. 9.11(b)) or vertically (Fig. 9.11(c)) by a computer. In the other controlling mode, an experimenter can control the stage movement manually by using a joystick with the sample surface monitored by a CCD camera. The stage-moving velocity was variable in the range of 0–20 μm ms^{-1}. A sample surface position to the focus in the laser incident direction (z direction) was always manually optimized to give the maximum X-ray intensity. The X-ray intensity was measured by a Geiger-Mueller counter and X-ray emission spectra were taken by a Si(Li) solid-state detector. A beryllium foil was attached to the input of the solid-state detector. A 1 mm thick lead or brass-made aperture was set in front of the X-ray detectors to reduce the X-ray intensity and to prevent sum peak detection. An intake duct was also set near the laser focus to remove debris from the laser optical path. An experiment was performed under atmospheric pressure at room temperature.

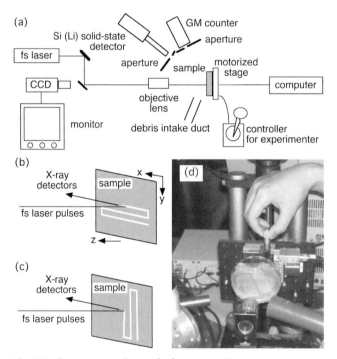

Fig. 9.11 The experimental setup for femtosecond-laser-induced X-ray emission spectroscopy of solid samples (a). Two different sample-moving directions for detection: horizontally (b) and vertically (c). The photograph of the experiment is shown in (d).

9.3.2
Effects of Air Plasma and Sample Self-absorption

Because the laser power at the focus in femtosecond-laser-induced X-rays generation is normally of the order of 10^{15} W cm^{-2} as described before, the air plasma
due to tight focusing of intense laser pulses may interfere with hard X-ray emission, which is one disadvantage for experiments under atmospheric pressure.
Figure 9.12 shows a counter plot of X-ray emission spectra as a function of relative
position of sample solution to the lens focus with different laser intensities [46].
In the case of higher laser intensity at 490 µJ per pulse, the X-ray emission intensity was the highest within the position region about 10 µm; precise positioning
of the sample solution to the lens focus is indispensable for higher X-ray intensity.
The optimum position of the solution for the highest X-ray intensity, in the case
of the higher laser intensity, was 13 µm closer to the lens than is the case for lower
laser intensity at 190 µJ per pulse. This may reflect that air plasma is formed during a laser pulse width in front of the solution surface when the laser intensity is
high enough, so that such air plasma can reflect or scatter the latter half of a laser
pulse and the coupling of laser pulses with solutions is ineffective. Of course,
introduction of an inert gas, like helium, into the laser focus can clear this effect
easily.

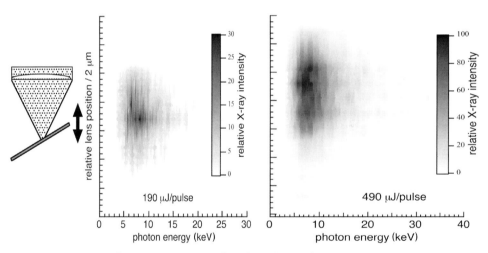

Fig. 9.12 Counter plots of X-ray emission spectra from distilled water obtained
by changing the position of the sample solution with respect to the lens focus,
for different laser intensities at 190 and 490 µJ per pulse.

Furthermore, due to the high power of laser pulses at the focus, laser ablation
can be induced. An inset in Fig. 9.13 shows a scanning electron micrograph of a
poly(vinyl chloride) plate irradiated by focused femtosecond laser pulses (530 µJ
per pulse) [55]. The sample-moving velocity was 6 µm ms^{-1}. It is clearly observed
that the sample surface was etched due to laser ablation. When it comes to femto-

second laser ablation of materials, generally speaking, sharp edge processing can be expected because there is little thermal diffusion during a short laser pulse width when it is compared with nanosecond laser ablation, as reported [57]. However, under X-ray generation conditions, other photons with a wide spectrum of wavelength from infrared, visible, and UV to EUV and soft X-ray, and also high-energy electrons, are generated densely at the same time. These result in a sample surface modification by laser irradiation. Thus, the condition may not be appropriate for laser material processing, although that is not the main subject of this chapter.

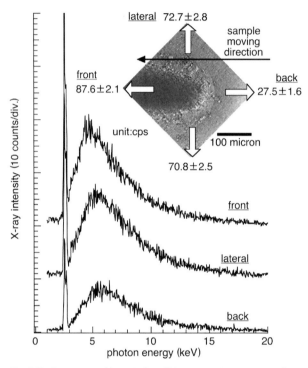

Fig. 9.13 Femtosecond laser-induced X-ray emission spectra and intensity from a poly(vinyl chloride) plate measured from different directions. The inset represents a scanning electron micrograph of a poly(vinyl chloride) plate irradiated by femtosecond laser pulses. The sample-moving velocity was 6 μm ms^{-1}. The solid-state detector used here was a Si(Li) detector.

As is also shown in Fig. 9.13, the X-ray intensity changed depending on the measurement direction [55]. The X-ray intensity measured from the forward-moving sample direction, was about 90 cps, while the intensity was measured to be about 30 and 70 cps when it was observed from the back side and the lateral direction, respectively. This was not affected by the laser polarization plane to the sample-moving direction. The X-ray emission spectra in Fig. 9.13 show that the X-ray intensity in the lower energy region was lower in the spectrum observed from the back side than in that observed from the front side. This indicates that X-rays emitted from the bottom of a trench made by laser ablation and X-rays was absorbed by the trench walls of the sample itself. Due to this effect, the X-ray intensity changes when samples move back and forward in the experimental setup shown in Fig. 9.11 (b). For measurements of the X-ray intensity, irrespective of the sample-moving direction, the experimental setup shown in Fig. 9.11 (c) is preferable.

A similar self-absorption effect was observed in cases of aqueous solutions. Figure 9.14 shows the X-ray intensity angle distribution of distilled water (a) and X-ray emission spectra of a cesium chloride aqueous solution (6.5 mol dm^{-3}) observed from the front and rear sides of solution film (b) [46]. Here the laser incident angle to the solution surface normal was 30°. The X-ray intensity angle distribution shows that X-rays are emitted homogenously except for angles around 120° and 300°. Also, in general, the X-ray intensity was higher at the front side of the solution film than at the rear side. These observations imply that X-rays are emitted from the solution surface and the X-rays are absorbed by the solution

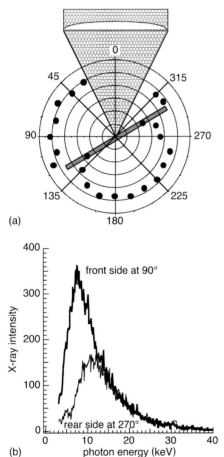

(a)

(b)

Fig. 9.14 X-ray intensity angle distribution of distilled water (a) and X-ray emission spectra from a cesium chloride aqueous solution (6.0 mol dm^{-1}) measured at the front (90°) and the rear (270°) sides of the solution film (b). The laser incident angle to the solution surface normal was 30°. The laser intensity was 490 μJ per pulse. The solid-state detector used here was a CdZnTe detector.

itself was observed from the rear side. Indeed, as shown in Fig. 9.14 (b), the X-ray emission spectrum observed from the rear side at 270° showed an intensity decrease in the lower energy range when compared with the spectrum observed from the font side at 90°, which proved the solution self-absorption of the X-rays. A further quantitative analysis of these X-ray emission spectra confirms that X-ray emitted from a point inside the solution within a 1 μm depth [46].

This self-absorption effect occurs markedly in metals rather than in transparent materials that are composed of relatively light atoms, since the X-ray absorption coefficients of metals are much higher than those of light atoms. In other words, X-ray emission induced by laser irradiation can be used from the back side of samples [58] and this method may be employed in various applications that require debris-free circumstances.

9.3.3
Multi-photon Absorption and Effects of the Addition of Electrolytes

In the case of metals, there are free electrons from the start of all processes leading to X-ray emission, while in the case of transparent materials, such as water, this is not the case. Here intense laser ionization, as the primary process leading to X-ray emission, should be briefly mentioned. The ionization potential of water has been reported to be ~ 6.5 eV by several groups [15–17, 59], while V. Y. Sukhonosov insisted that the energy for electron transition to the conduction band was 9.18 eV and the absorption in the lower energy region was due to interband electron transitions corresponding to exciton absorption [60]. This means that four to five photons at an energy of 1.59 eV (wavelength = 780 nm) are necessary at least for the photo-ionization of water. Three-photon absorption cross-section of water was reported to be 0.9×10^{-31} cm^4 W^{-2} in the power region of femtosecond laser pulses [50 fs, 400 nm (3.1 eV)] at 10^{11} W cm^{-2} [61]. When the total energy of multi-photons (2 eV, 620 nm) reaches the excited electronic state, which extends from ~ 7.4 to ~ 9.4 eV, a resonance effect becomes dominant and photo-ionization is promoted [62]. Furthermore, in the case of gas-phase water molecules, it was reported that the ac Stark effect, due to the intense laser field, induced an upward shift and broadening of the electronic states, when those states became sufficiently dressed with photons [63]. On the other hand, the probability of ionization of a water molecule after two-photon absorption of 266 nm (4.66 eV) light was reported to be 0.30, while the probabilities of dissociation and non-radiative relaxation were 0.26 and 0.44, respectively [64].

Once photoelectrons are ejected, they are forced to oscillate by the electronic field of the intense laser, collide with surrounding atoms and ions, and induce inner-shell ionization. A resulting phenomenon is Auger electron ejection and the probability of which $(1-\phi_K)$ can be estimated indirectly by Eq. (3). Figure 9.9 shows that the Auger electron ejection probability of lighter atoms is higher than that of heavier atoms, for instance, 0.99 for oxygen and 0.55 for copper [54]. In this sense, lighter atoms may be initiators for all processes leading to X-ray emission. Finally, avalanche ionization or cascade ionization is induced, resulting in breakdown. The threshold power of laser-induced (580 nm, 2.14 eV, 100 fs) break-down of water, was reported to be between 3.06×10^{12} and 1.11×10^{13} W cm^{-2} [15, 16, 65]. Although there is a difference in laser wavelength of between 580 nm and 780 nm, from the profile of the integrated laser power in Fig. 9.5, the laser-induced breakdown of water may be triggered at the former half or at the rising edge of a laser pulse with a total power of 10^{15} W cm^{-2}.

After the final stage of hard X-ray emission, relatively high-energy electrons, but which cannot induce hard X-ray emission, remain and induce another ionization and EUV emission. As a result, solvated electrons [66] may be formed at a late stage. Although there is a report on the solvated electron formation induced by intense laser pulses (1.3–3.3×10^{12} W cm^{-2}, 400 nm, 50 fs) showing that the life-time of solvated electrons decreases as the laser power increases [67], the laser power in this study was much lower than the power for hard X-ray emission (10^{14}–10^{15} W cm^{-2}). Late formation of solvated electrons has not yet been con-firmed spectroscopically, due to the bright luminescence of plasma and laser abla-tion. However, it may contribute to X-ray emission under the double-pulse excita-tion condition, which is described in Section 9.3.5.

On the basis of the discussion above, effects of the addition of electrolytes can be considered by comparison with the case of water. Similar to the case of two different electron temperatures, depending on laser intensity as in Fig. 9.3 (b), the electron temperatures increased as functions of the concentration and the atomic numbers of electrolytes [44]. As for the primary process of multi-photon ioniza-tion, it may be more effective if compared with water, since the ionization poten-tial of aqueous solutions of electrolytes decreases [59]. Once conductive electrons are ejected, electron impact ionization is induced more effectively, compared with the case of water, because the electron impact ionization cross-sections are larger in heavier atoms than in lighter atoms, for instance, 4.39×10^{-17} cm^2 for oxygen, 1.44×10^{-16} cm^2 for chlorine, and 8.66×10^{-16} cm^2 for rubidium for a conductive electron energy of 25 eV [50]. In other words, the ionization rate is higher in aque-ous solutions of electrolytes than in water. As a result, electron density easily reaches the critical density in the rising edge of a laser pulse and the remaining laser power can be effectively used for resonance absorption. This may be the rea-son why electron temperatures increase as a function of the concentration and the atomic number of the electrolytes.

9.3.4
Multi-shot Effects on Solid Materials

The X-ray intensity was also dependent on the sample-moving velocity. Kutzner et al. reported that the intensity of X-rays from a commercially available ferric audio-cassette tape target, when irradiated by femtosecond laser pulses (25 fs, 780 nm, 480 μJ per pulse, 1 kHz), varied with the tape speed in the range of 20–200 μm ms^{-1} [68]. The tape speed, where the X-ray intensity started to increase, coincided with the speed at which the focus spots of successive pulses started to overlap geometrically on the tape. Therefore, they insisted that the increase in X-ray yield was caused by the multiple laser irradiations on the target. Hatanaka et al. also reported that the intensity of the X-rays from transparent materials such as a polymer plate and a glass plate, varies with the sample-moving velocity and the optimum velocity also changes as the laser intensity changes [55]. Figure 9.15 shows the X-ray intensity dependences on the sample-moving velocity as a function of the laser intensity. Here the X-ray intensity was measured by the experimental setup shown in Fig. 9.11 (c). The sample was a poly(vinyl chloride) plate and a color glass filter (Toshiba, G54). When the laser intensity was as low as 37.5 μJ per pulse, the X-ray intensity was the highest, when the sample-moving velocity was at 4 μm ms^{-1}. The X-ray intensity decreased when the sample-moving velocity was higher or lower; there is an optimum sample-moving velocity for the highest X-ray intensity. As the laser intensity increased, the optimum sample-moving velocity increased; 6 and 8 μm ms^{-1} for 300 and 390 μJ per pulse, respectively. Furthermore, even when the laser intensity was the same, the optimum sample-moving velocity of the color glass filter changed to 1 μm ms^{-1}.

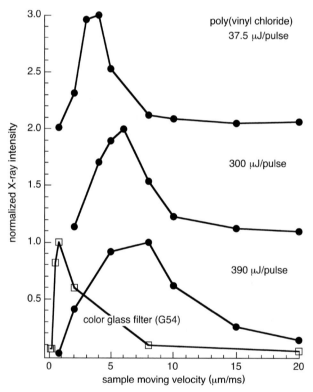

Fig. 9.15 X-ray intensity as a function of the sample-moving velocity with different laser intensities. Samples were a poly(vinyl chloride) plate and a color glass filter (Toshiba, G54).

Figure 9.16 shows electron micrographs of a poly(vinyl chloride) plate irradiated by femtosecond laser pulses (150 μJ per pulse) with different sample-moving velocities [55]. From the micrographs of samples irradiated with the velocity of 20 μm ms^{-1}, the laser spot size can be estimated to be about 15 μm, which means that a laser pulse irradiates a fresh sample surface every time. On the other hand, in the case of a lower velocity at 0.8 μm ms^{-1}, the width of laser-etched trench was 50 μm and sample surfaces were completely removed due to laser ablation. The number of multiple laser irradiations onto the same sample position can be calculated to be more than 18, from values of the laser spot size, the laser repetition rate (1 kHz), and the sample-moving velocity. This strongly indicates that the X-ray intensity decrease in the lower sample-moving velocity is due to sample depletion induced by successive laser ablation. Based on the above consideration, the reason for the optimum velocity to shift to the higher when the laser intensity increased is that the laser ablation rate (etch depth per single laser irradiation) is higher. Therefore, the sample should move faster, otherwise the sample would be removed. Also, the ablation rate of glass materials may be lower than that of poly-

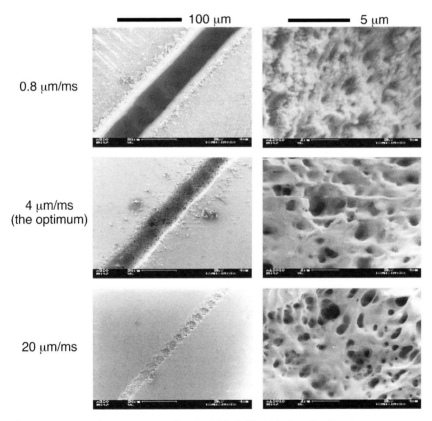

Fig. 9.16 Scanning electron micrographs of poly(vinyl chloride) plates irradiated by femtosecond laser pulses with different sample-moving velocities. The laser intensity was 150 μJ per pulse.

mer plates, so that the optimum velocity for the glass material shifted to lower, even if the laser intensity was the same.

The consideration described above has explained only the reason for the decrease in the X-ray intensity when the sample-moving velocity is low. If there is no other factor, the X-ray intensity should be constant when the velocity is higher because sample surfaces are always fresh and flat. However, this is not the case. It is well known that multiple laser irradiation onto material surfaces causes characteristic surface modification [69]. As is also shown in Fig. 9.16, sample surfaces of poly(vinyl chloride) plates were structured by multiple laser irradiation even under X-ray generation conditions [55]. This indicates that sample surfaces, which are initially flat and smooth, become structured because of multiple laser irradiation. Intense laser–matter coupling with structured sample surfaces can be different from that with flat and smooth surfaces. Indeed, Boyd et al. reported the local laser-field (1064 nm, 6 ns, 1×10^6 W cm^{-2}) enhancement on rough surfaces of metals, semimetals, and semiconductors with the use of optical second-harmonic generation [70]. Also, Stockman et al. reported a theory on the enhanced second-harmonic generation by metal surfaces with nanoscale roughness [71]. Even for X-ray generation, there are papers about the intensity enhancement of X-rays from rough surfaces. The group of Falcone et al. reported that metal (gold, silicon, and aluminium) surfaces consisting of grating structures and clusters, absorbed greater than 90% of the incident high-intensity laser light and the intensity of X-rays from such a sample surface increased, while flat surfaces absorbed only 10% of the incident laser light [72]. They calculated the absorption power of cluster-like structured gold particles by taking Mie scattering into account and confirmed a shift of the plasma resonance towards the visible region about 2.5 eV, when compared with a flat sample surface. They also reported that the intensity of soft X-rays from an aluminium target with colloidal surface irradiated by a femtosecond UV laser pulse (248 nm, 700 fs, 8×10^{15} W cm^{-2}) was enhanced when compared with the case of targets with a polished surface [73]. Not only where these intensity increases but also higher ionic states of aluminium (Al^{9+} and Al^{10+}) were observed, which indicates that the plasma produced on such rough surfaces is hotter than that on flat surfaces. They pointed out an important difference in colloidal sample surfaces. Since surface structures are of the order of 10 nm, which is much smaller than the skin depth (100 nm), the whole volume of colloids can be heated. When the whole volume is heated, the main energy loss process, which is present for flat surfaces, namely nonlinear heat conduction into adjacent cold bulk by electrons, does not work. Kulcsar et al. also reported that the intensity of X-ray emission from a nickel target sample with velvet-like nanostructures irradiated by a picosecond laser (1 ps, 1054 nm, 1×10^{17} W cm^{-2}) was 50 times higher than that from a flat sample [74]. They attributed the intensity increase partly to the enhanced laser–surface coupling. Fresnel reflection on the structured sample surface is much reduced because of the sample surface anisotropy. As a result, more light couples into the bulk of the material where the light is strongly absorbed. Nishikawa et al. also reported soft X-ray emission from a porous silicon target irradiated by femtosecond laser pulses (130 fs, 400 nm, 10 Hz,

10^{11}–10^{15} W cm^{-2}) [75]. Furthermore, they reported a study on the enhancement mechanism by using a nano-hole-array alumina target irradiated by femtosecond laser pulses (100 fs, 790 nm, 10 Hz, 1.4×10^{16} W cm^{-2}) [76]. It was found that the highest soft X-ray (5–20 nm, 60–250 eV) intensity was obtained with a nano-hole-array target with a 500 nm hole interval and a 450 nm hole diameter. They pointed out that the large surface area and the nanostructure wall enlarged the region of interaction with laser pulses, and that plasma collision at the nanometer spaces caused X-ray emission enhancement. Rajeev et al. reported that the intensity of X-ray emission in the region of 30–300 keV from copper samples, when irradiated by femtosecond laser pulses (100 fs, 806 nm, 10^{15}–10^{16} W cm^{-2}) increased when the sample surfaces were nanostructured [77]. They measured X-ray emission spectra and found that, not only the X-ray intensity, but also the electron temperature increased with nanostructured samples and the rough surface over-rode the role of laser polarization, while the laser polarization affected X-ray emission spectra of samples with flat surfaces. Later, they calculated the local enhancement of the laser field intensity due to surface nanostructures on the basis of the theory of surface plasmons and found that there was an optimum size of nanostructure when the laser field was the most intensified [78]. Hatanaka et al. also observed the X-ray intensity enhancement with plates of brass and poly(vinyl chloride) ground by sandpaper [55]. Also, as in the case of poly(vinyl chloride) plates with different sample-moving velocities (a different number of laser irradiations), Hironaka et al. reported that the intensity of X-ray emission from copper flat targets, when irradiated by femtosecond laser pulses (42 fs, 780 nm, $\sim 3 \times 10^{17}$ W cm^{-2}) was enhanced by about 100 times by multiple laser irradiations [79]. The X-ray intensity in the range 3–6 keV was a function of the number of laser shots and the maximum X-ray intensity was obtained in the fourth shot. They advocated the following two points for an X-ray intensity increase. Successive laser shots directed, not on the flat surface, but on the laser-ablated surface by previous laser shots. As a result, the laser focusing, the incident angle, and the polarization are changed. Such surface roughness can enhance laser absorption. Furthermore, when the plasma is produced in the cavity made by previous laser shots, the plasma can be confined to a small space and the collision of fast electrons with the solid material surface increases. Such an increase in the collisions may also enhance X-ray generation.

Based on the discussions described above, possible explanations for the X-ray intensity dependence on sample-moving velocities as shown in Fig. 9.15 are summarized in Fig. 9.17. In the case of a higher sample-moving velocity, successive laser pulses always irradiate fresh flat sample surfaces and never irradiate the previous laser pulse-produced structured sample surface. Of course, this results in X-ray emission with the same intensity as is the case for flat sample surfaces. On the other hand, in the case of a lower sample-moving velocity, successive laser pulses irradiate the same position of the sample surface many times. Since laser thresholds for ablation are much lower than those for X-ray emission, once the sample surface is ablated, the laser focus is out of the plane of samples for X-ray generation because the optimal laser focus position to the sample surface is about

10 μm long, as shown in Fig. 9.12. Then X-rays are not generated, because of sample depletion at the focus. Finally, in the case of an appropriate sample-moving velocity, successive laser pulses partly irradiate the same position that the previous laser pulse irradiated. However, the bottom of the ditch produced by previous laser pulses is still in the region of the laser focus for X-ray generation, which can induce X-ray emission and the bottom surface is structured, due to multiple laser irradiation. This results in an X-ray intensity increase, because of the optimal surface roughness.

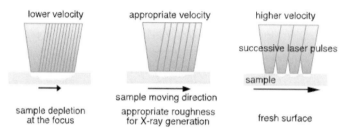

Fig. 9.17 A conceptual diagram of optimum sample-moving velocity for X-ray emission from solid samples when irradiated by successive laser pulses.

In the case of glasses, another possible mechanism for enhancing X-ray intensity, can be considered. As discussed in Section 9.3.2, emission of UV, EUV, and high-energy electrons was induced densely at the same time under X-ray emission conditions. Such high-energy quanta surely induce formation of color-centers or, in other words, defects [80]. As a result, glass materials are no longer transparent and have absorption at the laser wavelength. Such one-photon absorption may also contribute to the X-ray intensity increase, though this has not yet been confirmed experimentally.

9.3.5
Pre-pulse Irradiation Effects on Aqueous Solutions

There are reports on X-ray intensity increase under double-pulse laser irradiation conditions. Some reports on EUV and soft X-ray emission from transparent solid materials irradiated by double pulse laser have been already reviewed in Section 9.2.1. Here, some other reports on X-ray emission induced by double-pulsed laser are briefly summarized. Kuhlke et al. reported that X-ray (> 1 keV) intensity from a tungsten target irradiated by two successive femtosecond UV laser pulses (308 nm, 200–300 fs, 2.2 and 2.3 mJ per pulse, ~ 10^{17} W cm^{-2}) increased eight times at a delay time of about 30–50 ps [81]. They attributed the X-ray intensity enhancement to that the plasma generated by the first pulse expanded and absorbed the second pulse effectively. Tom and Wood reported that soft X-ray (17–35 nm, 35–73 eV) intensity from a tantalum target irradiated by two consecutive laser pulses (100 fs, 2.6×10^{13} W cm^{-2}) increased up to four times with the delay time of 80 ps and the X-ray intensity was the same until the delay time at

200 ps at the earliest [82]. They also attributed the intensity increase to the pre-
formed plasma expansion and its effective absorption of the second laser pulse.
Nakano et al. reported the intensity increase of soft X-ray from a 300 nm thick
aluminium layer when irradiated by a pair of femtosecond laser pulses (130 fs,
800 nm, 10^{15} W cm^{-2}) [83]. The soft X-ray intensity increased monotonically as the
delay time between the two pulses increased from 10 ps to 2 ns at the longest. The
maximum enhancement factor was obtained as one hundred at a delay time of
2 ns. They also observed that the soft X-ray pulse width increased as the delay
time increased. Pelletier et al. also reported soft X-ray emission from a tantalum
irradiated by a main pulse (400 fs, 527 nm, 5×10^{17} W cm^{-2}) with a pre-pulse
(550 fs, 1053 nm, 10^{16} W cm^{-2}) [84]. The conversion efficiency from laser to X-ray
increased in the range 0.8–1.2 keV from 0.16% with a delay time of 9 ps to 0.4%
with a delay time of 16 ps. Kutzner et al. reported a few hundredfold intensity
increase of X-ray emission from ferric audio-cassette tapes when irradiated by a
main pulse (25 fs, 3×10^{15} W cm^{-2}) with a pre-pulse (2×10^{12} W cm^{-2}) [85].

   Hatanaka et al. have observed an X-ray intensity increase in double-pulse laser
irradiation onto aqueous solutions [86]. Figure 9.18 (a) shows the X-ray intensity
as a function of the delay time between s-polarized pre-pulses and p-polarized
main pulses in the range from −10 to 40 ps. Here the distance between the X-ray
detector and the laser focus was 10 cm. Laser intensities of the pre- and main
pulses were 60 and 300 μJ per pulse, respectively. Also, the time dependence of
the specular reflection intensity of the main pulse, which is normalized to the
pulse without a pre-pulse, is shown in Fig. 9.18 (b). Under these experimental
conditions, the X-ray was a single pulse because the X-ray intensity by s-polarized
light was quite low as shown in Fig. 9.2. Two different maxima were observed at 4
and 13 ps. The X-ray intensity without the pre-pulse was only 30 counts under the

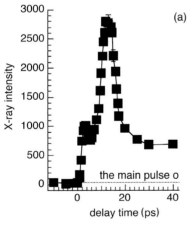
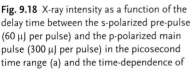
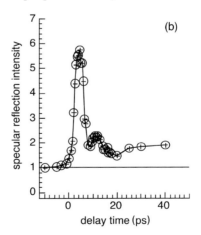

Fig. 9.18 X-ray intensity as a function of the
delay time between the s-polarized pre-pulse
(60 μJ per pulse) and the p-polarized main
pulse (300 μJ per pulse) in the picosecond
time range (a) and the time-dependence of
the specular reflection intensity of the main
pulse, normalized by the reflection intensity
without the pre-pulse (b). The distance for
X-ray intensity measurements was 10 cm.

experimental condition, so that the X-ray intensity was enhanced up to 90 times. In the same way the specular reflection intensity rose late and increased by up to almost 6 and 2 times at 4 and 13 ps, respectively.

The delayed increase of the specular reflection intensity was also observed by the group of Fotakis et al. They reported transient reflectivity changes induced by a femtosecond KrF excimer laser (500 fs, 248 nm, 10^{13} W cm^{-2}) of solutions such as poly-silicone oil, methyl-methacrylate, styrene, and water [87] and polymers such as poly(methyl methacrylate), polyethylene, poly(ethylene terephthalate), and polyimide [88] and the observed delayed (1–2 ps) rises in the reflectivity of up to 2.25 times. The refractive index, which is linked to the reflectivity of a plasma, is a function of the electron density, not of the electron energy or temperature.

$$n_p = [1 - (\omega_p/\omega_L)^2]^{0.5} \tag{4}$$

where n_p, ω_p, and ω_L represent the refractive index of a plasma, the plasma frequency, and the light frequency, respectively. Furthermore, the plasma frequency is related to the electron density as:

$$\omega_p = (e^2 n_e/\varepsilon_0 m_e)^{0.5} \tag{5}$$

where e, n_e, ε_0, and m_e represent the elementary electric charge, the electron density of a plasma, the dielectric constant, and the electron mass, respectively. On the other hand, according to a reference on electron-impact ionization cross-section for all electron shells [50], the cross-sections of electrons with higher energy are smaller than those of electrons with lower energy, as introduced in Section 9.2.2, which indicates that lower energy electrons can ionize atoms more than higher energy electrons in this energy region. Aqueous solution surfaces are originally insulators. However, the surface has a metal-like condition due to plasma formation by the pre-pulse irradiation. Since the s-polarized pre-pulse laser power (~ 10^{14} W cm^{-2}) was high enough for plasma generation (though it was not high enough for hard X-ray emission), the conductive electron energy could be high, of the order of a hundred eV, at least. Even after the pre-pulse had passed, conductive electrons with high energy collided with the surrounding atoms and ions in aqueous solution, inducing further impact ionization. As a result, the electron density may increase, though this is a competing process with plasma expansion and electron–cation recombination.

During such processes, the main pulse arrives late at the pre-formed plasma. Compared with the original solution surface, absorption of the main laser pulse by the pre-formed plasma is much higher, which results in a higher X-ray intensity. On the other hand, due to decay processes of the pre-formed plasma, such as expansion and recombination, absorption of the main laser pulse by the pre-formed plasma decreases and returns to that of the original solution surface. This may be the reason for the decrease in the X-ray intensity and specular reflection intensity after the peak at 4 ps. However, the leading edge of the main pulse also works as an ionization source, so that another increase in the specular reflection intensity was observed at 13 ps.

Similar changes in the X-ray intensity were observed in the nanosecond region of the delay time. Figure 9.19 (a) shows the X-ray intensity as a function of the delay time in the range 0–15 ns with different pre-pulse laser intensities at 20 and 60 μJ per pulse [86]. Here the distance between the X-ray detector and the laser focus was 30 cm. Obviously, there were two X-ray intensity peaks observed at 2.5 ns and 5 ns when the pre-pulse intensity was lower at 20 μJ per pulse. Furthermore, those peaks shifted to earlier times at 0.45 ns and 4 ns when the laser intensity was higher at 60 μJ per pulse. X-ray emission spectroscopy confirmed that the X-ray intensity in the range 8.5–9.5 keV was enhanced at least 800 times and the

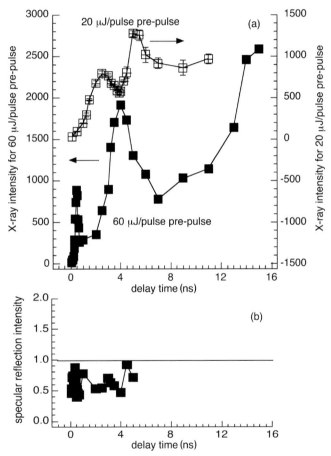

Fig. 9.19 X-ray intensity as a function of the delay time between the s-polarized pre-pulse and the p-polarized main pulse (300 μJ per pulse) in the nanosecond time range with different pre-pulse laser intensities at 20 and 60 μJ per pulse (a) and a time-dependence of the specular reflection intensity of the main pulse normalized by the intensity without the pre-pulse (b). The distance for X-ray intensity measurements was 30 cm.

electron temperature also increased when compared with the case of single-pulse irradiation without the pre-pulse. Additionally, the specular reflection intensity of the main pulse decreased as shown in Fig. 9.19 (b) and there was little difference observed in the reflection intensity between the two delay times at 0.45 ns and 4 ns. On the other hand, for X-ray intensity as a function of laser incident angle, under double-pulse laser irradiation conditions, as shown in Fig. 9.20, a clear difference between the two delay times was observed. The optimum incident angles for X-ray intensity shifted towards smaller angles at 35° and 30° for delay times of 0.45 ns and 4 ns when compared with the case of single-pulse irradiation (60°) as shown in Fig. 9.4.

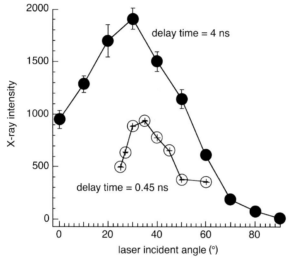

Fig. 9.20 X-ray intensity as a function of laser incident angle with a different time delay between the s-polarized pre-pulse (20 and 60 μJ per pulse) and the p-polarized main pulse (300 μJ per pulse). The distance for X-ray intensity measurements was 30 cm.

These results may indicate the following two points. First, laser ablation of the sample solution surface was induced in this time range, since the pre-pulse laser power at the focus was up to 10^{14} W cm^{-2} and the solution surface turned out to be transiently rough, which resulted in a decrease in the specular reflection intensity. Second, such transient surface roughness would grow as a function of time. Actually, Hatanaka et al. previously reported that sample surfaces of organic liquids such as toluene and benzyl chloride irradiated by femtosecond UV laser pulses (248 nm, 300 fs, 10^{11} W cm^{-2}) became rough at around the sub-nanosecond time region and the transient surface roughness grew from the order of a few tens of nanometers at sub-nanoseconds to the order of a few hundred nanometers at a few nanoseconds [89]. The different optimum laser incident angles shown in Fig. 9.20 suggest that such transient surface roughness was produced even under the 10^{14} W cm^{-2} pre-pulse irradiation to aqueous solutions. As discussed in Section 9.3.3, this nanometer-scaled surface roughness enhances the X-ray intensity. On the basis of discussions so far, the X-ray intensity peaks observed at 0.45 ns and 2.5 ns with pre-pulse laser intensities at 20 and 60 µJ per pulse may be due to this X-ray intensity enhancement by transient surface roughness. As introduced in the previous section, Rejeev et al. found that there was an optimum scale of nanostructures when the laser field was the most intensified [78]. This study suggests that transient surface nanostructures with optimum sizes where the laser field was mostly intensified were produced at the solution surface at a delay time of 0.45 ns when the pre-pulse intensity was 60 µJ per pulse. As time passed, such transient surface roughness grew and, as a result, X-ray intensity started decreasing.

As transient surface roughness grew more with the passage of time, a different effect of surface roughness, which is relatively macroscopic, can be considered. If the size of the transient surface roughness reaches the range of sub-micrometers to micrometers, the solution surface turns out to be like a two-dimensional array of convex and concave lenses because of the spatial modulation of the transparent solution surface. This may result in local focusing of the incident main laser pulses. Indeed, Pinnick et al. performed experiments on CO_2 laser ablation with 10–60 µm droplets of water, ethanol, and other solutions and observed that the rear surfaces of ethanol droplets were ablated due to a so-called ball lens effect [90]. In the case of water, which undergoes absorption at the wavelength of a CO_2 laser, the front side of the droplet was ablated. Based on the discussions above and the reference, the X-ray intensity peak observed at 4 ns when the pre-pulse intensity was 60 µJ per pulse can be due to local optical focusing of the main pulse and this effect can be one characteristic of transparent materials. All the processes considered so far are summarized pictorially in Fig. 9.21.

Further increase in the X-ray intensity at a later delay time after 8 ns was observed when the pre-pulse intensity was 60 µJ per pulse as shown in Fig. 9.19 (a) [86]. In this time range, the plasma is expected to decay and conductive electron energy may be low enough to be captured by the surrounding water molecules. As discussed in the Section 9.3.3, this results in solvated electron formation. As is well known, the absorption band of a solvated electron is in the wavelength region of the excitation laser wavelength at 780 nm [91]. This one-photon absorption of the main pulse by solvated electrons after the decay of plasma, produced by the pre-pulse irradiation, may cause the X-ray intensity increase in the late stage under double-pulse irradiation conditions.

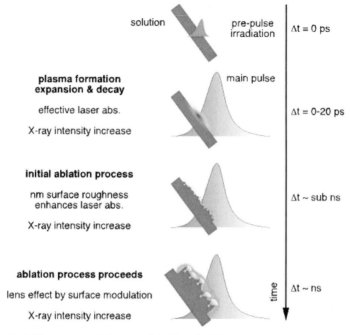

Fig. 9.21 A conceptual diagram of double-pulsed femtosecond laser-induced X-ray emission, using aqueous solutions.

9.4
Possible Applications

9.4.1
X-ray Imaging

One of the characteristics of femtosecond laser-based X-rays is that the X-ray source size is small at around 10 µm. The size will be made smaller by using objective lenses with high numerical apertures, after the optimization of laser excitation conditions, to obtain the highest X-ray intensity with the smaller laser intensity, which will result in higher spatial resolution. This ideal point source of X-rays can be used for imaging techniques. Sjogren et al. reported X-ray emission from tantalum targets when irradiated by femtosecond laser pulses (25 fs, 780 nm, 1 kHz) and applied X-ray pulses to the transmission imaging of rats [92]. However, this kind of a metal target limits the versatility of laser-induced X-ray sources, because the target needs to be constantly moving to avoid the depletion of material and, additionally, it gives rise to the surrounding pollution with ablation debris involving metal and metal oxides. These limitations of versatility can be mitigated by the use of an aqueous solution as a target material. This is because a liquid jet can easily supply harmless target material with a constant flow in any required position as an X-ray point source. As one example, X-ray transmission images of an IC chip, an insect body, and a pepper berry by using X-ray pulses from aqueous solutions are shown in Fig. 9.22. Due to the development of hollow fiber technologies, one will be able to generate X-rays anywhere in fine tubes and pipes, such as blood vessels and the esophagus in the human body, or in cooling pipes in atomic power plants. The technology for endoscopes and catheters has been well established and keeps progressing, so that X-ray generation using solutions like a normal saline solution is one of the most promising applications in the medical and industrial fields. Also, in the case of aqueous solutions, X-ray emission spectra are relatively broad. Thus, if one uses appropriate filters to select the X-ray wavelength, high-contrast images of X-rays can be obtained for selected elements contained within objects.

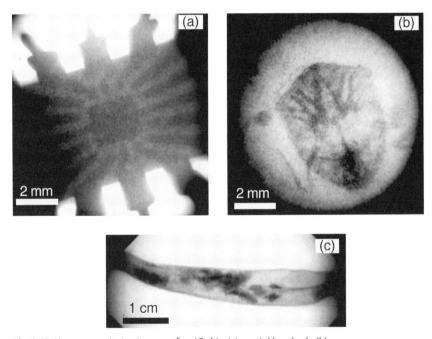

Fig. 9.22 X-ray transmission images of an IC chip (a); a stinkbug body (b); and a red pepper berry (c).

9.4.2
Elemental Analysis by X-ray Emission Spectroscopy

The experimental results described above indicate that any solid materials can be samples for femtosecond-laser-induced X-ray emission spectroscopy. Various solid samples have been used for X-ray emission spectroscopy, in addition to transparent materials, for example, audio-cassette tapes [43], alloys like brass [51], environmental samples like manganese nodules, naked filter papers and filter papers immersed once in solution, as well as sea foods (tuna, octopus, and scallops) [56]. For each sample, characteristic X-ray peaks were observed. Furthermore, Fukushima et al. performed experiments with natural rocks as one sample that elements distribute inhomogeneously [56]. The lightest element that can be detected by this method is silicon because of the X-ray absorption effects of air and the beryllium window. Furthermore, not only flat surface samples but other shaped samples can also be targets. For instance, Fukushima et al. used a Baccarat glass as shown in the photograph in Fig. 9.23 [56]. Here the glass was set on a ball mill rotator. As is well known, a crystal glass contains lead oxide and potassium oxide and their concentrations of Baccarat glass are ~ 30% and more than 10%, respectively. Such components were clearly observed, in addition to the silicon Kα line, potassium Kα at 3.31 keV, lead Mα at 2.35 keV, lead Lα at 10.5 keV, and lead Lβ at 12.6 keV.

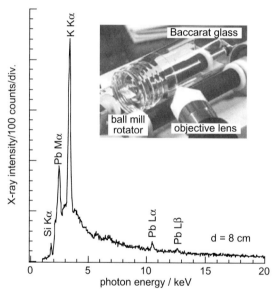

Fig. 9.23 An X-ray emission spectrum from Baccarat glass. The laser intensity was 350 μJ per pulse and the objective lens used here was Mitutoyo, M plan Apo 20× (NA = 0.42). The inset represents a photograph of experiment.

This versatility of the femtosecond-laser-induced X-ray emission spectroscopy to any solid samples may indicate its potential for application as an elemental analysis applicable under atmospheric pressure. Here, a comparative study can be made with conventional methods such as laser-induced plasma emission spectroscopy and electron probe micro-analysis (EPMA, JEOL, JSM-6500FT, EX-23000BU) where electron-induced X-ray emission spectra are measured. Figure 9.24 shows a plasma emission spectrum in the visible wavelength region (a) and an EPMA spectrum (b) of a color glass filter (Toshiba, B46). A plasma emission spectrum for air is also shown as a reference. Spectra shown in the figure are not corrected by the sensitivity the spectrometer used. Apparently, a lot of sharp lines are not detected in the plasma emission spectrum because of the intense background emission. For elemental analyses, detectors should be time-gated. Only the D line of sodium, which cannot be observed by the femtosecond-laser-induced X-ray emission spectroscopy, was clearly observed at 589 nm. Similar studies on elemental analyses of aqueous solutions by using laser-plasma emission spectra were also reported elsewhere [93]. In the EPMA spectrum, light elements such as oxygen, sodium, and aluminium were detected clearly, because EPMA is in vacuo. On the other hand, trace elements such as cobalt and arsenic were not detected. Generally speaking, emission spectra in lower energy ranges are essentially complex because a variety of radiative transitions are related. Although there are some difficulties in elemental analyses in X-ray fluorescence spectra, such as spectral interference and interelement effects [94], an elemental analysis in the hard X-ray region is comparatively much easier for the assignment of elements.

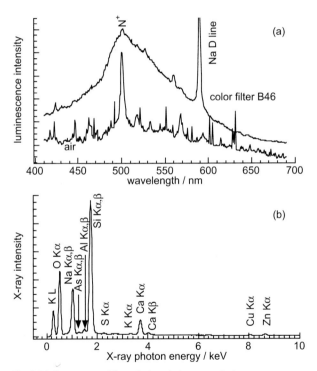

Fig. 9.24 Femtosecond laser-induced plasma emission spectra from air and a color glass filter (Toshiba, B46) (a); and an X-ray emission spectrum from a color glass filter (Toshiba, B46) in electron probe microanalysis (b).

9.4.3
Ultra-fast X-ray Absorption Spectroscopy

Another advantage of femtosecond laser-based X-ray is the temporally short-pulse length. The pulse width of the X-rays generated by femtosecond laser pulses is reported to be shorter than a few picoseconds at the longest [95]. Ultrafast X-ray pulses can be applied to various time-resolved measurements, for instance, X-ray diffraction [8] and X-ray absorption fine spectroscopy [9, 96]. Lee et al. recently reported time-resolved X-ray absorption fine spectroscopy at the iron K absorption edge of $Fe(CN)_6^{4-}$ solvated in water, when irradiated by femtosecond laser pulses (25 fs, 400 nm, 2 kHz) [9]. They generated X-ray pulses by using a metal target and the characteristic X-ray obtained was used for the calibration of the photon energy-axis. At this point, aqueous solutions have high versatility to any elements since one can add any ions to the aqueous solution as a photon energy-axis calibrator. Hatanaka et al. tried X-ray absorption near edge structure (XANES) measurements for various compounds by using X-ray pulses from a cesium chloride aqueous solution as a probe [46]. A XANES of copper thin foil, for instance, is shown in Fig. 9.25. A spectrum, obtained by a synchrotron radiation facility, is also shown in the figure. Fine structures at the copper K absorption edge were resolved and one small peak at 8980 eV was also observed which is assigned to the 1s → 4p transition. This experimental technique will play an important role in the clarification and understanding of molecular structures in excited states.

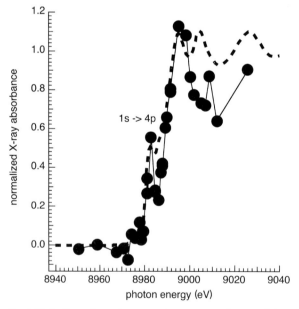

Fig. 9.25 An X-ray absorption near-edge structure of copper foil, measured by X-ray pulses from a cesium chloride aqueous solution under the focused femtosecond laser irradiation condition. The dotted line represents a spectrum obtained at a synchrotron radiation facility.

9.5
Summary

In this chapter, we have reviewed and summarized studies from the viewpoint of photon emission of EUV, soft X-rays, and hard X-rays. Interaction mechanisms between intense laser pulses of the order of from 10^{11} to 10^{17} W cm^{-2} and transparent materials such as glasses, polymers and solutions, have been discussed. There are quite a few groups studying hard X-ray generation using transparent materials, such as aqueous solutions. However, such studies using solutions as a model for transparent materials, will surely contribute to the clarification of the mechanisms of the interaction between an intense laser field and transparent materials and also to developments of applications for these X-ray sources.

References

1 T.P. Hughes, *Plasmas and Laser Light*, Adam Hilger, Bristol and Boston, 1975.

2 W.L. Kruer, *The Physics of Laser Plasma Interactions*, Addison-Wesley Publishing Company, California, 1988.

3 A. Rubenchik and S. Witkowski, eds., Physics of Laser Plasma in *Handbook of Plasma Physics*, vol. 3, 1991.

4 D. Attwood, *Soft X-rays and Extreme Ultraviolet Radiation*, Principles and Applications, Cambridge University Press, Cambridge, 1999.

5 I.C.E. Turcu and J.B. Dance, *X-rays from Laser Plasma*, Generation and Applications, John Wiley & Sons, Chichester, 1999.

6 K. Tsuji, J. Injuk, R.V. Grieken, eds., *X-ray Spectrometry: Recent Technological Advances*, John Wiley & Sons, Chichester, 2004.

7 A. Heuberger, J. Vac. Sci. Technol., B6, 107 (1988): E. A. Dobsiz, Emerging Lithographic Technologies IV, Proc. SPIE, 3997, 1 (2000).

8 J.R. Helliwell and P.M. Rentzepis, *Time-resolved Diffraction*, Oxford Science Publications, Oxford, 1997: M. Bargheer, N. Zhavoronkov, Y. Gritsai, J. C. Woo, D. S. Kim, M. Woerner, and T. Elsaesser, Science, 306, 1771 (2004) and references therein.

9 T. Lee, Y. Jiang, C.G. Rose-Petruck, F. Benesch, J. Chem. Phys., 122, 084506 (2005).

10 J. Yu, Z. Jiang, J.C. Kieffer, and A. Krol, IEEE J. Selected Topics in Quant. Electro., 5, 911 (1999): J. C. Kieffer, A. Krol, Z. Jiang, C. C. Chamberlain, E. Scalzetti, and Z. Ichalalene, Appl. Phys., B74, S75 (2002): F. Dorchies, L. M. Chen, Z. Ichalalene, Z. Jiang, J. C. Kieffer, C. C. Chamberlain, and A. Krol, J. Phys. IV, 108, 147 (2003).

11 G.A. Askar'yan, A.M. Prokhorov, G.F. Chanturiya, and G.P. Shipulo, Sov. Phys. JETP, 17, 1463 (1963).

12 R.G. Brewer and K.E. Rieckhoff, Phys. Rev. Lett., 13, 334 (1964).

13 P.A. Barnes and K.E. Rieckhoff, Appl. Phys. Lett., 13, 282 (1968).

14 C.E. Bell and J.A. Landt, Appl. Phys. Lett., 10, 46 (1967).

15 P.K. Kennedy, D.X. Hammer, and B.A. Rockwell, Prog. Quant. Electr., 21, 155 (1997).

16 C.H. Fan, J. Sun, and J.P. Longtin, J. Appl. Phys., 91, 2530 (2002): C. H. fan, J. Sun, and J.P. Longtin, J. Heat Transfer, 124, 275 (2002).

17 J. Noack and A. Vogel, IEEE. J. Quant. Elctr., 35, 1156 (1999).

18 C.B. Schaffer, N. Nishimura, E.N. Glezer, A. M.–T. Kim, and E. Mazur, Opt. Exp., 196 (2002).

19 C. Pepin, D. Houde, H. Remita, T. Goulet, and J.–P. Jay-Gerin, Phys. Rev. Lett., 69, 3389 (1992): F. Gobert, S. Pommeret, G. Vigneron, S. Buguet, R. Hai-

dar, J.–C. Mialocq, I. Lampre, and
M. Mostafavi, Res. Chem. Intermed.,
27, 901 (2001).

20 H. Yui, Y. Yoneda, T. Kitamori, and
T. Sawada, Phys. Rev. Lett., 82, 4110
(1999): H. Yui and T. Sawada, Phys. Rev.
Lett., 85, 3512 (2000).

21 R. M. Wood, *Laser Damage in Optical
Materials*, Adam Hilger, Bristol and Boston (1986).

22 S. Preuss, M. Spath, Y. Zhang, and
M. Stuke, Appl. Phys. Lett., 62, 3049
(1993).

23 C. B. Schaffer, A. Brodeur, and
E. Mazur, Meas. Sci. Technol., 12, 1784
(2001).

24 F. Jin and M. Richardson, Appl. Opt.,
34, 5750 (1995): M. Richardson,
D. Torres, C. DePriest, F. Jin, and
G. Shimkaveg, Opt. Commun., 145, 109
(1998): C. Keyser, R. Bernath, M. Al-
Rabban, and M. Richardson, Jpn. J.
Appl. Phys., 41, 4070 (2002): C. Keyser,
G. Schriever, M. Richardson, and
E. Turcu, Appl. Phys., A77, 217 (2003).

25 C. Rajyaguru, T. Higashiguchi,
M. Koga, W. Sasaki, and S. Kubodera,
Appl. Phys., B79, 669 (2004).

26 C. Rajyaguru, T. Higashiguchi,
M. Koga, K. Kawasaki, M. Hamada,
N. Dojyo, W. Sasaki, and S. Kubodera,
Appl. Phys., B80, 409 (2005).

27 U. Vogt, H. Stiel, I. Will, P. V. Nickles,
W. Sandner, M. Wieland, and T. Wil-
hein, Appl. Phys. Lett., 79, 2336 (2001):
U. Vogt, H. Stiel, I. Will, M. Wieland,
T. Wilhein, P. V. Nickles, W. Sandner,
SPIE Proc., 4343, 87 (2001).

28 S. Dusterer, H. Schwoerer, W. Ziegler,
C. Ziener, and R. Sauerbrey, Appl.
Phys., B73, 693 (2001).

29 B. A. M. Hansson, L. Rymell, M. Ber-
glund, H. M. Hertz, Microelectron.
Eng., 53, 667 (2000): B. A. M. Hansson,
M. Berglund, O. Hemberg, and H. M.
Hertz, J. Appl. Phys., 95, 4432 (2004):
B. A. M. Hansson, O. Hemberg, H. M.
Hertz, M. Berglund, H.–J. Choi,
B. Jacobsson, E. Janin, S. Mosesson,
L. Rymell, J. Thoresen, and M. Wilner,
Rev. Sci. Instrum., 75, 2122 (2004).

30 L. Rymell and H. M. Hertz, Opt Commun., 103, 105 (1993): L. Malmqvist,

L. Rymell, M. Berglund,H. M. Hertz,
Rev. Sci. Instrum., 67, 4150 (1996).

31 L. Rymell and H. M. Hertz, Rev. Sci.
Instrum., 66, 4916 (1995).

32 L. Rymell, M. Berglund, and
H. M. Hertz, Appl. Phys. Lett., 66, 2625
(1995).

33 L. Rymell, L. Malmqvist, M. Berglund,
H. M. Hertz, Microelectron. Eng., 46,
453 (1999): M. Berglund, L. Rymell,
H. M. Hertz, T. Wilhein, Rev. Sci.
Instrum., 69, 2361 (1998).

34 L. Malmqvist, L. Rymell, and
H. M. Hertz, Appl. Phys. Lett., 68, 2627
(1996): L. Malmqvist, A. L. Bogdanov,
L. Montelius, H. M. Hertz, Microelec-
tron. Eng., 35, 535 (1997): L. Malmqvist,
A. L. Bogdanov, L. Montelius, and
H. M. Hertz, J. Vac. Sci. Technol., B15,
814 (1997).

35 M. Berglund, L. Rymell, and
H. M. Hertz, Appl. Phys. Lett., 69, 1683
(1996).

36 P. Dunne, G. O'Sullivan, and
D. O'Reilly, Appl. Phys. Lett., 76, 34
(2000).

37 H. Nakano, T. Nishikawa, and
N. Uesugi, Appl. Phys. Lett., 70, 16
(1997).

38 H. Nakano, T. Nishikawa, and
N. Uesugi, Appl. Phys. Lett., 72, 2208
(1998).

39 R. J. Tompkins, I. P. Mercer, M. Fett-
weis, C. J. Barnett, D. R. Klug, G. Porter,
I. Clark, S. Jackson, P. Matousek,
A. W. Parker, and M. Towrie, Rev. Sci.
Instrum., 69, 3113 (1998).

40 T. D. Donnelly, M. Rust, I. Weiner,
M. Allen, R. A. Smith, C. A. Steinke,
S. Wilks, J. Zweiback, T. E. Cowan, and
T. Ditmire, J. Phys., B34, L313 (2001).

41 K. Hatanaka, T. Miura, and H. Fukumura, Appl. Phys. Lett., 80, 3925 (2002).

42 K. Hatanaka, T. Miura, H. Ono, Y. Watanabe, and H. Fukumura, Science of
Super-Strong Field Interactions, American Institute of Physics Conference Proceedings, p.260, Ed. K. Nakajima and
M. Deguchi (2002).

43 K. Hatanaka, T. Miura, H. Odaka,
H. Ono, and H. Fukumura, Bunseki
Kagaku, 52, 373 (2003).

44 K. Hatanaka, T. Miura, and H. Fukumura, Chem. Phys., 299, 265 (2004).

45 K. Hatanaka, T. Ida, S. Matsushima, and H. Fukumura, in prep.

46 K. Hatanaka, S. Matsuhima, H. Ono, and H. Fukumura, in prep.

47 B.L. Henke, E.M. Gullikson, and J.C. Davis, Atomic Data and Nuclear Data Tables, 54, 181 (1993): http://www-cxro.lbl.gov/optical constants/

48 X-RAY DATA BOOKLET, Lawrence Berkeley National Laboratory, University of California (2001).

49 K. W. Hill, *et al.*, Nucl. Fusion, 26, 1131 (1986).

50 P. L. Bartlett and A. T. Stelbovics, Atomic Data and Nuclear Data Tables, 86, 235 (2004).

51 K. Hatanaka, K. Yomogihata, H. Ono, H. Fukumura, Appl. Surf. Sci., (2005), in press.

52 D. L. Walters and C. P. Bhalla, Phys. Rev., A3, 1919 (1971).

53 J. P. Santos, F. Parente, Y.-K. Kim, J. Phys., B36, 4211 (2003).

54 W. Bambynek, B. Crasemann, R. W. Fink, H.-U. Freund, H. Mark, C. D. Swift, R. E. Price, and P. Venugopala Rao, Rev. Mod. Phys., 44, 716–813 (1972).

55 K. Hatanaka, K. Yomogihata, and H. Fukumrua, in prep.

56 M. Fukushima, K. Hatanaka, H. Ono, S. Matsushima, and D. Fukumura, Abstract in Euro-Mediterranean Symposium on Laser-Induced Breakdown Spectroscopy (EMSLIBS II), 2003.

57 B. N. Chichkov, C. Momma, S. Nolte, F. von Alvensleben, and A. Tunnermann, Appl. Phys., A63, 109 (1996).

58 S. Juodkazis and H. Misawa, private communication.

59 D. N. Nikogosyn, A. A. Oraevsky, and V. I. Rupasov, Chem. Phys., 77, 131 (1983): F. Williams, S. P. Varma, and S. Hillenius, J. Chem. Phys., 64, 1549 (1976): D. Grand, A. Bernas, and E. Amouyal, Chem. Phys., 44, 73 (1979).

60 V. Ya. Sukhonosov, High Energy Chemistry, 32, 71 (1998).

61 R. Naskrecki, M. Menard, P. van der Meulen, G. Vigneron, and S. Pommeret, Opt. Commun., 153, 32 (1998).

62 C. Pepin, D. Houde, H. Remita, T. Goulet, and J.-P. Jay-Gerin, Phys. Rev. Lett., 69, 3389 (1992).

63 M. V. Fedorov and A. E. Kazakov, Prog. Quantum Electron., 13, 1 (1989).

64 D. N. Nikogosyn, A. A. Oraevsky, and V. I. Rupasov, Chem. Phys., 77, 131 (1983).

65 J. Noack, D. X. Hammer, G. D. Noojin, B. A. Rockwell, and A. Vogel, J. Appl. Phys., 83, 7488 (1998): D. X. Hammer, R. J. Thomas, G. D. Noojin, B. A. Rockwell, P. K. Kennedy, and W. P. Roach, IEEE. J. Quantum Electron., QE32, 670 (1996).

66 A. Mozumder, *Fundamentals of Radiation Chemistry*, Academic Press, San Diego, 1999.

67 F. Gobert, S. Pommeret, G. Vigneron, S. Buguet, R. Haidar, J.-C. Mialocq, I. Lampre, and M. Mostafavi, Res. Chem. Intermed., 27, 901 (2001).

68 J. Kutzner, M. Silies, T. Witting, G. Tsilimis, H. Zacharias, Appl. Phys. B78, 949 (2004).

69 S. Ono, S. Nakaoka, J. Wang, H. Niino, and A. Yabe, Appl. Surf. Sci., 127–129, 821 (1998): J. Kruger, W. Kautek, M. Lenzner, S. Sartania, C. Spielmann, and F. Krausz, Appl. Surf. Sci., 127–129, 892 (1998): A. Bensauoula, C. Boney, R. Pillai, G. A. Shafeev, A. V. Simakin, and D. Starikov., Appl. Phys., A79, 973 (2004).

70 G. T. Boyd, Th. Rasing, J. R. R. Leite, and Y. R. Shen, Phys. Rev., B30, 519 (1984).

71 M. I. Stockman, D. J. Bergman, C. Anceau, S. Brasselet, and J. Zyss, Phys. Rev. Lett., 92, 057402 (2004).

72 M. M. Murnane, H. C. Kapteyn, S. P. Gordon, J. Bokor, E. N. Glytsis, R. W. Falcone, Appl. Phys. Lett., 62, 1068 (1993): M. M. Murnane, H. C. Kapteyn, S. P. Gordon, and R. W. Falcone, Appl. Phys., B58, 261 (1994): S. P. Gordon, T. Donnelly, A. Sullivan, H. Hamster, and R. W. Falcone, Opt. Lett., 19, 484 (1994).

73 C. Wulker, W. Theobald, D. R. Gnass, F. P. Schafer, J. S. Bakos, R. Sauerbrey, S. P. Gordon, and R. W. Falcone, Appl. Phys. Lett., 68, 1338 (1996).

74 G. Kulcsar, D. AlMawlawi, F. W. Budnik, P. R. Herman, M. Moskovits, L. Zhao, and R. S. Marjoribanks, Phys. Rev. Lett., 84, 5149 (2000).

75 T. Nishikawa, H. Nakano, H. Ahn, N. Uesugi, and T. Serikawa, Appl. Phys. Lett., 70, 1653 (1997).

76 T. Nishikawa, H. Nakano, K. Oguri, N. Uesugi, K. Nishio, and H. Masuda, J. Appl. Phys., 96, 7537 (2004).

77 P. P. Rajeev, S. Banerjee, A. S. Sandhu, R. C. Issac, L. C. Tribedi, and G. R. Kumar, Phys. Rev., A65, 052903 (2002): P. P. Rajeev and G. R. Kumar, Opt. Commun., 222, 9 (2003).

78 P. P. Rajeev, P. Taneja, P. Ayyub, A. S. Sandhu, and G. R. Kumar, Phys. Rev. Lett., 90, 115002 (2003): P. P. Rajeev, P. Ayyub, S. Bagchi, and G. R. Kumar, Opt. Lett., 29, 2662 (2004). H. Raether, *Surface Plasmons on Smooth and Rough Surfaces and on Gratings*, Springer, Berlin, 1988.

79 Y. Hironaka, Y. Fujimoto, K. G. Nakamura, K. Kondo, and M. Yoshida, Appl. Phys. Lett., 74, 1645 (1999).

80 J. Xiongwei, Q. Jianrong, Z. Congshan, K. Hirao, and G. Fuxi, Opt. Mater., 20, 183 (2002).

81 D. Kuhlke, U. Herpers, and D. von der Linde, Appl. Phys. Lett., 50, 1785 (1987).

82 H. W. K. Tom and O. R. Wood, II, Appl. Phys. Lett., 54, 517 (1989).

83 H. Nakano, T. Nishikawa, H. Ahn, and N. Uesugi, Appl. Phys. Lett., 69, 2992 (1996).

84 J. F. Pelletier, M. Chaker, and J. C. Kieffer, J. Appl. Phys., 81, 5980 (1997).

85 J. Kutzner, M. Silies, T. Witting, G. Tsilimis, H. Zacharias, Appl. Phys. B78, 949 (2004).

86 K. Hatanaka, H. Ono, and H. Fukumura, in prep.

87 B. Hopp, Z. Toth, K. Gai, A. Mechler, Zs. Bor, S. D. Moustaizis, S. Georgiou, and C. Fotakis, Appl. Phys., A69, S191 (1991).

88 Zs. Bor, B. Racz, G. Szabo, D. Xenakis, C. Kalpouzos, and C. Fotakis, Appl. Phys., A60, 365 (1995).

89 K. Hatanaka, T. Itoh, T. Asahi, N. Ichinose, S. Kawanishi, T. Sasuga, H. Fukumura, H. Masuhara, Appl. Phys. Lett., 73, 3498 (1998): K. Hatanaka, Y. Tsuboi, H. Fukumura, H. Masuhara, J. Phys. Chem., B106, 3049 (2002).

90 R. G. Pinnick, A. Biswas, R. L. Armstrong, S. G. Jennings, J. D. Pendleton, and G. Ferbandez, Appl. Opt., 29, 918 (1990).

91 E. J. Hart and J. W. Boag, J. Am. Chem. Soc., 84, 4090 (1962): A. Rogers, Ed., *Radiation Chemistry: Principles and Applications*, VCH, New York, 1987.

92 F. Albert, A. Sjogren, C.–G. Wahlstrom, S. Svanberg, C. Olsson, and H. Merdji, J. Phys. IV, 11, 429 (2001): A. Sjogren, H. Haebst, C.–G. Wahlstrom, S. Svanberg, and C. Olsson, Rev. Sci. Instrum., 74, 2300 (2003).

93 J. H. Eickmans, W, -F. Hsieh, and R. K. Chang, Appl. Opt., 26, 3721 (1987): D. A. Cremers, L. J. Radziemski, and T. R. Loree, Appl. Spcctrosc., 38, 721 (1984): W. F. Ho, C. W. Ng, and N. H. Cheung, Appl. Spectrosc., 51, 87 (1997): C. W. Ng, W. F. Ho, and N. H. Cheung, Appl. Spectrosc., 51, 976 (1997), L. M. Berman and P. J. Wolf, Appl. Spectrosc., 52, 438 (1998).

94 K. L. Williams, *Introduction to X-ray Spectrometry*, Allen & Unwin, London, 1987.

95 M. Yoshida, Y. Fujimoto, Y. Hironaka, K. G. Nakamura, K. Kondo, M. Ohtani, and H. Tsunemi, Appl. Phys. Lett., 73, 2393 (1998).

96 C. Bressler and M. Chergui, Chem. Rev., 104, 1781 (2004).

10

Femtosecond Laser Microfabrication of Photonic Crystals

Vygantas Mizeikis, Shigeki Matsuo, Saulius Juodkazis, and Hiroaki Misawa

Abstract

The evolution of modern photonic technologies depends on the possibilities of obtaining large-scale photonic crystals cheaply and efficiently. Photonic crystals [1, 2] are periodic dielectric structures which are expected to play an important role in optics and optoelectronics due to their unique capability of controlling the emission and propagation of light via photonic band gap (PBG) and stop-gap effects. A comprehensive summary of the properties of various classes of PBG materials and their potential capabilities can be found in the literature, for example, books [3–6]. According to common knowledge, the wavelengths at which PBGs or stop-gaps open are close to the period of the dielectric lattice. At the same time, the most desirable spectral region for opto-electronic devices, including those based on photonic crystals, is in the visible and near-infrared wavelength range. Given this requirement, fabrication of structures periodic in one, two or three dimensions, and comprising many lattice periods, is not a trivial task.

This challenging task requires one to be able perform a complicated microstructuring of various materials with a spatial resolution better than 1 µm. Although planar semiconductor processing techniques, borrowed from microelectronics, can provide such resolution and are successfully used to built high-quality photonic crystals working in the above mentioned spectral ranges [7–9], the tediousness of the planar approach, especially when used for the fabrication of three-dimensional photonic crystals, has prompted a search for alternative fabrication strategies that are more flexible and cost-effective. Fabrication techniques based on laser microstructuring of materials via dielectric breakdown or other kinds of permanent photomodification are among the most promising candidates for succesful implementation of these novel strategies. In this respect, techniques that employ ultrashort (picosecond or femtosecond) laser pulses emerge as particularly strong candidates due to their strong nonlinearity, high efficiency and non-thermal character of the photomodification [10, 11]. In this section we describe microfabrication of photonic crystal structures using femtosecond laser radiation. At first we shall briefly outline the main physical principles that underlie femtosecond laser

3D Laser Microfabrication. Principles and Applications.
Edited by H. Misawa and S. Juodkazis
Copyright © 2006 WILEY-VCH Verlag GmbH & Co. KGaA, Weinheim
ISBN: 3-527-31055-X

microfabrication, in Section 10.1. Later, the studies and the results achieved using this technique will be outlined in Sections 10.2 and 10.3.

10.1
Microfabrication of Photonic Crystals by Ultrafast Lasers

Microfabrication of materials using lasers is a field widely explored and increasingly applied in modern technologies. Fabrication of photonic crystal structures is just one of the many applications existing within this field. The most important principles that underlie laser microfabrication of bulk dielectric materials are illustrated schematically in Fig. 10.1. The simplest arrangement of the laser microfabrication is depicted in Fig. 10.1(a). The method is based on the irradiation of materials by intense optical waves that can be derived from both continuous-wave and pulsed lasers. Pulsed lasers providing ultrashort (picosecond or femto-second) pulses have the advantage of high peak power, that helps achieve photo-modification without undesired thermal effects. The high peak power density is needed in order to induce two-photon absorption (TPA) [12] or multi-photon absorption (MPA) [13] in materials that have negligible one-photon absorption at the laser wavelength. TPA and MPA are nonlinear processes that are dependent on the local power density. The laser beam is focused by a positive lens into the bulk of the material and propagates toward the focal region without losses due to linear or nonlinear absorption. With an appropriately chosen laser intensity, it is possible to achieve conditions when nonlinear absorption is induced in the focal region only, where the local radiation power density exceeds some threshold value. The absorption leads to the generation of nonequilibrium charge carriers in the area localized within the focal spot. What happens in the absorbing area at subsequent stages depends on the properties of the material used. The properties of the photomodified region and its surroundings are shown in Fig. 10.1(b). Usually some kind of physical or chemical process is induced which changes the physical properties or phase of the material. For the fabrication of photonic crystal structures, photomodification processes which lead to instantaneous or subsequent modification of the dielectric properties are the most important. The simplest example of an instantaneous permanent photomodification occuring under irradiation by laser pulses of various length is the light-induced damage [14–17], which obliterates the material through dielectric breakdown, creating a void filled by air or gaseous products of the breakdown. In such case, the contrast of refractive index, (the ratio between the maximum and minimum values of the refractive index, (n), is created at the void boundary between the nondamaged material ($n_2 > 1$) and the void ($n_1 \approx 1$). Another example of the instantaneous modification is photosolidification of the liquid photo-curring polymeric resin [18]. The modified region of the initial material, which is a liquid, becomes solid, and after the removal of the unexposed liquid, the refractive index contrast between the solidified material ($n_1 > 1$) and the surroundings ($n_2 \approx 1$) is created. The subsequent modification is achieved in photoresists, where the modified regions are initially latent, but become clearly visible after subsequent development [19]. Both voids and solid features can be fabricated, depending on the kind of photoresist used (negative or positive), and refractive index contrast occurs between the photoresist ($n_{1,2} > 1$) and the surrounding medium, usually air ($n_{2,1} \approx 1$). Figure 10.1(c) illustrates another important element of the laser microfabrication process. The focal spot of the laser is translated inside the material along the desired path with high precision and accuracy of several nanometers is achievable (via the use of piezo-electric transducer (PZT)-controled translation stages). Translation of the sample or the laser beam thus facilitates the drawing of complex-shaped patterns, including the periodic ones, inside the material. As a result, dielectric features, for example, similar to those shown in Fig. 10.1(d) can be permanently recorded. The method described in Fig. 10.1 also known as direct laser writing (DLW), is used extensively in our laboratories for various studies involving fabrication of photonic crystal structures [10, 20–27] and 3D optical memories [28]. It must be noted that, beside DLW, there also exists another microfabrication technique which allows one to obtain spatially-modulated optical radiation patterns by using multiple-beam interference [29]. This technique will be described in Section 10.3.

It is appropriate to discuss here the requirements for the fabrication process that are specific for photonic crystals. Ideally, the fabrication should produce structures with: 1) good long-range periodic ordering; 2) intentionally engineered nonperiodic defects; 3) high dielectric contrast of at least $n = 2$ to $n = 1$ and 4) controllable feature size. The preceeding discussion makes it clear that laser

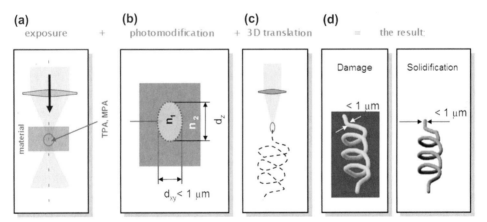

Fig. 10.1 The laser microfabrication procedure. The focused laser beam penetrates without absorption into the bulk of the material until it reaches the focal region where it invokes TPA or MPA (a). This region is elongated in the beam's propagation direction and is marked by the ellipse. Tight focusing and nonlinear absorption helps to reduce the size of the photomodified region, which may become somewhat shorter than the wavelength of the fabricating laser, and in most practical cases is smaller than 1 μm. TPA or MPA modifies the structural, physical, chemical or other properties of the material leading to a pronounced dielectric contrast (b). The smooth or step-like translation of the focal spot is used (c) to record extended dielectric features (solid or voids) or their arrays (d).

microfabrication can satisfy requirements (1) and (2). In fact, its capabilities far exceed those of other known techniques, especially when three-dimensional structures are needed. Requirement (3) is the one most often not fulfilled because most of the successfully used initial materials (glasses, resins, photoresists) have a low refractive index of about 1.5–1.7. Materials that have a refractive index $n > 2$, usually crystalline or amorphous semiconductors as well as some inorganic glasses, can be fabricated using lasers. However, high refractive index mismatch between the material, surrounding medium (e.g., air or immersion oil) and the focusing lens, leads to strong optical abberations which lower the quality of beam focusing and limit the highest achievable resolution [30]. At the same time it is known that a dielectric contrast of about 2 to 1 is required to be observable PBG in most kinds of periodic lattices [3, 31]. This means that photonic crystals obtained by laser microfabrication will most likely not have complete PBGs due to their low refractive index contrast. In these circumstances the contrast can be increased by subsequent infiltration of the low-contrast templates by other materials which have higher refractive indexes. Such an approach was succesfully used with artificial opal structures obtained by self-organized sedimentation of microspheres from colloidal suspensions [32, 33], which also have insufficient refractive index contrast. The artificial opals have been successfully infiltrated by Si, InP, or other materials [34, 35]. The requirement (4) arises due to the need to scale the period and other parameters of the periodic lattice, such that PBGs and stop-gaps atthe desired wavelengths are obtained. According to Maxwell's scaling law [3], these wavelengths are proportional to the lattice parameters. Usually, the PBGs and stop-gaps open at wavelengths close to (but usually longer than) the lattice period of the photonic crystal. The wavelength regions of highest interest to modern optoelectronics and optical information processing are in the visible (a wavelength of 350–750 nm) and near-infrared telecommunications (a wavelength of 1.1–1.7 μm) ranges. Hence lattices with periods smaller than 1 μm are required. Since several interfaces with dielectric contrast must be present within the lattice period, the fabricated feature size must be somewhat smaller than 1 μm. The highest resolution achievable in any optical recording is fundamentally limited by diffraction to approximately the wavelength of the recording radiation. Most lasers used for the fabrication operate at visible and near-infrared (NIR) wavelengths. Although it would seem that these circumstances prevent high resolution patterning of materials, nonlinearity of the absorption allows one to obtain spatial absorption profiles which are sharper than those of the initial light intensity distribution. As will be shown below, the resolution can thus be increased beyond the diffractive resolution limit of the optical focusing system.

10.1.1
Nonlinear Absorption of Spatially Nonuniform Laser Fields

Linear absorption causes the intensity, I of the laser beam propagating along the direction z to decay according to the Lambert–Beer law $I(z) = I_0 e^{-az}$, where a_0 is the intensity-independent absorption coefficient. Thus, linear absorption blocks the optical access from outside into the bulk of the material making it impossible to achieve the conditions depicted in Fig. 10.1. In the presence of nonlinear absorption, the absorption coefficient is intensity-dependent and is expressed as a sum of contributions from the linear and nonlinear absorption processes: $a(I) = a_0 + a_{NL}(I)$. When the laser photon energy is tuned below the fundamental electronic band gap of the material, the linear absorption vanishes. Although at low excitation power densities the nonlinear absorption is negligible, it may become significant at higher excitation levels. For the focused laser beam, its power density in the focal spot region may reach a level sufficient for inducing substantial nonlinear absorption and subsequent photomodification of the localized region in the bulk of material.

To provide the illustration, some numerical analysis may prove useful. Figure 10.2 shows the calculated laser intensity distribution near the focal spot of a lens which has a high numerical aperture (NA = 1.35), close to that of the best microscope oil-immersion lenses used for laser microfabrication. This distribution represents so-called point-spread function (PSF). The calculation is based on the scalar Debye theory which neglects polarization effects (for this, full vectorial calculation is required) and is performed following the expression for three-dimensional PSF [36] in the cylindrical coordinate system (arguments r, ψ, z):

$$E_{sc}(r, \psi, z) = \frac{2\pi i}{\lambda} \int_0^a P(\theta) \sin(\theta) J_0(kr \sin(\theta)) e^{-ikz \cos(\theta)} d\theta \qquad (1)$$

where $J_0(x)$ is the zero-order Bessel function of the first kind, a is the half-angle, $P(\theta)$ is the apodization function, and $k = 2\pi/\lambda$ is the wave vector defined by the wavelength λ. For the simulation, the Helmholtz apodization function $P(\theta) = \cos(\theta)^{-3/2}$, which corresponds to a uniformly illuminated objective aperture (the pupil function of unity) is employed. This condition is appropriate for the so-called perfect imaging case, where there are no distortions along all three dimensions. The above expression allows one to simulate various distributions of light intensity over the pupil as well as to introduce the aberration function [36], and the effects of spherical aberrations on the focusing of femtosecond pulses as described in [30]. The perspective and cross-sectional views of the intensity distribution at the focal region is shown in Fig. 10.2(a). Three lateral (lying in the x-y plane) cross-sections of this distribution taken at different longitudinal positions (along the z axis) are shown in Fig. 10.2(b–d). The laser field distribution is smooth, but when visualized using a constant intensity surface as is done in Fig. 10.2(a), it consists of an ellipsoidal core region, inside which the power density is highest, surrounded by several ellipsoidal and ring-like shells. Multi-photon absorption processes depend on the power density as $\sim I^n$ where n denotes the number of photons required to complete the absorptive transitions. For example, TPA is characterized by the quadratic dependence on the irradiation intensity. Thus, the absorption distribution usually has a somewhat sharper spatial profile than has the laser field distribution. Although absorption is in principle thresholdless, its nonlinear character, combined with a variety of subsequent processes, usually results in threshold-like photomodification onset in the regions where the irradiating power density exceeds the threshold value. Given all these circumstances, the incident laser power can be adjusted to produce the photomodification only within a small region inside the core. For example, the lateral diameter of the region where the local intensity exceeds the $1/e$ level is already smaller than the laser wavelength. Thus, it becomes possible to surpass the resolution limited by diffraction of the laser beam.

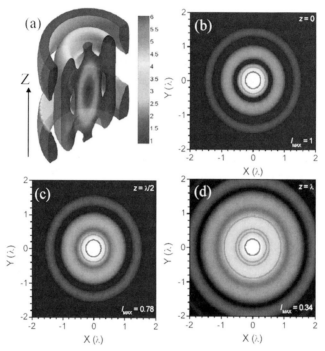

Fig. 10.2 3D intensity distribution ($I = |E_{sc}|^2$ Eq. (1) at the focus of objective lens NA = 1.35 calculated by the scalar Debye theory. (a) 3D view with the cross-section width of 2.88λ and height 4λ. (b–d) The intensity normalized lateral cross-sections of PSF at the center of focus ($z = 0$) and shifted by $\lambda/2$ and λ, respectively. Contours in (b–d) mark $1/e^2$ and FWHM levels by intensity; the intensity scale is logarithmic.

10.1.2
Mechanisms of Photomodification

The region where nonlinear absorption is induced may undergo a variety of transformations resulting in transient or permanent photomodification. Transient photomodification occurs at moderate instantaneous power densities because of the interaction between the molecules and atoms of the condensed medium and the electric field of the laser radiation. The transient photomodification can be revealed by changes in the absorption coefficient or refractive index and disappears when all optically induced excitations relax to the ground state after the laser field is turned off. This transient regime is widely exploited in ultrafast laser spectroscopy. Permanent photomodification occurs at laser power densities exceeding a threshold value, and remains recorded in the material forever after the exciting field is gone. Dynamics and consequences of the photomodification depend on the laser radiation and material used. Below we shall provide a brief overview of several kinds of photomodification, which are the most important for laser microfabrication of photonic crystals.

Laser-induced damage in solid dielectrics is perhaps the most common consequence of intense laser field action on the materials. The permanent photomodification results from the destruction of the material by the generation of an electron–hole plasma which produces even more plasma absorption and leads to mass-density modification and dielectric breakdown. Generation of the free carrier plasma is usually treated as a consequence of the following processes: multiphoton ionization causing the excitation of electrons into the conduction band, electron–electron collisional ionization due to Joule heating, and plasma energy transfer to the lattice. Femtosecond laser-induced breakdown in glasses has been widely studied in the literature [10, 11, 15, 37, 38]. These and other investigations have indicated that, for pulses shorter than approximately 100 fs, the role of multi-photon ionization is to supply the seed electrons for the subsequent avalanche ionization. The electronic subsystem acquires excess energy faster than it can be transferred to the ionic subsystem. Therefore, electron temperatures from a few to tens of electron volts (1 eV = 11 600 K) can be reached during the laser pulse, while the ion subsystem remains relatively cold. The electron–ion energy exchange, which usually takes place after the laser pulse is gone, heats ionic lattice. When the laser beam is focused, breakdown occurs as a "microexplosion" inside the material [11] and leaves an empty void. Explosive breakdown is often accompanied by shock-wave generation which may create a shell of densified material having a higher refractive index (compared to nondamaged regions) surrounding the void which has a low refractive index. Thus, periodic structures consisting of isolated voids or extended channels can be generated optically with little or no post-processing required. However, in these circumstances, the shape of the focal region must be controlled as precisely as possible during the recording. Due to its explosive nature, laser-induced damage may also generate a considerable number of cracks and other randomly scattered unwanted micro- and nano-features.

Figure 10.3 shows two kinds of damage-induced voids in solids. The images were taken using scanning electron microscopy (SEM). The first of them (Fig. 10.3(a)) shows an isolated void recorded by a single laser pulse in a sapphire slab focused by an oil-immersion microscope objective lens (magnification of 100 ×, NA = 1.35) approximately 15 μm below the surface. Voids embedded in the bulk of the samples cannot be straightforwardly visualized by SEM. For this purpose the slab was broken along the line of voids, and voids located on freshly formed edges were then etched in HF aqueous solution in order to remove the debris resulting from the laser damaging and breakage. Although the void seen in

Fig. 10.3(a) is the product of laser-induced damage *and* etching rather than damage alone, it nevertheless illustrates the shape, size, and morphology of the laser-microfabricated voids. The void has the shape of the raindrop with a sharp tip oriented along the pulse propagation direction, this shape is most likely the result of aberrations due to the refractive index mismatch when focusing into sapphire ($n \approx 1.7$) is done using oil-immersion optics designed for materials with lower $n \approx 1.5$. Another kind of void, which has turned out to be particular useful for the buildup of photonic crystals is shown in Fig. 10.3(b). Empty microchannels were recorded by scanning the focal spot inside the film of polymeric photoresist polymethylmethacrylate (PMMA) at a constant velocity, which ensured substantial spatial overlap between adjacent irradiated spots at a pulse repetition rate of 1 kHz. The recording was based on the microexplosion mechanism [39, 40]. For visualization, the film was broken and its open edges inspected under SEM. The voids have slightly ellipsoidal cross-sections with major and minor diameters of about 350 nm and 290 nm, respectively. Organic glasses usually have much lower mass density and and mechanical rigidity compared with nonorganic (e.g., silica) glasses and therefore provide much better opportunities for visualizing the densification in the region surrounding the microexposion sites [41].

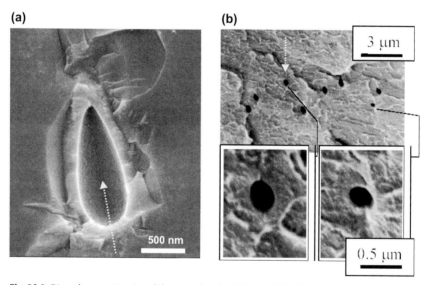

Fig. 10.3 Direct laser writing in solids: an isolated void in sapphire (a); extended linear voids in PMMA (b), cross-sectional views.

Photochemical processes in organic materials like liquid polymers or solid polymer photoresists provide the opportunity to create dielectric features using lower power densities without explosive damaging. Epoxy-based liquid resins are widely used in everyday life as standard two-part systems that, after mixing, cure to brittle solids. In these systems the epoxy cures by reacting with another compound called "hardener". In specially tailored resins, exposure to light (with wavelength usually in the ultraviolet spectral region) can also play the role of hardener. The photosensitive resins are doped with photoinitiator (or photoacid generator) molecules that absorb the ultraviolet photons creating free radicals. The free radicals connect with the molecules of the resins and monomers, and they, in turn, cross-link with each other, forming chains of molecules that build polymerized solid material. The solid resin typically has a refractive index of about 1.5. By exposing the resin to spatially periodic illumination, periodic dielectric structures can be recorded. The unexposed parts of the resin, which remains liquid, can afterwards be removed from the solidified framework by washing. Photocuring resins are very convenient systems for the laser fabrication of various microstructures. Uniformly pre-cured (by single-photon exposure to ultraviolet radiation) solid resins can serve as an initial material for laser-induced damaging experiments [41–44]. Solidified by laser irradiation from the liquid phase, they provide very high spatial resolution. For example, by using multi-photon absorption in a commercially available urethane acrylate-based resin a resolution of 120 nm, for isolated features, was achieved [19]. Application of liquid resins for laser fabrication of photonic crystals is described in Section 10.2.

Like liquid resins, various photoresists are also usable for laser microfabrication. Being designed mostly for lithography by electron-beam writing or projection imaging using single-photon absorption, these materials are usually photosensitive at ultraviolet wavelengths and hence are suitable for exposure by visible or infrared lasers via TPA or MPA. Photosensitivity mechanisms in organic photoresists are similar to those in photocuring resins. Unlike the resins, photoresists are solid before and after the exposure. Therefore the photomodified regions are embedded in a solid matrix of unexposed material, which helps prevent their unwanted mechanical deformation during the recording. In negative photoresists, the exposed regions are rendered resistant to subsequent development, while the unexposed parts are removed, leaving networks of solid features with a refractive index of 1.5–1.6 in air. In positive photoresists the development removes the optically exposed regions.

A photoresist which turned out to be highly suitable for the fabrication of photonic crystals and their templates, is an epoxy-based photoresist SU-8, designed for the fabrication of high-aspect-ratio micromechanical structures in ultrathick films. SU-8 has low intrinsic absorption at wavelengths longer than 360 nm, and is capable of resolution of a few tens of nanometers. Polymerized SU-8 possesses a dense network of cross-links and provides a high solubility contrast between the strongly and weakly exposed regions. Photopolymerization in SU-8 does not occur immediately after the exposure; for this, the resist must be thermally baked after exposure. SU-8 is currently commercially available from several companies worldwide, including Microchem Corp. A summary of the main properties and potential applications of SU-8 can be found at their web site: http://www.microchem.com/products/su_eight.htm.

Figure 10.4 illustrates two of the simplest kinds of features that can be recorded in SU-8. Figure 10.4(a) shows arrays of volume elements (also called voxels) recorded on the interface between the SU-8 film and the glass substrate on which

(a)

(b)

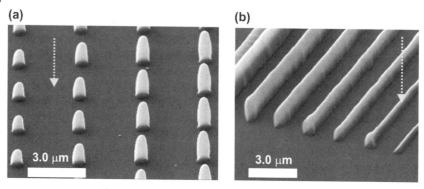

Fig. 10.4 Simplest features recorded by direct laser writing in SU-8. Arrays of isolated voxels (a) arranged into a square lattice with period of 3 μm. In the recording of each column of voxels the focal spot position along the z-axis increases by 100 nm (from left to right). Arrays of equidistant (separation of 3 μm) linear features (b), recorded by scanning the focal spot position along the y-axis in steps of 150 nm; in the recording of each line, the focal spot position along the z axis was decreased by 100 nm (from left to right).

it was coated. The voxels are therefore firmly attached to the substrate. Since it is not very easy to detect the SU-8/glass interface before recording, different columns of voxels were recorded with different positions of the focal spot above (and below) the interface. The voxels which were not attached to the substrate were washed away during the subsequent development, and no recording was done if the focal spot region was fully buried in the substrate. In qualitative agreement with analysis presented in Fig. 10.2, the voxels have ellipsoidal shapes. Their lateral and axial diameters are about 0.6 μm and 1.4 μm, respectively. When individual voxels are allowed to overlap, they form extended linear features, which are elliptical cylinders similar to those shown in Fig. 10.4(b). From these cylinders extended periodic structures can be constructed.

Thermal effects are widely used in laser machining applications where they induce melting and vaporization of solids. Usually these applications employ longer laser pulses in the nanosecond range. Thermal effects are not particularly useful for the microfabrication of photonic crystals because the heating spreads away from the optically modified region inducing undesired thermal modification and compromising the spatial resolution of the recording. Nevertheless, since thermal effects, to some extent, accompany most photoexcitation events, their possible role deserves some attention. Thermal dynamics occurring during and after photoexcitation can be subdivided into several stages. Each stage has its own characteristic timescale and spectral signatures. These signatures can be identified from the spectra of the secondary electromagnetic emission originating from the photoexcited region.

The primary product of irradiation by a high-power laser is the hot electron–hole plasma. The electron temperature is the fastest to be established after the excitation. This typically occurs within a timescale of few femtoseconds. The hot plasma yields quasi-thermal emission similar to the black-body radiation described by Planck's formula [45]

$$I_\nu = \frac{2\pi k_B^3 T^3}{c^2 h^2} \frac{x^3}{e^3 - 1}, \text{ where } x = \frac{h\nu}{k_B T} \tag{2}$$

where $\nu = c/\lambda$ is the optical frequency corresponding to the speed of light, c, and the wavelength, λ; T is the temperature and k_B is the Boltzman constant. In the case of ultrashort pulses (< 1 ps) the nonequilibrium electrons can possess excess energy of few electronvolts (1 eV is equivalent to 11 605 K). The nonequilibrium temperature, T_e, of hot plasma is much higher than that of the core ions, T_i. Depending on the irradiance, $T_e > T_i$ remains for times comparable to or up to several times longer than the laser pulse.

Equation (2), known as Wien's law, describes the emission with spectral peak at the frequency $\nu_{max} = 2.82 k_B T/h$. Thus, a peak at ultraviolet wavelengths corresponds to a hot plasma temperature of about 10^4 K. For example, 308 nm wavelength of the Ce^{3+} absorption band in PTR glass, corresponds to a temperature of 15 500 K and an excess energy of 1.42 eV. The "white-light" plasma emission relaxes within 1–100 ps after excitation and might be attractive as an internal light source, which can be switched on locally inside solid, liquid or gaseous dielectrics. This internally generated broadband radiation may contribute to the optical recording. Although most observations report a continuum generation in glasses and liquids, photoresists can also exhibit continuum generation. For example, when working with DLW in photoresist SU-8, we have observed substantial broadband radiation which was easily detectable using a CCD camera. Such generation may be responsible for the additional "internal" microfabrication via one-photon absorption of the broadband radiation, and can also facilitate the readout of three-dimensional optical memories.

The atomic and ionic core subsystem of the material establishes the temperature on a somewhat longer, picosecond timescale. The relation between the number density of atoms and ions, N_+, and the absolute temperature, T, assuming that Boltzman distribution has already been established, is described by the Saha equation:

$$\frac{N_e N_+}{N_a} = A \frac{g_+}{g_a} T^{3/2} e^{I/k_B T} \tag{3}$$

where $A = 2(2\pi m k_B/h^2)^{3/2} \simeq 6.04 \times 10^{21}$ cm^{-3} eV$^{-3/2}$, N_a and N_e are the atoms and electrons densities, respectively, I is the ionization potential, k_B and h are the Boltzmann and Planck constants; g_a with g_+ are the occupation factors for the ground and excited/ionized states, respectively. The temperature of ionic and atomic subsystem can be determined by measuring the intensity of the luminescence or Raman emission lines. When melting or vaporization temperatures are reached, thermal damage takes place in the material.

10.2
Photonic Crystals Obtained by Direct Laser Writing

The basic principles of the DLW technique have already been described in Section 10.1. A wide range of materials can be used for DLW, for example, silica or other inorganic or organic glasses (by damaging), liquid photopolymerizeableresins (by photosolidification), or photoresists. The optical setup for 3D DLW experiments used in our own experiments is schematically shown in Fig. 10.5. The laser source is a Spitfire or Hurricane X system (Spectra-Physics) which provide femtosecond pulses with duration $\tau_{pulse} = 130$ fs at the central wavelength $\lambda_{pulse} = 800$ nm. The fabrication is performed in an optical microscope (Olympus IX71) equipped with oil-immersion objective lenses (magnification 60× and 100×, numerical apertures NA = 1.4 and 1.35, respectively). With both objective lenses the diffraction-limited beam diameter at $1/e^2$ level is $d = 1.22\frac{\lambda}{NA} \approx 720$ nm). 3D drawing is accomplished by mounting the samples on a piezoelectric transducer-controlled 3D translation stage (Physik Instrumente PZ48E) which has a maximum positioning range of up to 50 μm and an accuracy of several nanometers. The stage motion is controlled by a custom-made software running on a personal computer. During fabrication the samples were translated along the predefined path, along which the desired locations were sequentially irradiated by the tightly focused laser beam. This method allowed one to record periodic structures which consist

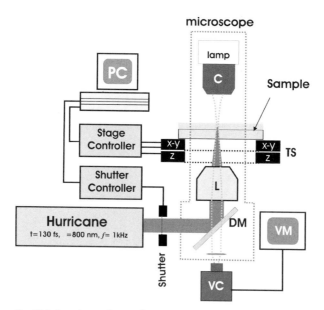

Fig. 10.5 Experimental setup for 3D two-photon and multi-photon lithography by DLW. Hurricane is the pulsed femtosecond laser system, DM is a dielectric mirror with R ≈ 100% at the laser wavelength, L is the microscope objective lens, TS is the 3D translation stage on which the sample is mounted, C is the condenser, VC and VM is the video camera and video monitor, respectively.

both of isolated voxelsor of extended (e.g., linear) features. The lattice symmetry, size of the sample and the presence of structural defects can be defined from the software.

10.2.1
Fabrication by Optical Damage in Inorganic Glasses

The recording of periodic structures in solids by laser-induced damaging is perhaps the most straightforward method of obtaining photonic crystal structures. We have used dry (OH concentration < 10 ppm) vitreous silica, v-SiO$_2$ (ED brand from Nippon Silica Glass Co.) as the starting material. The PZT coordinates were scanned at the low speed of 16 μm s^{-1} which ensured a 16 nm spacing between the adjacent exposed sites (at a 1 kHz laser repetition rate), i.e., much smaller than the laser wavelength and the size of the focal spot. Since the damage instantaneously produces the refractive index modulation, the drawing can be monitored *in situ* by a CCD camera and a video monitor (see Fig. 10.5). In glasses the most critical parameter for microfabrication is the light-induced damage threshold (LIDT). The single-shot LIDT of 7 J cm^{-2} was determined experimentally.

Among various 2D geometries of the photonic crystals, the geometry described by a 2D triangular point lattice is one of the most promising because, with sufficient index contrast, it allows one to achieve a complete PBG for both transverse electric (TE) and transverse magnetic (TM) polarizations [3]. 2D photonic lattices were recorded by scanning the position of the damage spot along the lines arranged in the triangular pattern [46]. Figure 10.6 (a) shows the image of the PhC structure with a 2D triangular lattice. The image is reconstructed from optical micrographs of the top and the side walls of the structure. The structure exhibits good long-range periodicity. The bright cylinders seen in the image were confirmed to be hollow by AFM measurements. The lattice constant $a = 1.2$ μm, and the lateral size of the fabrication on the *x-y* plane is 40 × 40 μm^2. The transmission spectrum of the structure was measured with an FTIR spectrometer and is shown for both polarizations in Fig. 10.6 (b). A transmission dip of about 10% occurs for TE polarization at $\lambda = 2.45$ μm, while for TM polarization it occurs at $\lambda = 2.52$ μm and is somewhat less pronounced. These findings indicate that a photonic stop-gap exists in the structure in the infrared spectral range.

3D photonic crystals can be also easily microfabricated in glass using a similar approach. An example of a 3D PhC recorded in silica is shown in Fig. 10.6 (c) [23, 47], where a single (111) plane of an fcc lattice is shown. The entire crystal was recorded by layer-by-layer stacking the (111) planes. Optical characterization of the 3D PhC structure (Fig. 10.6 (d)) reveals a minimum in the transmittance around 3490 cm^{-1} (2.9 μm). The calculated transmission spectrum (shown in the same plot) reproduces the wavelength 2.87 μm (3490 cm^{-1}) of the experimental transmission dip, by taking the void radius $r = 125$ nm, and the refractive index modulation $\Delta n = 1.45$. The value of r measured by AFM was larger than 125 nm used in the calculations. This difference may arise due to the lateral size of the voxel being altered by polishing. In addition, the AFM data must be deconvoluted using a tip profile, which is not known precisely.

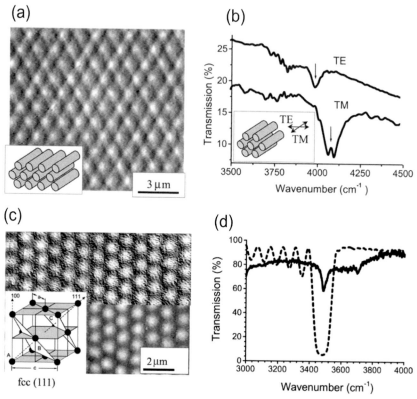

Fig. 10.6 2D triangular PhC lattice, cross-sectional image with rod alignment scheme shown in the inset (a), and transmission spectra of the sample measured by FTIR spectrometer (b) [46]. The (111) plane of the 3D fcc photonic crystal with the shadowed lower half showing the image of a neighboring atomic plane (c). Calculated (dashed line) and measured (solid line) transmission spectra (d) [23,47].

As can be seen, despite looking good under the optical microscope, photonic crystal structures in glass exhibit only quite weak signatures of the photonic stop-gaps. The two tentative reasons behind these observations are low refractive index contrast and the presence of significant scattering by damage-induced disorder. In order to identify these reasons more clearly we have recorded the simplest PhC structure which is a 1D diffraction grating. The grating allowed us to evaluate the refractive index modulation induced by the laser damage by measuring its diffraction efficiency (the ratio of the diffracted and incident light intensities). Assuming a sinusoidal grating profile, we have estimated the refractive index modulation $\Delta n > 10^{-2}$ [48]. This estimate shows that the modulation amplitude is much lower than the maximum achievable value. Hence, it can be concluded that the microfabricated voids are not completely empty and most likely are filled with byproducts of the laser-induced damaging processes. Light scattering may decrease the PhC efficiency in the spectral region of the PBG as well. In the microfabricated glass areas, random scattering may consume up to 49% of the probing light intensity in structures fabricated at laser intensities approaching $3.5 \times$ LIDT at 800 nm. To reduce the scattering, fabrication and post-fabrication treatment of the sample needs to be optimized. For example, annealing may increase the performance of the PhCs by reducing the scattering and defect-related absorption. Indeed, waveguides fabricated by multi-shot irradiation of glass [49] show $\Delta n \approx 10^{-2}$. In the case of multi-shot irradiation, annealing is performed locally by the subsequent laser pulses, butthermal annealing of the entire sample at temperatures of 700–900 °C is also possible. Finally, there is a very promising possibility of using etching of the recorded structures in KOH ar other highly potent etchants. In such cases the material partially damaged by the laser is removed from irradiated areas by etching, which is known to attack the irradiated areas much more actively than the unexposed ones.

10.2.2
Fabrication by Optical Damage in Organic Glasses

Organic glassy materials may often provide a more suitable platform for laser microfabrication of photonic crystals. Since most organic glasses are less dense and hard than their inorganic counterparts, they allow easier relaxation of the mechanical strains and tensions as well as a release of gaseous products of laser-induced damage. Thus, possible destructive consequences of these relaxation events are minimized and extended structures of higher quality are obtained. These circumstances, together with significant densification, contribute to overall higher refractive index contrast and help to avoid collapse of the structures.

Below we shall describe in some detail the DLW of photonic crystal structures in PMMA. This organic material, applied in planar semiconductor processing technologies as a photoresist, with some additives is also widely used in everyday life as organic glass ("plexiglass"). The commercial availability and low cost makes PMMA a promising candidate for photonic applications. The PMMA samples for DLW were prepared by casting drops of PMMA formalin solution on the glass substrate (microscope slide glass) and allowing them to dry for a few days. During the fabrication the laser was focused into the PMMA layer through the substrate. This arrangement prevented direct contact between the immersion oil and PMMA.

As a first step toward photonic crystal fabrication, the recording of extended microchannel voids was investigated, having in mind the possibility of building the so-called woodpile structures (to be discussed below) by fabricating arrays of microchannels. As will be demonstrated later, periodic structures displaying signatures of stop gaps can be formed in PMMA from isolated voids, but even stronger signatures are obtained in the extended linear void structures. By smoothly scanning the focal spot inside the PMMA film at a constant velocity which ensured substantial spatial overlap between the adjacent irradiation spots (at the pulse repetition rate 1 kHz) allowed one to obtain empty channels smaller than 0.5 μm in diameter [39, 40]. The channels, recorded by 10 nJ pulses were indeed empty as was evidenced by their permeability to the luminescent dye solution which was photo-excited by a 543 nm wavelength laser and visualized in a confocal microscope. The smallest diameter of channels, shown earlier in Fig. 10.3(b) was determined to be 290 nm, while their average diameter was approximately 350 nm. These values are small and thus make the channel voids attractive as elements suitable for the building of photonic crystal structures.

For extended structures the recording of the so-called woodpile architecture was chosen [50]. Woodpile structure has a face-centered cubic (fcc) of face-centered tetragonal (fct) point lattice and an "atomic" basis consisting of two perpendicular dielectric rods, which results in a diamond-like structure, capable of opening a wide PBG. The woodpile architecture and definition of its parameters are shown in Fig. 10.7. Woodpiles can be formed conveniently by stacking layers of uni-formly spaced dielectric rods along the z axis direction and hence are easy to build in a layer-by-layer manner. The DLW of woodpile structures was performed by smoothly scanning the focal spot position along the channel lines in small steps, $\Delta l = 0.05$–0.2 µm, with single or multiple-shot exposure at each step. The neigh-boring exposed spots overlapped strongly leading to the formation of void chan-nels. The structures were recorded starting from the deepest layers (from the PMMA-substrate interface) toward the shallowest ones. Figure 10.8(a) shows top-view image of the woodpile structure (see caption for the parameters) in PMMA, taken by a confocal laser microscope (LSM). Because linear voids are strongly reflective, only the two topmost layers can be seen clearly in the image. Although the image cannot provide detailed insight into the long-range order at larger depths, it is apparently sufficient for the formation of photonic stop-gaps with ob-servable spectral signatures. Figure 10.8(b) shows transmission and reflection spectra of the same sample measured along the z axis direction (perpendicular to the image plane). The measurements were performed using a Fourier-Transform Infrared (FT-IR) spectroscope (Valor III, Jasco) coupled with infrared microscope (Micro 20, Jasco). The most apparent feature of both spectra is the presence of spectrally matched transmission dip and reflection peak near the wavelength of $\lambda = 4.1$ µm. The matching indicates that optical attenuation observed at this wave-length is due to the rejection of the incident radiation by the structure and is not related to absorption. Such behavior is typical for PBG. However, since the magni-tude of the transmission dip is quite low, the photonic stop gap (a gap which exist only along a particular direction) should be held responsible for this observation. Low magnitudes of the dips and peaks may also result from the following factors: 1) low refractive index contrast, about 1.6 to 1; 2) scattering by random defects; 3) angular spreading and elimination of normally incident rays of the light beam in the infrared Cassegrainian microscope objective used for probing the sample, which therefore measures transmission and reflection properties along multiple directions.

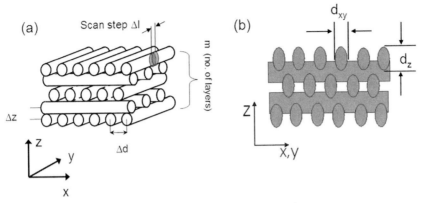

Fig. 10.7 Woodpile structure and its main parameters: Δd distance between the rods, Δz distance between the layers, m number of layers. (a) Schematic side-view of the woodpile structure consisting of rods having elliptical cross-sectional shapes, the ellipses are elongated in the beam focusing direction, their lateral and longitudinal diameters are d_{xy} and d_z, respectively (b).

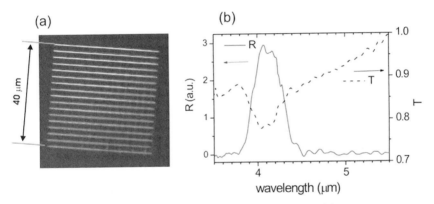

Fig. 10.8 Top-view LSM image of woodpile photonic crystal in PMMA with lattice parameters $\Delta d = 3$ μm, m, $\Delta z = 1.5$ μm, $m = 20$, recorded with 1.2 nJ single-shot exposure with step $\Delta l = 200$ nm (a), infrared transmission and reflection spectra of this sample (b).

Another feature typical of periodic photonic structures that exhibit PBG or stop-gaps is Maxwell's scaling behavior. This behavior implies that the PBG (and also stop-gap or other features related to the photonic band dispersion) wavelength is proportional to the lattice period (in-depth explanation of Maxwell's scaling can be found in [3]). Thus, in photonic crystals with all lattice parameters proportionally scaled up or down, the PBG or stop-gap wavelengths will be scaled up or down accordingly. Figure 10.9 illustrates this scaling by presenting the reflection spectra of three different woodpile structures with proportionally scaled lattice parameters. In these plots several reflectivity peaks, which most likely originate from the fundamental and higher order stop-gaps, can be seen. Higher-order stop gaps are an evidence of the good structural quality achieved, because higher photonic bands are more susceptible to disorder. In addition, they can be exploited for achieving photonic band-gap effects in structures with larger structural parameters [43]. Their central wavelengths decrease when the lattice parameters are scaled down. During the fabrication lattice parameters Δd and Δz can be controlled most directly. The diameters of the voids d_{xy} and d_z (see Fig. 10.7(b) for the definitions) are controlled indirectly by adjusting the laser pulse energy and the scanning step size, Δl. Thus, it is not easy to achieve strictly proportional scaling of *all* lattice parameters. Nevertheless, the above observations are in qualitative agreement with Maxwell's theoretical scaling behavior.

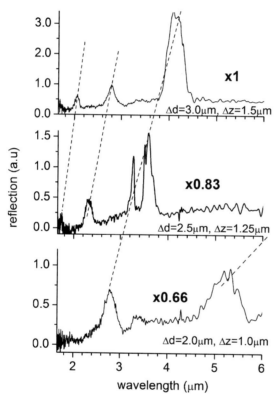

Fig. 10.9 Maxwell's scaling in PMMA structures. The plots show reflection spectra of three samples which have lattice parameters proportionally scaled down (the parameters and scaling factors of 1, 0.83 and 0.66 are indicated in the plots). The numbered dashed lines are guides to the eye that emphasize the scaling of the peaks' wavelengths. In the middle plot, the two peaks seen between 3.0 and 4.0 μm actually originate from a single-peak split at the center by the PMMA intrinsic absorption band near the 3.4 μm wavelength.

As was pointed out earlier, PMMA also supports structures composed of isolated voxels. This is illustrated by the images and data in Fig. 10.10 showing the diamond structure with an fcc point lattice having period $a = 3.3$ μm. The LSM images in Fig. 10.10(a,b) show that the structural quality of the sample is quite high, although there is also a significant disorder present, which is the most likely reason for the relative weakness of the photonic stop-gap manifestations in the absorption and reflection spectra shown in Fig. 10(c). The major disadvantage of photonic structures composed of isolated voxels is their closed architecture, which is unsuitable for the refractive index contrast enhancement by infiltration with other materials. However, such structures may be still of interest in applications that do not require full PBG, for example, photonic crystal superprisms and collimators.

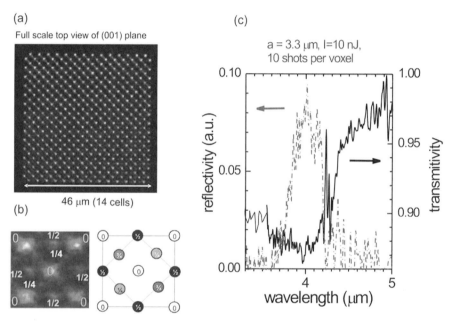

Fig. 10.10 A photonic crystal with diamond structure in PMMA, Full-scale LSM image (a), magnified portion of LSM image with positions of "atoms" belonging to different atomic planes along the z axis (marked by numbers in the units of the cubic cell) together with the schematic explanation (b), optical transmission and reflection spectra of the sample (c).

10.2.3
Lithography by Two-photon Solidification in Photo-curing Resins

In contrast to laser damaging, which leaves empty regions in the solid, the polymerization of resins under optical illumination provides the possibility of obtaining solidified regions in the initial material. Unsolidified regions not exposed to optical illumination can be removed in post-processing. The densified volumes of the resin typically have refractive index of about 1.5. There are two possibile methods of recording. The first of them is the DLW, while the second one is to use periodic patterns of light interference [29], to be discussed later.

Photonic crystal structures having woodpile geometry were recorded in acrylic acid ester-based Nopcocure800 (from San Nopco). This resin is strongly absorbing $(a > 10^3 \text{ cm}^{-1})$ at ultraviolet wavelengths shorter than $\lambda \approx 300$ nm. Thus, second harmonic pulses of a Ti:Sapphire laser system having wavelength of 400 nm are weakly absorbed ($a = 1.4 \text{ cm}^{-1}$ at 400 nm), and photopolymerization occurs due to the TPA. The recording procedure is very similar to that used in DLW by laser damaging, i.e., closely located spots are illuminated in a sequence, and ultimately form periodic patterns of photopolymerized material. After the fabrication, unexposed resin is removed by washing the structure in acetone. It should be noted that, in order to obtain mechanically stable self-supporting structures, the solidified areas should be interconnected. This circumstance imposes some restriction on the available resin filling ratio. Figure 10.11 shows SEM image of a woodpile structure and infrared transmission spectra of several structures with different lattice parameters, indicating the Maxwell's scaling behavior [24].

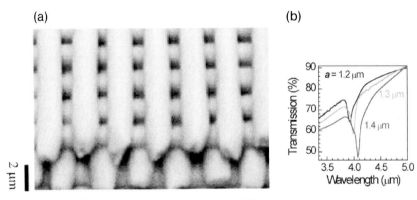

Fig. 10.11 Woodpile structures fabricated in Nopcocure800 resin by DLW (a), transmission spectra of the samples with different in-plane rod distance Δd = 1.4, 1.3, 1.2 µm (b) [24].

Next, planar defect states were introduced into the resin-based woodpile structures. The top inset in Fig. 10.12 shows a schematic view of a woodpile structure which, in the middle, contains a defected layer with every second rod missing. Such removal of rods allows one to disrupt the crystal periodicity without significant loss of mechanical stability. For the light incident perpendicular to the woodpile layers, the defected PhC may be regarded as a couple of planar photonic mirrors, separated by the spacer, i.e., it is essentially a planar microcavity. Figure 10.12 compares transmission spectra of the sample with the defect (solid line) with that of uniform sample (dashed line). The defected sample clearly exhibits a transmission peak at the middle of the woodpile stop-gap, thus indicating the presence of

defect states in the structure. The peaks were centered at 3.801 μm (for TM polarization) and at 3.838 μm (for TE polarization), had Lorentzian line shapes and
almost identical amplitudes. There was a slight displacement between the peaks
of different polarizations due to the anisotropic nature of the defect. From the
peaks' spectral width, microcavity Q-factor of about 85 was determined. Despite
the fact that the refractive index of the polymerized resin used in the PhC mirrors
is insufficient for achieving higher Q-factors, the structure may be considered as a
step toward the realization of polymer-based photonic crystal elements [25].

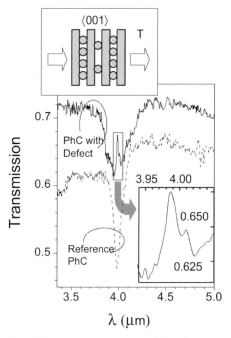

Fig. 10.12 Transmission spectra of the reference structure without
defects (dashed line, offset by –0.1 for clarity), and the structure with
defect. The top inset illustrates the formation of the microcavity and
light propagation during transmission measurements, the bottom
inset shows a detailed view of the defect peak [25].

10.2.4
Lithography in Organic Photoresists

Photoresists are designed to protect the surface of dielectrics and semiconductors
during planar processing, which may involve reactive ion bombardment, etching,
and other rigorous treatment. Therefore, they can withstand a wide range of
chemical, thermal, mechanical and other influences and are potentially interesting materials for the fabrication of photonic crystals. Recently, use of SU-8, an
ultra-thick negative polymeric photoresist that is much more robust than resins,

attracted a lot of attention. Optical exposure induces polymer cross-linking in SU-8, rendering the material insoluble to a wet development process, which reveals the latent photomodified regions by dissolving ad removing the unexposed parts of the resist. Since SU-8 is solid before and after optical exposure, the absence of an instantaneous liquid-to-solid transition during the DLW minimizes the local changes in the refractive index, thus creating stable recording conditions. It also permits the fabrication of areas behind the already fabricated features, thus the order in which fabrication is performed becomes unimportant. SU-8 is designed for lithographic fabrication of micro-mechanical systems and is therefore very robust mechanically. Although the refractive index of SU-8 ($n \approx 1.6$) is too low for the formation of PBG, very stable templates of photonic crystals can be fabricated. Recently, wodpile structures having photonic stop-gaps at telecommunication wavelengths were reported [51], and templates for a novel class of spiral structures with potentially wide PBGs were fabricated in SU-8 [27].

10.2.4.1 Structures with Woodpile Architecture

We have fabricated various structures from the initial samples which were prepared as films of formulation 50 SU-8 (NANO, Microchem), spin-coated to a 50 µm thickness. Single-photon absorption in SU-8 is negligible at the laser wavelength of 800 nm, but becomes dominant at the wavelength < 400 nm. Therefore, two-photon absorption is responsible for photomodification. Focusing the laser pulses by high NA = 1.35–1.40 objective lenses, pulse energies as low as 0.2 nJ were required for fabrication, thus confirming that output of an unamplified Ti:Sapphire laser would suffice for the fabrication of SU-8 as demonstrated earlier [52].

Figure 10.13(a,b) shows scanning electron microscopy (SEM) images of the woodpile sample recorded in SU-8. The recording was done by translating the sample in small Δl =80 nm steps along the rod lines with single-pulse exposure on each step. The pulse energy was I = 0.55 nJ. The images demonstrate that the sample is a perfect parallelepiped with dimensions of $(48 \times 48 \times 21)$µm. Its size in the x-y plane is limited only by the range of the translation stage, while along the z axis it is limited to about 40 µm by the requirement to maintain uniform, aberation-free focusing. The structures retain their shapes even after being dislodged from the substrates, and have not degraded within at least several months after fabrication. The individual rods have smooth surfaces and elliptical cross-sections with diameters of 0.5 µm (x-y plane) and 1.3 µm (z axis) adjustable within a certain range by changing the pulse energy. Elongation in the focusing direction by a factor of about 2.6 is due to the ellipsoidal shape of the focal region and forces an increase in the distance Δz in order to avoid an unacceptably high overlap between the neighboring layers tantamount to monolithic SU-8. For the sample shown in Fig. 10.13 (a,b), $\Delta z/\delta l = 1/\sqrt{(2)}$, which corresponds to stretching of a face-centered cubic unit cell by a factor of 2. The elongation still allows decrease the lattice parameters to decrease even further, as is illustrated by the structure in Fig. 10.13 (c,d) which has lattice parameters Δd = 1.2 µm, Δz = 0.85 µm, and con-

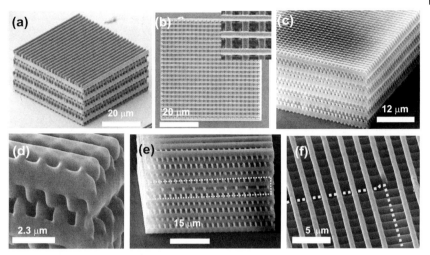

Fig. 10.13 SEM images of various woodpile photonic crystal structures in SU-8, the sample with parameters $\Delta d = 2.0\ \mu m$, $\Delta z = 1.4\ \mu m$, $m = 14$, recorded with scan step $\Delta l = 0.08\ \mu m$ with $l = 0.55 nJ$ pulses focused by $60 \times NA = 1.4$ objective (a,b). Another sample with lattice parameters scaled down proportionally to $\Delta d = 1.2\ \mu m$, $\Delta z = 0.85\ \mu m$, $m = 22$, other parameters are similar (c,d). Demonstration of a planar defect (emphasized by the rectangle) formed by a layer with every second rod missing (e). Demonstration of linear defect with 90° bend; the tentative light-flow direction is indicated by the dotted line (f). For easier visualization, the defect is formed at the top of the structure [26].

tains 22 layers of rods. Despite the apparently close packing of rods, this, as well as other structures with similar parameters, display pronounced signatures of the photonic stop-gaps in their transmission and reflection spectra.

Figure 10.13 (e) demonstrates the possibility of creating a planar defect similar to that engineered in the resin-based woodpile (see above). Another interesting possibility is to create a bent linear waveguide and this is demonstrated in Fig. 10.13 (f). The missing halves of two rods in the two neighboring top layers are connected to form a waveguide with a 90° bending angle as proposed in the literature [53, 54]. Removal of every second rod from the middle layer can produce a planar microcavity similar to that fabricated in resin (see above). Although the refractive index of SU-8 is too low to achieve significant waveguiding or microcavity effects, this example illustrates that defects with the required geometry can be easily created by simply shutting off the laser beam for appropriate time intervals during the recording. This would be impossible to achieve by recording with multiple beam interference fields, which can generate periodic patterns only (see the next section).

Figure 10.14(c) shows the transmission and reflection spectra from Fig. 10.13(a,b), measured along the z axis. Due to the lower sensitivity of the FT-IR setup in transmission mode at longer wavelengths, the transmission was measured in a narrower spectral range than was the reflectivity. Two major high reflectance regions can be seen centered at $\lambda = 7.0$ and $3.6\ \mu m$. The shorter-wavelength peak has a spectrally matched dip in transmission. The shapes and relative amplitudes of both reflectivity peaks are well reproduced by the theoretical spectrum obtained by the transfer-matrix calculations [55, 56], shown in the same figure. The model structure consisted of elliptical rods with $n = 1.6$, major and minor axes of 1.3 and 0.5 μm, respectively, and other parameters (Δl, Δz, m) matching those of the sample. The good overall agreement between the experiments and calculations and the peak's central wavelength ratio of approximately 2 to 1, implies that they represent photonic stop-gaps of different orders. Next, Maxwell's scaling behavior was examined by comparing the reflection spectra of several samples (including the samples shown in Fig. 10.13 (a–d)) with proportionally scaled lattice parameters. The parameters of the sample shown in Fig. 10.13 (a,b) are regarded as a reference, while other structures have parameters scaled down by a certain scaling parameter. Figure 10.15 shows the measuredreflectivities of three other structures with different lattice scaling factors. The positions of the major reflectivity peaks in these samples exhibit a blue shift when the scaling factor decreased, which is agreement with Maxwell's scaling behavior [3]. This behavior is summarized in

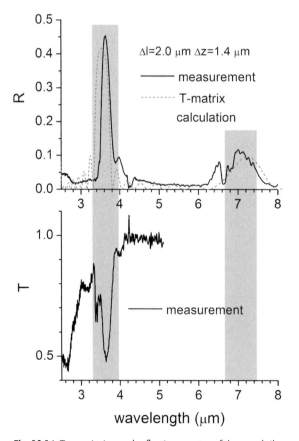

Fig. 10.14 Transmission and reflection spectra of the woodpile sample shown in Fig. 10.13 (a,b).

the inset to the figure, where peak's positions are plotted against the lattice scaling factor. Their linear dependencies constitute a clear evidence that photonic band dispersion is present in the samples [26].

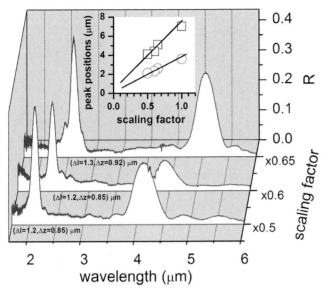

Fig. 10.15 Maxwell's scaling behavior in woodpile structures with different lattice parameters described by the proportional scaling factor, which is normalized to the parameters of the structure with $\Delta d = 2.0$ μm, $\Delta z = 1.4$ μm, $m = 14$ shown in Fig. 10.13 (a,b). The inset shows the spectral positions of the reflectivity peaks versus the scaling factor.

The shortest wavelength of a reliably detected stop-band, 2.11 μm, is already fairly close to the optical communications spectral region (1.1–1.6 μm). In the future it may become possible to reach this spectral region by further scaling down theparameters of the woodpile structure. It must be noted at this point that woodpile structures with stop-gaps at optical communication wavelengths have already been reported [51]. However, these woodpile structures, recorded with about twice as high a resolution, seem to be strongly affected by polymer shrinkage, which is counteracted by building monolithic SU-8 walls around the structures. Although the walls prevent structural deformation, significant mechanical strain is likely to buid up in the structures, and it is not clear if they would retain their shapes if some extended structural defects were incorporated into them during the fabrication. If strain is indeed present, it may also make infiltration more difficult, since strained SU-8 templates are easier to be destroyed by various mechanical or thermal disturbances. In free-standing structures without walls, the distortions usually lead to the characteristic pyramid-like shapes of the initially tetragonally-shaped structures, i.e., the shrinking increases with distance from the substrate, and side edges of the structures become nonparallel. Our woodpile structures, though having larger lattice parameters, are free-standing and are simultaneously nearly free of the polymer shrinkage-related distortions. We have found that the top edges of our structures are just about 0.5% shorter than the bottom edges, and nonparallelism of the vertical side edges is about 0.5°. Thus, although the high resolution of recording and stop-gaps at short wavelengths are important prerequisites for the building of functional photonic crystals and their templates, their overal stability, distortion-free character, and capability of sustaining structural defects are also of importance.

10.2.4.2 Structures with Spiral Architecture

The woodpile structures discussed above were recorded in a layer-by-layer manner, starting from the layers nearest to the substrate. This procedure essentially emulates the woodpile buildup process implemented using the planar lithography of semiconductors. It is worth noting that the versatility of the DLW technique when applied to SU-8 extends beyond this capability. Since during and after the exposure SU-8 is optically transparent, the exposed areas create no obstacles or optical distortions and recording in SU-8 can be done in arbitrary order.

The intense theoretical search for photonic architectures having strong PBG properties, has produced a number of interesting candidate structures that would be difficult to build layer-by-layer. One of them is the recently elaborated 3D square spiral architecture [57, 58], intended for fabrication using the glancing angle deposition (GLAD) method, which uses growth of silicon spirals on pre-patterned substrates. The shape of an individual square spiral and the 3D periodic structure consisting of spirals is shown in Fig. 10.16(a) along with definitions of its characteristic parameters (L is the length of the spiral arms, and c is the vertical pitch of the spiral). The spirals need not necessarily have a square shape; a circular spiral similar to the one shown in Fig. 10.16(b) is a generic symmetry-breaking element that helps to open a sizeable PBG [59]. The extended square spiral structures are generated by centering the spirals on the nodes of a two-dimensional square latticewith period a as illustrated in Fig. 10.16(c). Optimization of the PBG properties might require adjacent spirals that are strongly intertwined and mutually phase-shifted. Although the simplest square spiral structures were successfully fabricated from silicon by the GLAD technique [60–62], their more-complex variants (e.g., containing defects and phase-shifted spirals) can only be obtained by DLW.

(a)   (b)   (c)

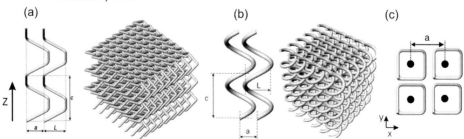

Fig. 10.16 Schematic explanation of spiral architecture and its parameters.

Figure 10.17(a–c) shows a scanning electron microscopy (SEM) image of a sample with design parameters a = 1.8 μm, L = 2.7 μm, and c = 3.04 μm (L = 1.5 a, c = 1.67 a), fabricated with pulse energy I = 0.6 nJ. The dimensions of the structure (48 × 48 × 30)μm are limited on the x-y plane only by the available range of the translation stage. Along the z axis, the sample size is limited by the need to ensure aberration-free focusing at all depths. With the 100× NA = 1.35 lens used, a maximum height of 40 μm was achieved without loss of structural uniformity. According to the existing classification, the fabricated structure belongs to the category of [001]-diamond:5, which is obtained by extruding the spirals from points on the [001] plane of the diamond lattice to their fifth-nearest neighbors [58]. Hence, the spirals are strongly interlaced, in contrast to the [001]-diamond:1 structure, which was earlier realized in silicon by the GLAD technique. The structure comprises ten spiral periods along its entire height, i.e., more than were obtained with the GLAD [60]. Although the image in Fig. 10.17(a) seems to indicate that the vertical edges of the free-standing structure are slightly nonparallel, which, as discussed above, signifies polymer shrinkage, this impression is caused largely by SEM imaging distortions. The top- and side-view SEM images were examined in detail for this and other fabricated structures, and only minor distor-

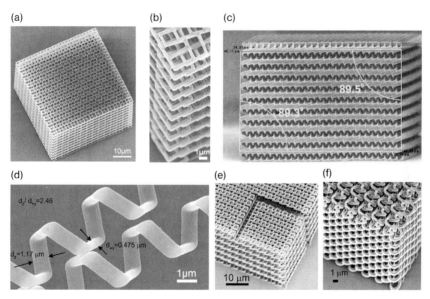

Fig. 10.17 SEM images of square spiral structures recorded by the DLW technique in SU-8. The sample with parameters a = 1.8 μm, L = 2.7 μm, and c = 3.04 μm, fabricated with pulse energy I = 0.6 nJ (a). Enlarged image of its vertical edge region showing 10 spiral periods in the vertical direction (b). Estimates of the shrinkage-related distortions from the side-view image of the previous sample (d). Image of individual spirals separated from the structure with evaluation of their parameters (d). Square spiral structure with two L-shaped waveguides on the walls by missing parts of the spirals, lattice parameters are the same (e). Circular spiral structure with parameters (a = 1.8 μm, L = 2.7 μm, c = 3.6 μm) with 180° phase shift between the adjacent spirals (f).

tions were revealed. As Fig. 10.17(c) illustrates, the combined nonparallelism of the sample's vertical edges is about 1.2°. While being virtually distortion-free in the *x-y* plane, our structures show stronger deformation in the *z* axis, where uniformexpansion of about 3% was observed. Thus, the structure in Fig. 10.17(a,b) has $c = 3.14$ μm instead of the design value of $c = 3.04$ μm. This expansion can be easily compensated for by reducing the design value of c.

The individual spirals are mechanically rigid and preserve their shape even when dislodged from the parent structures. Figure 10.17(d) shows a detailed view of the individual spirals, on which smooth turning points and low-roughness surfaces are evident. The cross-sectional area of the spiral arms derives its shape from the focal region, which is an ellipsoid, elongated along the *z* axis. For the sample shown in Fig. 10.17(a,b), the minor and major diameters of the ellipsoids are 0.475 m and 1.17 m, respectively. Their ratio, about 2.6, is determined by the two-photon point-spread function. The elliptical cross-section and low refractive index of SU-8 ($n \approx 1.6$) are the major deviations of the recorded structures from the ideal models [57, 58].

Figure 10.17(e) shows a square spiral structure with a linear defect, created by missing parts of the spirals, that are turning at a 90° angle. The defect was achieved by simply closing the laser beam during fabrication. For easier inspection, the defect was fabricated on top of the structure. At present, there are no theoretical suggestions regarding which defect topology is the most suitable for spiral structures. Therefore, this sample should be regarded as an illustration that sustainable defects can be built into our samples, rather than a demonstration of the truly functional waveguide.

Figure 10.17(f) shows a circular spiral structure similar to that suggested in an earlier theoretical study [59]. Notice the 180° phase shift between the adjacent spirals, which would be impossible to achieve by the GLAD technique.

Next, we examine the optical properties of spiral structures without intentional defects. Figure 10.18 shows reflectivity and transmission spectra, measured along the *z* axis in the sample with $a = 1.5$ μm, $L = 2.25$ μm, and $c = 3.75$ μm. In the plot, the wavelength interval 2.7–3.6 μm is omitted because it contains some bands of intrinsic SU-8 absorption, which suppress photonic stop-gaps. Spectrally matching pairs of transmission dip and reflection peaks centered, at 2.4 and 4.7 μm wavelengths can be seen. This behavior indicates a fundamental and a higher order stop-gap, similar to that observed in woodpile structures. The peaks and dips also exhibit Maxwell's scaling; this is illustrated by the plots presented in Fig. 10.19. The spectrally matched transmission and reflection features, as well as their Maxwell scaling, are clear manifestations of photonic stop-gaps.

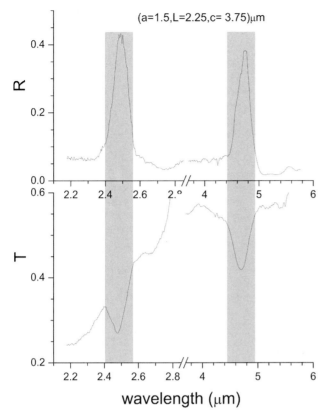

Fig. 10.18 Optical transmission and reflection spectra of the square spiral structure with parameters $a = 1.5\ \mu m$, $L = 2.25\ \mu m$, and $c = 3.75\ \mu m$. The spectral ranges of the tentative photonic stop-gaps are emphasized by the gray rectangles.

A deeper insight into the origin of the stop-gaps can be gained by examining the photonic band diagram shown in Fig. 10.20, which was calculated for the sample with lattice period $a = 1.5\ \mu m$. Along the Γ–Z direction, which coincides with the direction of the optical measurements, two fairly narrow but noticeable stop-bands (the dark rectangles within Γ–Z) are present, one between bands 4 and 5 and the other between bands 8 and 9. The spectral ranges, emphasized by the gray rectangles in Fig. 10.19, are translated into the normalized frequency ($f = \omega a/(2\pi c_l)$, where ω is the frequency and c_l is the speed of light), and are shown in Fig. 10.20 by similar gray rectangles. It can be seen that the observed transmission and reflection features are spectrally close to the two aforementioned stop-gaps, which therefore should be responsible for the experimental observations. The lower stop-gap between bands 4 and 5 is the fundamental gap that may develop into a full PBG, provided that the refractive index contrast is sufficient, and lattice parameters are properly chosen [57, 58]. The presence of an upper stop-gap can be regarded as evidence of the good structural quality of the samples, since higher bands are usually more susceptible to the disorder. The theoretical stop-bands are spectrally narrow due to the SU-8 low refractive index and also due to the ellipsoidal cross-sections of the spiral arms, which force one to choose a verticalspiral pitch ($c = 2.49a$) that is larger than the optimum value ($c \approx 1.6a$) suggested in the literature. Hence, the signatures of the stop-gaps in transmission and reflection spectra are relatively weak. During the measurements, they become further suppressed due to the limited numerical aperture of the microscope objective, which causes deviations from the z axis for the propagation of incident and collected radiation. Nevertheless, these stop-gaps already indicate the presence of long-range order, which is one of the key requirements for the photonic crystal templates.

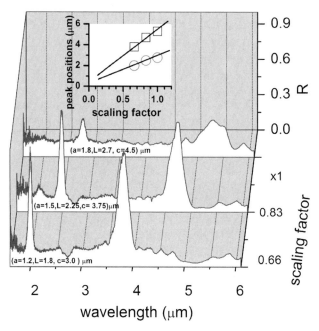

Fig. 10.19 Maxwell's scaling behavior in square spiral structures with proportionally scaled lattice parameters. The reflectivities of three samples having the same normalized lattice parameters ($L = 1.5a$, $c = 2.48a$), but different lattice periods, $a = 1.2, 1.5, 1.8$ µm are shown. The scaling factor is defined relative to the largest structure ($a = 1.8$ µm). The inset shows the spectral positions of the peaks versus the scaling factor.

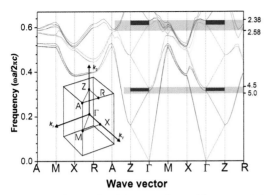

Fig. 10.20 Photonic band diagram of a model square spiral structure with parameters $L = 1.5a$, $c = 2.49a$, $d_{xy} = 0.34a$, $d_z = 0.87a$, and $n = 1.6$. The gray-shaded boxes emphasize the same spectral ranges (given in micrometers by the numbers on the right) as in Fig. 10.19. The dark boxes emphasize stop-gaps along the Γ–Z direction. The inset shows the positions of the high symmetry points of the first Brillouin zone.

Practical infiltration of 3D SU-8 templates (with woodpile, spiral, or other architecture) by other materials having higher refractive indexes still needs to be studied. There are several tentative possibilities. One of them is sol-gel infiltration with TiO_2 ($n = 2.2$–2.6) or other nanoparticles into the air voids, followed by the removal of the template (for example, SU-8 can be ashed at temperatures above 600 °C with very little residue. The thermal and chemical robustness of SU-8 also permits the use of certain chemical vapor deposition (CVD) or electrodeposition processes, provided that they are performed at temperatures below the softening temperature of cross-linked SU-8 (about 350 °C). Single infiltration and template removal will produce an inverse spiral structure. Alternatively, the double-templating approach can be used. First, a CVD infiltration of silica is performed by the low-temperature CVD. Then, the SU-8 templateis thermally removed and the remaining secondary silica template is infiltrated by a high refractive index semiconductor. Afterwards, the silica template is removed by chemical etching, and a direct structure is obtained.

10.3
Lithography by Multiple-beam Interference

10.3.1
Generation of Periodic Light Intensity Patterns

Direct writing with tightly focused laser beams creates periodic or nonperiodic structures in a sequence of point-by-point recording events. As an alternative to this approach, periodic field patterns generated by the interference of two or multiple coherent laser beams overlapping at nonzero mutual angles, can be used to record an entire periodic structure at once. The periodic patterns produced by the interference of multiple laser beams depend on the mutual alignment of the beams' directions and their phases. These patterns can be visualized and studied using numerical or analytical calculations. The relation between the interfering beams' parameters and the structure of the interference field is widely studied in the literature. In fact, it was demonstrated that in three dimensions interference patterns with symmetries of all fourteen Bravais point lattices can be generated by the interference of four beams [63]. As a starting point, here we address the simplest case of two interfering beams, denoted by numbers *1* and *2*, having the same angular frequency, ω, but different wave-vectors, k_1 and k_2 and phases, ϕ_1 and ϕ_2. This situation is depicted in Fig. 10.21.

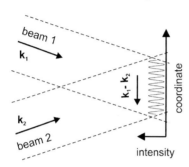

Fig. 10.21 Interference of two coherent plane waves with wave-vectors k_1 and k_2 represented by the beams 1 and 2. In the region of their spatial overlap the electric field intensity becomes spatially modulated as described by Eq. (6) along the direction parallel to the difference vector $k_1 - k_2$.

The electric field $E_{1,2}$ of each beam can be, to a first approximation, described by a plane wave as follows:

$$E_{1,2}(\mathbf{r}_{1,2}, t) = \mathbf{E}^0_{1,2} \cos \left(\mathbf{k}_{1,2} \cdot \mathbf{r}_{1,2} - \omega_{1,2} t + \phi_{1,2} \right) \tag{4}$$

where the index 1 or 2 is used to denote the particular wave. The electric field interference pattern is then obtained by summing the electric fields of the form (4), and the intensity pattern is obtained by time-averaging the squared sum-field $\mathbf{E}^2$. For simplicity let us assume that both waves are scalar (i.e., polarization is ignored) and the field amplitudes in both beams are equal to E_0. Then, the field can be expressed as

$$
\begin{aligned}
E(\mathbf{r}, t) &= E_1 + E_2 \\
&= E_0 \cos \left(\mathbf{k}_1 \cdot \mathbf{r} - \omega t + \phi_1 \right) + E_0 \cos \left(\mathbf{k}_2 \cdot \mathbf{r} - \omega t + \phi_2 \right) \\
&= 2 E_0 \cos \left(\frac{\mathbf{k}_1 + \mathbf{k}_2}{2} \cdot \mathbf{r} - \omega t + \frac{\phi_1 + \phi_2}{2} \right) \\
&\quad \times \cos \left(\frac{\mathbf{k}_1 - \mathbf{k}_2}{2} + \frac{\phi_1 - \phi_2}{2} \right)
\end{aligned}
\tag{5}
$$

The intensity field $I(\mathbf{r})$ is proportional to the square of the electric field:

$$I(\mathbf{r}) \propto \left\langle E(\mathbf{r}, t)^2 \right\rangle$$

$$\propto E_0^2 \{1 + \cos[(\mathbf{k}_1 - \mathbf{k}_2) \cdot \mathbf{r} + \phi_1 - \phi_2]\} \tag{6}$$

The intensity pattern has periodic sinusoidal dependence on the coordinate along the direction defined by the vector $\Delta\mathbf{k} = 3D(\mathbf{k}_1 - \mathbf{k}_2)$. The period of the pattern, Λ, can be expressed as:

$$\Lambda = \frac{\lambda}{2 \sin\left[\frac{\widehat{(\mathbf{k}_1, \mathbf{k}_2)}}{2}\right]} = \frac{\pi}{|\mathbf{k}_{1,2}| \sin\left[\frac{\widehat{(\mathbf{k}_1, \mathbf{k}_2)}}{2}\right]} \tag{7}$$

where λ is the wavelength of the interfering waves. Hence, the period is inversely proportional to the grating vector of the one-dimensional pattern $|\Delta\mathbf{k}|$. It is easy to see that the assumption of unequal intensities of the two beams would slightly alter the Eq. (6) which would then contain a spatially uniform "dc" background component $E_1^2 + E_2^2$ and an oscillating sinusoidal component with amplitude $E_1 E_2$. The spatial modulation of the intensity pattern will therefore become weaker. Varying the the phase difference of the two beams would result in the spatial shift of the periodic pattern.

The same approach can be used to calculate 2D and 3D periodic patterns formed by the interference of multiple waves. In the case of three and four-beam interference, three and six grating (or difference) vectors can be defined. However, only two and three of them, respectively, are linearly independent. The independent wave-vectors also are the primitive translation vectors of the reciprocal lattice. Let us consider four interfering beams. The wave-vectors of the beams can be defined using the unit vectors $d\mathbf{x}, d\mathbf{y}, d\mathbf{z}$ of Cartesian coordinate system. For example, if the directions of the four beams are

$$\mathbf{k}_1 = \left[\frac{3}{2}; \frac{3}{2}; \frac{3}{2}\right], \ \mathbf{k}_2 = \left[\frac{5}{2}; \frac{\overline{1}}{2}; \frac{\overline{1}}{2}\right], \ \mathbf{k}_3 = \left[\frac{\overline{1}}{2}; \frac{5}{2}; \frac{\overline{1}}{2}\right], \ \mathbf{k}_4 = \left[\frac{\overline{1}}{2}; \frac{\overline{1}}{2}; \frac{5}{2}\right] \tag{8}$$

and their electric fields are $E_{1,2,3} = E_0 \cos(\mathbf{k}_i \cdot \mathbf{r} - \omega t)$, the resulting intensity pattern, $I(\mathbf{r})$, can be expressed as

$$I(\mathbf{r}) \propto E_0^2 \{2 + \cos(2\pi x) + \cos(2\pi y) + \cos(2\pi z) + \\ \cos(2\pi(y-z)) + \cos(2\pi(x-z)) + \cos(2\pi(x-y))\} \tag{9}$$

The pattern is periodic along the x, y, and z axes. The 3D intensity distribution calculated according to the above equations is shown in Fig. 10.22(a). Its point lattice is defined by the primitive translation vectors $d\mathbf{x}, d\mathbf{y}, d\mathbf{z}$, and has has face-centered cubic (fcc) symmetry. In the reciprocal space, the primitive translation vectors are the difference vectors $\mathbf{k}_1 - \mathbf{k}_4 = 2\pi[1\ 0\ 0]$, $\mathbf{k}_2 - \mathbf{k}_4 = 2\pi[0\ 1\ 0]$, and $\mathbf{k}_3 - \mathbf{k}_4 = 2\pi[0\ 0\ 1]$, and the point lattice has a body-centered cubic (bcc) symmetry. As is well known from crystallography, symmetries of the direct and inverse lattices are related to each other. Thus the desired lattice symmetry can be chosen by selecting the beams directions that yield the appropriate difference vectors.

Similar analysis can be repeated for other arrangements of the recording beams. Figure 10.22(b) shows the result for four beams with directions

$$\mathbf{k}_1 = [\bar{1}; 0; \bar{2}],\ \mathbf{k}_2 = [2; 0; \bar{1}],\ \mathbf{k}_3 = [0; 1; \bar{2}],\ \mathbf{k}_4 = [0; \bar{1}; \bar{2}] \tag{10}$$

and amplitudes the same as before. The resulting periodic intensity pattern has a body-centered cubic (bcc) point lattice. Rearranging the beams also allows one to obtain a simple cubic (sc) lattice [29]. Systematic studies of the beam arrangements required for obtaining of various 2D and 3D lattices can be found in the literature [63–66].

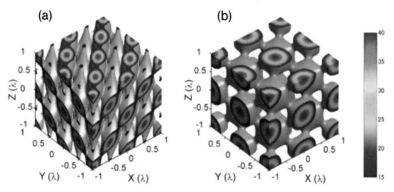

Fig. 10.22 Constant-intensity surface plot representation of the periodic structures with face-centered cubic (a) and body-centered cubic (b) lattices generated by four-beam interference. The coordinates are given in the units of the laser wavelength.

It should be stressed here that the above discussion has ignored the vectorial nature of the electromagnetic waves, and also has assumed the intensities of all beams to be equal. Their phases were also neglected. With these simplifications, the calculations yield correct lattice symmetry, but cannot predict the correct shape of the "atomic" basis element associated with each node of the point lattice. More realistic calculations must account for the vectorial nature (polarization) of the fields and their different amplitudes. For example, by adjusting the wave vectors, polarizations, and intensities of the beams it is possible to obtain a diamond structure which has a fcc unit cell and a "two-atom" basis [67, 68].

10.3.2
Practical Implementation of Multiple-beam Interference Lithography

The key element of every multi-beam interference setup is the method by which multiple coherent laser beams are obtained from a single laser beam. The simplest technique for accomplishing this task is well known and is widely used in laser physics and spectroscopy: the laser beam is split into two components using a transmission/reflection type beam-splitter. Each of these components can be subsequently split into two components again until the required number of the beams is obtained. Then, the mutual temporal delays of the beams should be adjusted to be shorter than the laser coherence length (for continuous-wave lasers) or the pulse length (for pulsed lasers), and all beams should be steered toward their common interference plane in accordance with the required lattice type and the basis shape. Since by using this method all beams can be well separated in space, it is easy to insert other elements into the paths of the beams for the control of their intensities, polarizations and other desired properties. However, the experimental setup might become quite crowded and tricky to align, especially when using ulrashort laser pulses. For example, the spatial length of a 100 fs pulse is only 30 μm, and adjustment of each optical path length is needed with micrometer precision. Nevertheless, such an experimental setup enabled the recording of 3D structures with fcc lattice [29, 67–69].

In spite of these achievements, a recently proposed alternative method of obtaining multiple laser beams using diffraction gratings is becoming widely used for the multi-beam interference lithography. This technique employs the diffractive beam-splitter (DBS), which can be collected by stacking together several 2D diffraction gratings oriented at different angles. As the laser beam passes through the first grating, it splits into the transmitted (zero-order) and diffracted (first and higher order) components. These components are subsequently passed through the second grating where they diffract again. Two identical gratings with grooves oriented at a 90° angle will split a normally-incident input beam into multiple beamlets that will form a square pattern on the plane, perpendicular to the input beam's propagation direction. From these beams, only those required for the creation of the desired interference pattern are selected using a transmission mask. The unmasked beams can then be converged to overlap in space using a system of lenses. DBS is particularly convenient for experiments with ultrafast lasers because it automatically ensures zero temporal delay between all individual beams. Furthermore, due to the diffraction angle dependence on the laser wavelength, the overlapping beams create patterns whose period is wavelength-independent and uniquely defined by the period of the DBS gratings. Thus, intensity patterns with well-defined periodicity can be obtained even with multicolor lasers [70].

By using more than two gratings and by orienting them at different angles, even more beamlets can be obtained. In fact, instead of combining several one-dimensional gratings, a single plate with equivalent transmission function can be fabricated holographically. The setup that was used in our experiments is shown schematically in Fig. 10.23. The input beam of an amplified Ti:Sapphire laser system (the same system as used for direct laser writing experiments) enters the DBS and is split into several components. The DBS is placed at the focal plane of a positive lens, which transforms the diverging beams into parallel ones. The transmission mask placed in their path selects the beams required for the interference pattern creation. These beams may be optionally passed through the phase-retarding unit, which consists of variably-tilted glass plates. The relative phases of the beams can be adjusted by adjusting the tilt angles. For the assessment of the pattern quality, a small fraction of the selected beams' intensity is reflected by a glass plate and imaged by a lens on the CCD camera. The transmitted beams are then focused by a microscope lens, with a numerical aperture of 0.75, into the sample, where they overlap at nonzero mutual angles, creating the desired intensity patterns. According to Eq. (7), the highest spatial frequency in the pattern is determined by the maximum mutual angle at which two (or more) beams overlap. The theoretical resolution limit is the half-wavelength, which is achieved for counter-propagating beams. However, in most experiments based on the setup shown in Fig. 10.23, the beams converge on the sample at somewhat smaller angles and the resolution is lower. The samples used for the recording are thin films of photoresist SU-8 (formulation 25) spin-coated to the thickness of 20 μm on the cover glass substrate.

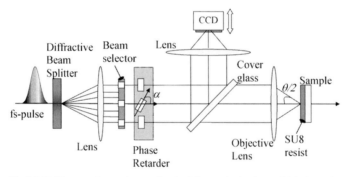

Fig. 10.23 The experimental setup for the lithography by the multiple-beam interference.

This experimental arrangement has allowed one to demonstrate easy implementation of multi-beam interference with a pulsed femtosecond laser [71–73]. Similar setups have also been used in other reported studies [74–80]. Chelnokov et al. [81] have demonstrated the possibility of controlling the lattice by varying the optical phases of the interfering beams. Nakata et al. [74] have pointed out another important advantage of the DBS: for ultrashort pulses overlapping at non-zero angles, the interference region is only limited by the pulses' temporal width, and is not affected by the diameter of the beams. This is especially important when the mutual angles of the beams increase. Without the DBS, significant temporal delay would develop near the edges of the overlap zone, thus preventing the formation of a periodic pattern and limiting the size of the interference region.

10.3.3
Lithographic Recording of Periodic Structures by Multiple-beam Interference

10.3.3.1 Two-dimensional Structures

At first we shall focus on the two-dimensional structures. By superimposing the two-dimensional interference pattern on a photoresist film, patterned films can be obtained and inspected by SEM. Consequently, it is easy to check whether or not the recorded patterns correspond to the intended ones. Three-dimensional structures, besides being more difficult to record, can be inspected by SEM only from the outside. Their inner structure must be visualized using other techniques, like laser-scanning confocal microscopy, which have lower resolution than SEM.

Figure 10.24(a) shows the beam configuration used for the recording. The beams 1–4 are depicted after the selection mask and the focusing lens. The beams are arranged symmetrically around the principal optical axis of the system and converge to that axis and on the sample with the angle $\theta/2$. Their phases can be controlled using glass slides which are inserted into the beams and can be tilted, thus inducing the phase shifts as shown in Fig. 10.24(b). The intensity patterns taken by the CCD camera are shown in the upper part of Fig. 10.24(c) for the case when all beams have the same phase, and in Fig. 10.24(d) for the case when a $\pi/2$ phase shift between the beam pairs 1,2 and 3,4 is induced. For comparison, the calculated intensity patterns for the same configuration are shown in the lower part of Fig. 10.24(c) and (d). As can be seen, these patterns have distinct square symmetries. The matching between the observed and calculated patterns is very close.

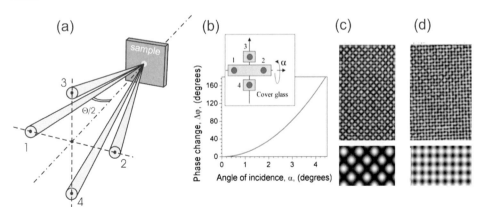

Fig. 10.24 Four-beam arrangement for the recording of two-dimensional structures (a). Phase control using tilted glass plates (b). Experimentally observed (top) and calculated (bottom) interference pattern for zero phase difference between the beams (c) The same when there is $\pi/2$ phase difference between the beam pairs 1,2 and 3,4. The four-beam scheme can be also supplemented with the fifth beam propagating along the principal optical axis of the setup.

Figure 10.25 shows SEM images of several structures recorded in SU-8. All patterns have a square lattice with period of about 3.0 ± 0.1 μm. This value compares favorably with 2.98 μm, which is expected from the calculations. During the experiments, the recording beams were allowed to enter the SU-8 film resist from both sides, i.e., from the free surface of SU-8, and from the SU-8 and cover–glass interface. In the latter case, back-reflection from the free SU-8 surface due to considerable refractive index mismatch ($n_{SU8} = 1.68$ to $n_{air} = 1$) is strong, and creates additional intensity modulation along the direction of the optical axis. As a result, the rods had a noticeable periodic modulation of their diameter with period of 240 nm. The period of the pattern inside the SU-8 film can be calculated using Snell's law, which for $\theta/2 = 10.9°$ incidence angle used, yields the internal reflection angle in SU-8 of approximately 6.8°. The interference between the incident and reflected beams thus yield $\Lambda_s = \lambda/(2n_{SU8}) \times \cos^{-1}(6.8°) \simeq 250$ nm modulation period, which is close to the previous observation. No modulations were observed when beams were incident from the free SU-8 surface. In these circumstances, back reflection into SU-8 was inhibited by the low refractive index mismatch at the SU-8/glass interface ($n_{SU8} = 1.68$ to $n_{glass} = 1.52$).

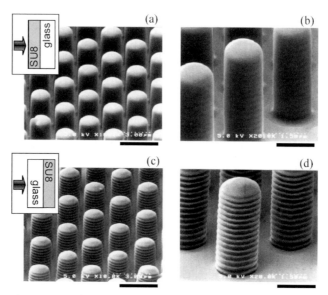

Fig. 10.25 Two-dimensional square patterns of cylindrical rods recorded in SU-8 with femtosecond laser pulses having 800 nm central wavelength and duration of approximately 250 fs (at the focus). The exposure time was 20 s, the total irradiation dose was 28 μJ. The exposure was directly to the SU-8 layer (a,b) and through the glass substrate (c,d). The scale bars are 3 μm (a,c) and 1.5 μm (b,d).

Two-dimensional structures can also be recorded using only one pair of beams (1,2 or 3,4 in Fig. 10.24(a)) in a two-step exposure with 90° degree rotation of the sample between the steps. The result is a superposition of two one-dimensional gratings which yields a rectangular arrangement of holes and is shown in Fig. 10.26. The structure is very similar to that expected from the calculations; for example, its lattice period is 1.0 μm whereas calculations predict the value of 1.02 μm. The structure exhibits a high degree of long-range order, and despite the fact that low refractive index of the photoresist precludes the occurrence of significant PBG effects, it may prove to be useful as a template for infiltration with materials of high refractive index.

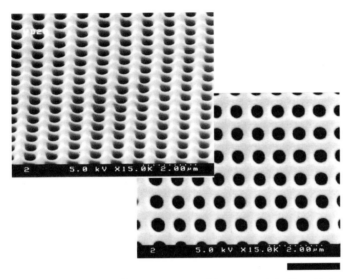

Fig. 10.26 Perspective and top-view images of the rectangularly aligned hole structure recorded by two consecutive exposures to the two-beam interference pattern. The scale bar is 2 μm.

The arrangement of the beams on the corners of a square allows one to record structures which possess relatively simple square symmetry. Increasing the number of beams and arranging them into more sophisticated patterns allows to perform more complex two-dimensional patterning. An example of such patterning is described in Fig. 10.27. The beams were centered on the corners of a hexagon as shown in the top half of Fig. 10.27(a). In the bottom half, the calculated interference pattern is shown. The structure recorded in SU-8 (see Fig. 27(b)) closely resembles this pattern. Figure 10.27(c) demonstrates the high resolution of the fabrication which is obvious from the narrow spikes in the optical density of the image in (b), measured along the diagonal dashed line. The spikes have half-widths of about 200 nm, i.e., the smallest dimensions of the features in the structure are close to $\lambda/4$. The high resolution is most likely achieved due to nonlinear absorption.

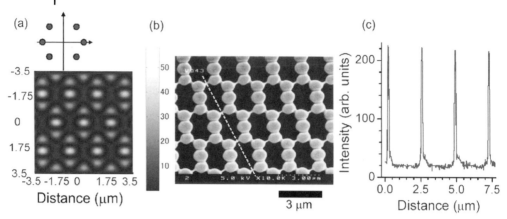

(a)

-3.5
-1.75
0
1.75
3.5
-3.5 -1.75 0   1.75 3.5
Distance (μm)

(b)

50
40
30
20
10

3 μm

(c)

200
100
0
0.0   2.5   5.0   7.5
Distance (μm)

Intensity (arb. units)

Fig. 10.27 Recording by six-beam interference. The beams, centered on the corners of a hexagon are expected to produce a complex interference pattern as inferred from the calculations (a), the structure recorded by this pattern in SU-8 (b), the intensity profile measured along the dashed line in (b) is shown in (c).

10.3.3.2 Three-dimensional Structures

Three-dimensional light intensity patterns can be achieved using the four-beam arrangement similar to that shown in Fig. 10.24(a) supplemented by the fifth central beam propagating along the principal optical axis. However, fabrication of three-dimensional structures always presents a challenge, since such structures after the fabrication must be self-supporting, and hence must comprise well-connected regions of photoresist. The connectivity is also required for the structure to withstand action of the capillary forces during the development and subsequent drying steps. This usually requires careful choice of the optimum pre- and post-processing conditions.

Figure 10.28 shows SEM images of a three-dimensional structure which was recorded with the five-beam arrangement described above. The recording side beams converged at angles of $\theta = 34°$, the total energy of the five pulses was 24 μJ, and the exposure time was 90 s at 1 kHz laser repetition rate. The calculations performed prior to the fabrication indicated that the structure will have a body-centered tetragonal (bct) lattice with period of 1.45 μm (in the x-y plane) and 8.1 μm (along the z axis). The SEM image of the topmost x-y plane of the sample shown in Fig. 10.28(a) reveals an ordered square array of solid features (emphasized by a dashed line in the figure), whose side length corresponds to the x-y plane bct lattice period of 1.4 μm, i.e., close to the calculated period. The side-view image of the cleaved edge of the same sample, taken at the 60° viewing angle (Fig. 10.28(b)) illustrates that the solid features are ellipsoids with major (long) axis aligned in the z axis direction (coincident with the principal optical axis of the setup). The lattice period along the z axis is 8.06 μm, close to the calculated value. The strong elongation of the bct unit cell is undesirable because it leads to the tendency of photonic stop-gaps opening at longer wavelengths, and from the

(a) (b)

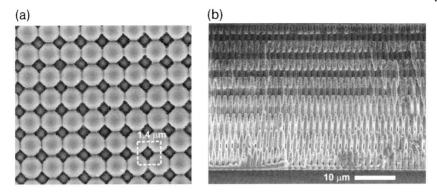

Fig. 10.28 SEM image of the body-centered tetragonal structure taken along the z axis direction (a), image of its cleaved edge (b), courtesy of T. Kondo.

applications point of view is the major disadvantage of the structures discussed here. The main reason responsible for the elongation is the relatively low beam convergence angle $\theta/2$. Although the angle can be increased by using focusing optics with higher numerical aperture, the fcc lattice, which is the ultimate limit of optimization, requires internal angles (inside the photoresist) $\theta/2$ that are not achievable due to the existence of the critical angle of total internal reflection. Miklyaev et al. [82] have used a specially designed prism mounted on top of the photoresist, which allowed entrance of the beams into photoresist at high internal angles without the refraction-related problems This approach was successfully used to fabricate fcc structures in SU-8 with a lattice constant of about 550 nm and photonic stop-gaps at visible wavelengths.

Experimentally measured optical transmission and reflection spectra of the sample, similar to the one discussed above but with slightly different lattice parameters (lattice period 2.0 µm in the x-y plane and 1.71 µm along the z axis) are shown in Fig. 10.29. The signatures of photonic band dispersion, though weak, can be unmistakably identified. Pairs of spectrally matched transmission dips and reflection peaks are seen at wavelengths of 5.0, 3.4 and 2.6 µm and are emphasized by vertical dashed lines with arrows. These pairs most likely represent photonic stop-gaps of different orders.

Images of another three-dimensional structure recorded in SU-8 using interference of five laser beams ($\theta/2 = 33°$, pulse energy used 17 µJ, exposure time 2.5 min) are shown in Figure 10.30. The two images shown in Figure 10.30 (a) and (b), taken in different parts of the structure, illustrate a quite common problem: the patterns appear to be different in the two areas. This is most likely due to the uncontrollable phase variations occurring in various regions of the cross-section of the interfering beams with diameters in the range of 50–200 µm). This example shows that, for the best results, phases of the recording beams must be controlled precisely. Numerical analysis also implies that an ideal bct structure obtainable without the phase control is self-supporting, provided that suitably high exposure levels are used. However, the structures also become impermeable, which inhibits their development and infiltration by other materials. At lower exposures the bct structure becomes disconnected and dislodges during development.

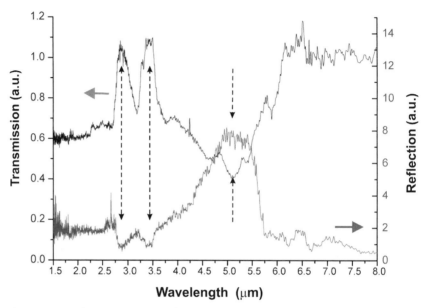

Fig. 10.29 Optical transmission and reflection spectra of the body-centered tetragonal structure (see text for details), the vertical dashed lines with arrows emphasize the photonic stop-gap regions. Courtesy of Dr. T. Kondo.

(a)   (b)

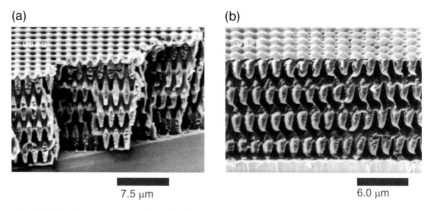

7.5 μm   6.0 μm

Fig. 10.30 Illustration of nonuniformity of the samples recorded by multi-beam interference, the images shown in (a) and (b) were taken in different parts of the sample with intended bct structure.

The need for phase control is strengthened even further by the possibilities it offers for achieving other types of structure with better photonic band-gap properties. For example, it can be shown that, when two non-axial beams (for instance beams 1 and 2 or 3 and 4 in Fig. 10.24) in the five-beam scheme, acquire a $\pi/2$ phase shift with respect to the other beams, the resulting structure is well-connected, self-supporting, and permeable. Moreover, it acquires the "two-atom" basis of a diamond structure. The topology of the experimentally obtained diamond-like structure is illustrated in Fig. 10.31. This structure is different from other SU-8 structures discussed so far because it was fabricated using the second harmonic wavelength (400 nm) of the Ti:Sapphire laser system. This was done in an attempt to improve the connectivity of the structure. However, one-photon absorption at this wavelength was also substantial, and it was necessary to control the laser pulse energy very carefully in order to avoid the complete unpatterned exposure of the material. Figure 10.31 presents the images acquired by the scanning laser microscopy in reflection mode. The top image in the figure is the reconstructed cross-sectional image of the sample in the x-z plane. The lines crossing this image mark z-positions at which the x-y plane images shown in the lower row were taken. These positions correspond to the atomic planes of the diamond structure along the z axis of a tetragonal diamond cell. The cross-sectional images demonstrate explicitly the diamond-like character of the structure obtained. This is emphasized by the circles that mark the locations of "photonic atoms" in the planes, and it is easy to recognize their diamond-like ordering.

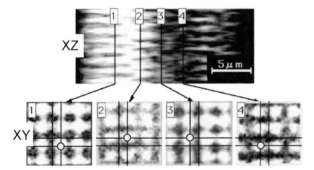

Fig. 10.31 Laser scanning microscopy images of the diamond-like structure recorded by five-beam interference using phase control. The recording laser wavelength is 400 nm. The image on the top depicts a cross-sectional view of the sample along the z axis direction (along the principal axis), and the lower shows cross-sectional images in the x-y plane taken at different z coordinates. These coordinates correspond to different atomic planes along the z axis of a tetragonal diamond cell.

10.4
Conclusions

Micro- and nano-optics is becoming increasingly important in a wide range of applications, including optoelectronics, communications, sensors, biomedical, data storage, and other technology-driven areas. Microstructuring of materials using lasers allows relatively easy, low-cost fabrication of various micro- and nanostructures applicable in these fields. In photonics, various kinds of laser fabrication and lithography techniques offer highly versatile, simple, and low-cost tools for the structuring of materials. These tools still need to be improved before they can be widely applied in practice. The possibility of obtaining three-dimensional periodic structures, if desired, with intentional defects, and simultaneously avoiding the restrictions and tediousness of the planar semiconductor processing approach, is very tempting. The main challenge faced by laser micro- and nanofabrication is the improvement of the resolution needed in order to bring the PBG and stop-gap wavelengths into the near-infrared and visible ranges, widely used in telecommunications and optoelectronics. These goals are pursued actively by research groups worldwide. Recent reports indicate clearly that 3D templates with adequate resolution are already obtainable bydirect laser drawing and multiple-beam interference techniques. It can be expected that in the very near future, reliable reports about photonic crystal structures with complete PBG at telecommunications wavelengths obtained from such laser-fabricated templates will be published.

References

1 E. Yablonovitch, "Inhibited spontaneous emission in solid-state physics and electronics," *Phys. Rev. Lett.* **58**, pp. 2059–62, 1987.

2 S. John, "Strong localization of photons in certain disordered dielectric superlattices," *Phys. Rev. Lett.* **58** (23), pp. 2486–9, 1987.

3 J. D. Joannopoulos, R. D. Meade, and J. N. Winn, *Photonic Crystals: Molding the Flow of Light*, Princeton University Press, Princeton, New Jersey, 1995.

4 S. Johnson and Joannopoulos J.D., *Photonic crystals: The Road From Theory to Practice*, Kluwer Academic Publishers, 2002.

5 S. Noda and T. Baba, eds., *A Roadmap on Photonic Crystals*, Kluwer Academic Publishers, 2003.

6 K. Sakoda, *Optical Properties of Photonic Crystals*, vol. 80 of *Springer Series in Optical Sciences*, Springer Verlag, 2001.

7 M. Qi, E. Lidorikis, P. Rakich, S. Johnson, J. Joannopoulos, E. Ippen, and H. Smith, "A three-dimensional optical photonic crystal with designed point defects.," *Nature* **429** (6991), pp. 538–42, 2004.

8 S. Ogawa, M. Imada, S. Yoshimoto, M. Okano, and S. Noda, "Control of light emission by 3D photonic crystals," *Science* **305** (5681), pp. 227–9, 2004.

9 S. Noda, K. Tomoda, N. Yamamoto, and A. Chutinan, "Full three-dimensional photonic bandgap crystals at near-infrared wavelengths," *Science* **289** (5479), pp. 604–6, 2000.

10 H. Misawa, H. Sun, S. Juodkazis, M. Watanabe, and S. Matsuo, "Microfabrication by femtosecond laser irradiation," in *Laser Applications in Microelectronic and Optoelectronic Manufacturing v*, H. Helvajian, K. Sugioka, M. C. Gower, and J. J. Dubowski, eds., pp. 246–260, SPIE, vol. 3933, 2000.

11 E. N. Glezer and E. Mazur, "Ultrafast-laser driven micro-explosions in transparent materials," *Appl. Phys. Lett.* **71**, pp. 882–4, 1997.

12 M. Goppert-Mayer, "Uber Elementarakte mit zwei Quantensprungen," *Ann. Phys.* **9**, pp. 273–94, 1931.

13 N. Tanno, K. Ohkawara, and H. Inaba, "Coherent transient multi-photon scattering in a resonant two-level system," *Phys. Rev. Lett.* **46**, pp. 1282–85, 1981.

14 B. Stuart, M. Feit, A. Rubenchik, B. Shore, and M. Perry, "Laser-induced damage in dielectrics with nanosecond to subpicosecond pulses," *Phys. Rev. Lett.* **74**, pp. 2248–51, 1995.

15 N. Bloembergen, "Laser-induced electric breakdown in solids," *IEEE Journ. Quant. Electron.* **10**, pp. 375–86, 1974.

16 B. Stuart, M. Feit, A. Rubenchik, S. Herman, B. Shore, and M. Perry, "Nanosecond-to-femtosecond laser induced breakdown in dielectrics," *Phys. Rev. B.* **53**, pp. 1749–61, 1996.

17 C. Carr, H. Radousky, and S. Demos, "Wavelength dependence of laser-induced damage: determining the damage initiation mechanisms.," *Phys Rev Lett* **91** (12), p. 127402, 2003.

18 B. Cumpston, S. Ananthavel, S. Barlow, D. Dyer, J. Ehrlich, L. Erskine, A. Heikal, S. Kuebler, I.-Y. Lee, D. Mccord-Maughon, J. Qin, H. Rockel, M. Rumi, X.-L. Wu, S. Marder, and J. Perry, "Two-photon polymerization initiators for three-dimensional optical data storage and microfabrication," *Nature* **398**, pp. 51–4, 1999.

19 S. Kawata, H.-B. Sun, T. Tanaka, and K. Takada, "Finer features for functional microdevices," *Nature* **412**, pp. 697–8, 2001.

20 H. Sun, S. Matsuo, and H. Misawa, "Three-dimensional photonic crystal structures achieved with two-photon-absorption photopolymerization of resin," *Appl. Phys. Lett.* **74**, pp. 786–8, 1999.

21 H. B. Sun, Y. Xu, S. Matsuo, and H. Misawa, "Micro-fabrication and characteristics of two-dimensional photonic crystal structures in vitreous silica," *Opt. Rev.* **6**, pp. 396–8, 1999.

22 H. B. Sun, Y. Liu, S. Juodkazis, K. Sun, J. Nishii, Y. Suzuki, S. Matsuo, and H. Misawa, "Photonic lattices achieved with high-power femtosecond laser microexplosion in transparent solid materials," in *Proc. SPIE*, e. a. X. Chen, ed., **3888**, pp. 131–42, 2000.

23 H. B. Sun, Y. Liu, K. Sun, S. Juodkazis, M. Watanabe, S. Matsuo, H. Misawa, and J. Nishii, "Inlayed "atom"-like three-dimensional photonic crystal structures created with femtosecond laser microfabrication," *Mat. Res. Soc. Symp. Proc.* **605**, pp. 85–90, 2000.

24 H. Sun, S. Matsuo, and H. Misawa, "Three-dimensional photonic crystal structures achieved with two-photon-absorption photopolymerization of resin," *Appl. Phys. Lett.* **74**, pp. 786–8, 1998.

25 H. Sun, V. Mizeikis, Y. Xu, S. Juodkazis, J.-Y. Ye, S. Matsuo, and H. Misawa, "Microcavities in polymeric photonic crystals," *Appl. Phys. Lett.* **79**, pp. 1–3, 2001.

26 V. Mizeikis, K. Seet, S. Juodkazis, and H. Misawa, "Three-dimensional woodpile photonic crystal templates for the infrared spectral range.," *Opt Lett* **29** (17), pp. 2061–3, 2004.

27 K. Seet, V. Mizeikis, S. Matsuo, S. Juodkazis, and H. Misawa, "Three-dimensional spiral-architecture photonic crystals obtained by direct laser writing," *Adv. Mater.* **17**, pp. 541–4, 2005.

28 M. Watanabe, S. Juodkazis, H.-B. Sun, S. Matsuo, H. Misawa, M. Miwa, and R. Kaneko, "Transmission and photoluminescence images of three-dimensional memory in vitreous silica," *Appl. Phys. Lett.* **74**, pp. 3957–9, 1999.

29 M. Campbell, D. Sharp, M. Harrison, R. Denning, and A. Turberfield, "Fabrication of photonic crystals for the visible spectrum by holographic lithography," *Nature* **404** (6773), pp. 53–6, 2000.

30 A. Marcinkevicius, V. Mizeikis, S. Juodkazis, S. Matsuo, and H. Misawa, "Effect of refractive index mismatch on laser microfabrication in silica glass," *Appl. Phys. A* **76**, pp. 257–60, 2003.

31 K. Ho, C. Chan, and C. Sokoulis, "Existence of photonic gaps in periodic

dielectric structures," *Phys. Rev. Lett.* **65**, p. 3152, 1990.

32 P. N. Pusey and W. van Megen, "Phase behaviour of concentrated suspensions of nearly hard colloidal spheres," *Nature* **320**, pp. 340–2, 1986.

33 H. Míguez, C. López, F. Meseguer, A. Blanco, L. Vásquez, and R. Mayoral, "Photonic crystal properties of packed submicrometer SiO_2 spheres," *Appl. Phys. Lett.* **71**, pp. 1148–50, 1996.

34 H. Míguez, A. Blanco, F. Meseguer, C. López, H. M. Yates, M. E. Pemble, V. Fornés, and A. Mifsud, "Bragg diffraction from indium phosphide infilled fcc silica colloidal crystals," *Phys. Rev. B* **59**, pp. 1563–6, 1999.

35 A. Blanco, E. Chomski, S. Grabtchak, M. Ibisate, S. John, S. Leonard, C. Lopez, F. Meseguer, H. Miguez, J. Mondia, G. Ozin, O. Toader, and H. van Driel, "Large-scale synthesis of a silicon photonic crystal with a complete three-dimensional bandgap near 1.5 micrometres," *Nature* **405** (6785), pp. 437–40, 2000.

36 M. Gu, *Advanced Optical Imaging Theory*, vol. 75 of *Springer Series in Optical Sciences*, Springer Verlag, 1999.

37 T. Apostolova and Y. Hahn, "Modelling of Laser-induced Breakdown in Dielectrics with Subpicosecond Pulses," *J. Appl. Phys.* **88**, pp. 1024–34, 2000.

38 J. Natoli, L. Gallais, H. Akhouayri, and C. Amra, "Laser-induced damage of materials in bulk, thin-film, and liquid forms.," *Appl Opt* **41** (16), pp. 3156–6, 2002.

39 K. Yamasaki, M. Watanabe, S. Juodkazis, S. Matsuo, and H. Misawa, "Three-dimensional drilling in polymer films by femtosecond laser pulse irradiation," in *Proc. 49th Spring Meeting, Japan Soc. of Appl. Phys.*, p. 1119, 2002.

40 K. Yamasaki, S. Juodkazis, S. Matsuo, and H. Misawa, "Three-dimensional microchannels in polymers: one step fabrication," *Appl. Phys. A* **77**, pp. 371–3, 2003.

41 M. J. Ventura, M. Straub, and M. Gu, "Void channel microstructures in resin solids as an efficient way to infrared photonic crystals," *Appl. Phys. Lett.* **82**, pp. 1649–51, 2003.

42 M. Straub and M. Gu, "Near-infrared photonic crystals with higher order bandgaps generated by two-photon polymerization," *Opt. Lett.* **27**, pp. 1824–6, 2002.

43 M. Straub, M. Ventura, and M. Gu, "Multiple higher-order stop gaps in infrared polymer photonic crystals.," *Phys Rev Lett* **91** (4), p. 043901, 2003.

44 G. Zhou, M. Ventura, M. Vanner, and M. Gu, "Fabrication and characterization of face-centered-cubic void dots photonic crystals in a solid polymer material," *Appl. Phys. Lett.* **86**, p. 011108, 2005.

45 G. Fowles, *Introduction to Modern Optics*, Dover Publications, New York, 1989.

46 H.-B. Sun, Y. Xu, S. Matsuo, and H. Misawa, "Microfabrication and characteristics of two-dimensonal photonic crystal structures in vitreous silica," *Optical Review* **6**, pp. 396–8, 1999.

47 H. Misawa, H.-B. Sun, S. Juodkazis, M. Watanabe, and S. Matsuo, "Microfabrication by femtosecond laser irradiation," in *Proc. SPIE*, **3933**, pp. 246–59, 2000.

48 M. Watanabe, S. Juodkazis, S. Matsuo, J. Nishii, and H. Misawa, "Crosstalk in photoluminescence readout of three-dimensional memory in vitreous silica by one- and two-photon excitation," *Jpn. J. Appl. Phys.* **39**, pp. 6763–7, 2000.

49 K. Miura, J. Qiu, H. Inouye, T. Mitsuyu, and K. Hirao, "Photowritten optical waveguides in various glasses with ultrashort pulse laser," *Appl. Phys. Lett.* **71**, pp. 80–2, 1997.

50 K. M. Ho, C. T. Chan, C. M. Soukoulis, R. Biswas, and M. Sigalas, "Photonic band gaps in three dimensions: New layer-by-layer periodic structures," *Solid State Commun.* **89**, pp. 413–6, 1994.

51 M. Deubel, G. von Freymann, M. Wegener, S. Pereira, K. Busch, and C. Soukoulis, "Direct laser writing of three-dimensional photonic-crystal templates for telecommunications.," *Nat. Mater.* **3** (7), pp. 444–7, 2004.

52 G. Witzgall, R. Vrijen, E. Yablonovitch, V. Doan, and B. Schwartz, "Single-shot two-photon exposure of commercial photoresist for the production of three-

dimensionalstructures," *Opt. Lett.* **23**, pp. 1745 –7, 1998.

53 S. Noda, A. Chutinan, and M. Imada, "Trapping and emission of photons by a single defect in a photonic bandgap structure," *Nature* **407**, pp. 608–10, 2000.

54 S. Noda, "Three-dimensional photonic crystals operating at optical wavelength region," *Physica B* **279**, pp. 142–9, 2000.

55 J. Pendry and A. Mackinnon, "Calculation of photon dispersion relations," *Phys. Rev. Lett* **69**, p. 2772, 1992.

56 J. B. Pendry, "Calculating photonic bandgap structure," *J. Phys. Cond. Matt.* **8**, pp. 1085–108, 1996.

57 O. Toader and S. John, "Proposed square spiral microfabrication architecture for large three-dimensional photonic band gap crystals.," *Science* **292** (5519), pp. 1133–5, 2001.

58 O. Toader and S. John, "Square spiral photonic crystals: robust architecture for microfabrication of materials with large three-dimensional photonic band gaps.," *Phys. Rev. E* **66**, p. 016610, 2002.

59 A. Chutinan and S. Noda, "Spiral three-dimensional photonic-band-gap structure," *Phys. Rev. B* **57**, pp. R2006–8, 1998.

60 S. Kennedy, M. Brett, O. Toader, and S. John, "Fabrication of tetragonal square spiral photonic crystals," *Nano Letters* **2**, pp. 59 –62, 2002.

61 S. Kennedy, M. Brett, H. Miguez, O. Toader, and S. John, "Optical properties of a three-dimensional silicon square spiral photonic crystal," *Photonics and Nanostructures* **1**, pp. 37–42, 2003.

62 M. Jensen and M. Brett, "Square spiral 3D photonic bandgap crystals at telecommunications frequencies," *Opt. Express* **13**, pp. 3348–54, 2005.

63 L. Cai, X. Yang, and Y. Wang, "All fourteen bravais lattices can be formed by interference of four noncoplanar beams," *Opt. Lett.* **27**, pp. 900–2, 2002.

64 L. Z. Cai, X. L. Yang, and Y. R. Wang, "Formation of a microfiber bundle by interference of three noncoplanar beams," *Opt. Lett.* **26**, pp. 1858–60, 2000.

65 L. Cai, X. Yang, and Y. Wang, "Formation of a microfiber bundle by interference of three noncoplanar beams," *Opt. Lett.* **26**, pp. 1858–60, 2001.

66 X. Yang and L. Cai, "Wave design of the interference of three noncoplanar beams for microfiber fabrication," *Opt. Commun.* **208**, pp. 293–7, 2002.

67 C. Ullal, M. Maldovan, E. Thomas, G. Chen, Y.-J. Han, and S. Yang, "Photonic crystals through holographic lithography: Simple cubic, diamond-like, and gyroid-like structures.," *Appl. Phys. Lett.* **84**, pp. 5434–36, 2004.

68 D. Sharp, A. Turberfield, and R. Denning, "Holographic photonic crystals with diamond symmetry," *Phys. Rev. B.* **68**, p. 205102, 2003.

69 X. Wang, J. Xu, H. Su, Z. Zeng, Y. Chen, H. Wang, Y. Pang, and W. Tam, "Three-dimensional photonic crystals fabricated by visible light holographic lithography," *Appl. Phys. Lett.* **82**, pp. 2212–14, 2003.

70 V. Berger, O.Gauthier-Lafaye, and E. Costard, "Photonic band gaps and holography," *J. Appl. Phys.* **82**, pp. 60–4, 1997.

71 T. Kondo, S. Matsuo, S. Juodkazis, and H. Misawa, "Femtosecond laser interference technique with diffractive beam splitter for fabrication of three-dimensional photonic crystals," *Appl. Phys. Lett.* **79**, pp. 725–7, 2001.

72 T. Kondo, S. Matsuo, S. Juodkazis, V. Mizeikis, and H. Misawa, "Multiphoton fabrication of periodic structures by multibeam interference of femtosecond pulses," *Appl. Phys. Lett.* **82**, pp. 2758–60, 2003.

73 T. Kondo, K. Yamasaki, S. Juodkazis, S. Matsuo, V. Mizeikis, and H. Misawa, "Three-dimensional microfabrication by femtosecond pulses in dielectrics," *Thin Solid Films* **453-454**, pp. 550–6, 2004.

74 Y. Nakata, T. Okada, and M. Maeda, "Fabrication of dot matrix, comb, and nanowire structures using laser ablation by interfered femtosecond laser beams," *Appl. Phys. Lett.* **81**, pp. 4239–41, 2002.

75 Y. Nakata, T. Okada, and M. Maeda, "Nano-sized hollow bump array generated by single femtosecond laser pulse,"

Japanese J. Appl. Physics **42**, pp. L1452–54, 2003.

76 H.-B. Sun, A. Nakamura, S. Shoji, X.-M. Duan, and S. Kawata, "Three-dimensional nanonetwork assembled in a photopolymerized rod array," *Adv. Mater.* **15**, pp. 2011–2014, 2003.

77 H. Segawa, S. Matsuo, and H. Misawa, "Fabrication of fine-pitch tio2-organic hybrid dot arrays using multi-photon absorption of femtosecond pulses," *Appl. Phys. A* , pp. 407–409, 2004.

78 I. Divliansky, A. Shishido, I.-C. Khoo, T. Mayer, D. Pena, S. Nishimura, C. Keating, and T. Mallouk, "Fabrication of two-dimensional photonic crystals using interference lithography and electrodeposition of CdSe," *Appl. Phys. Lett.* **79**, pp. 3392–94, 2001.

79 I. Divliansky, T. Mayer, K. Holliday, and V. Crespi, "Fabrication of three-dimensional polymer photonic crystal structures using single diffraction element interference lithography," *Appl. Phys. Lett.* **82**, pp. 1667–69, 2003.

80 G. Schneider, J. Murakowski, S. Venkataraman, and D. Prather, "Combination lithography for photonic-crystal circuits," *J. Vac. Sci. Technol. B* **22**, pp. 146–51, 2004.

81 A. Chelnokov, S. Rowson, J.-M. Lourtioz, V. Berger, and J.-Y. Courtois, "An optical drill for the fabrication of photonic crystals," *J. Opt. A: Pure Appl. Opt.* **1**, pp. L3–L6, 1999.

82 Y. V. Miklyaev, D. C. Meisel, A. Blanco, G. von Freymann, K. Busch, W. Koch, C. Enrich, M. Deubel, and M. Wegener, "Three-dimensional face-centered-cubic photonic crystal templates by laser holography: fabrication, optical characterization, and band-structure calculations," *Appl. Phys. Lett.* **82**, p. 1284, 2003.

11

Photophysical Processes that Lead to Ablation-free Microfabrication in Glass-ceramic Materials

Frank E. Livingston and Henry Helvajian

"Force without wisdom falls of its own weight"

[1]

Abstract

Glass-ceramics (GC) represent an important and versatile class of materials with an application base that continues to grow as multi-functional properties are integrated into these formerly passive materials. Ceramics have now replaced the printed circuit board in most miniaturized wireless electronics, and glass materials will likely lead the way for the development of all-photonic information service units of the future. Adding dopant compounds can alter the material properties of GC systems, and this feature provides the potential to tailor a specific functional property on either a local or a global scale. Given these features, it is surprising that glass-ceramic materials have not been utilized on a broader scale.

Ultra fast laser techniques have been successful in showcasing the potential of laser ablation processing of glass and ceramic materials, especially in fabricating precision microstructures. Although this is a notable achievement, the laser ablation process is an inherently slow serial approach when applied to manufacturing. The application of a pure and fleeting energy form, like a laser pulse, to generate plasmas for material ablation is also an inefficient process. However, the cooperation and interplay between specific matrix agents can help to facilitate a photophysical event. With these cooperative principles in mind, The Aerospace Corporation began, in 1994, to investigate a cost-effective and versatile approach to the processing of glass-ceramic materials. There is a subclass of glass-ceramic materials commonly labeled as photostructurable glass-ceramics, photocerams, or photositalls. These materials are manufactured in the glass phase and contain a photosensitizer that permits the controlled devitrification of the material following ultraviolet (UV) radiation. A key aspect of this photo-induced phase transformation process is that the resulting crystal is soluble in a dilute hydrofluoric acid (HF). The processing is similar in nature to a positive type photoresist; a pulsed UV laser can be used to pattern the regions that are to be removed. However, the photoceramic material differs from a photoresist material in that the entire material can act as a photoresist, not just the overlayer section, and that very high aspect ratio (30:1)

3D Laser Microfabrication. Principles and Applications.
Edited by H. Misawa and S. Juodkazis
Copyright © 2006 WILEY-VCH Verlag GmbH & Co. KGaA, Weinheim
ISBN: 3-527-31055-X

and millimeter-scale structures, can be fabricated. The primary advantage of the photostructurable material is that it permits the merging of two processing techniques: laser patterning and batch chemical etching. This combination establishes a processing approach whereby the laser's unique properties of directed energy and wavelength are used to advantage, without sacrificing overall processing speed.

This chapter summarizes our investigations concerning the photophysical processes that are related to the laser-induced excitation and subsequent devitrification of a commercially available photostructurable glass-ceramic. Optical spectroscopy and microscopic structural analysis techniques have been used to examine the deposition of laser energy and the utilization of this energy to form nanoparticles that induce the precipitation of a metastable crystalline phase. The etching rates of exposed samples have also been measured, and from these results we have established two laser processing models: one that describes the photophysical excitation process and the subsequent energy transformations, and the second that describes the efficacy of the chemical etching process. We also present representative microstructures that showcase the possibilities of laser microfabrication of glass-ceramic materials via cooperative photophysical events.

11.1
Introduction

Over the last decade, micro and nanofabrication technologies have progressed at remarkable rates. These material processing technologies, which were once regarded primarily for their utility in the manufacture of microelectronics, have recently gained attention due to their impact on the development of microsystems or microelectronmechanical systems (MEMS) and nanosystems. Based on the significant growth of these exciting new technologies and the need to optimize functionality at the quantum level, there must be a readily achievable means for true three-dimensional (3D) patterning and structure fabrication. The optimal approach would be to utilize a material that can be specifically "engineered" for 3D processing and which permits the formation of high-fidelity features with minimal collateral damage to the host substrate.

Materials in which the properties can be tailored by the variation of slight compositional changes nominally acquire a strategic commercial advantage, since the base material can be refined to high quality standards and new applications can be realized with "variations on the theme" [2]. By the inclusion of photo-initiator compounds, along with the ability to pattern lithographically, it then becomes feasible to locally alter a material property and thereby enable the cofabrication of a variety of functionalities (e.g., electronic, magnetic, optical, mechanical compliance) on a common substrate. Semiconductor-grade crystalline silicon is one example of a material that can now be produced nearly defect free. However, the electronic properties can be dramatically altered by the inclusion of a small amount of a dopant admixture. The combination of selective oxidation and patterned doping facilitated the ability to control current flow in two-dimensional patterns and spawned the microelectronics industry.

Materials in which the properties can be "engineered" through photolithographic possessing can facilitate significant technological developments. The promise is the capability to cofabricate a diverse array of operational subsystems on a common substrate. Each subsystem would derive its function from a specific property of the shared base material, yet enable the ensemble to perform as an integrated device with near seamless boundaries. An integrated system that is designed around a common base material will be crucial to the further miniaturization of the next-generation of instruments, and could further promote the integration of photonics (controlling light), bionics (controlling fluids) and wireless communications (controlling RF) technologies with microelectronics (controlling current) and MEMS (controlling inertial motion).

There are several approaches to the realization of a successful photo-activated engineered material. One approach is to employ a base material where the effects of compositional variations on the material properties are well understood and can be accurately predicted by calculations. For example, silicon is a material where technical innovation could produce a variant that has chemical and physical properties that are locally alterable by activation. A second material-manufacturing approach involves material growth by the sequential addition of functionalized units. Examples of such materials include the use of functionalized nanoparticles in the development of sintered complex glasses and novel metal alloys [3] and the use of functionalized inorganic-organic hybrid polymers (Ormisils [4] and Ormocer™[5]) in the fabrication of application-specific microstructures [6]. A third approach is to utilize the thermodynamic properties of doped glass-ceramics (GCs) [7] to induce local phase changes that result in desirable material attributes. The use of glass-ceramics as a multi-functional material is the subject of this paper.

GC materials are a unique material class which combines the special properties of sintered ceramics with the characteristics of glasses. Glass ceramics are manufactured in the amorphous homogenous glass state and can be transformed to a composite material via heat treatment and the subsequent controlled nucleation and crystallization of the glass-ceramic constituents. Since their invention nearly 60 years ago, the GC materials have been used in a wide range of scientific, industrial, and commercial applications. GC materials are particularly well suited for aerospace engineering, biotechnology, and photonics, due to their attractive chemical and physical properties (e.g., no porosity, optically transparent, high-temperature stability, limited shrinkage, corrosion resistance, and biocompatibility). A key advantage which has contributed to the widespread use and success of glass-ceramics, is the ability to alter the chemical and physical properties of the material through controlled variation of the properties of the amorphous glass and the incorporated crystalline ceramic phase. The physical property changes that have enabled new application areas include modifications to the material strength, density, thermal conductivity, maximum temperature of operation, electrical insulation and RF transmission, color and transparency in the optical wavelengths, and susceptibility to chemical etching. However, many new exciting applications of GC systems are yet possible with the likelihood of incorporating "intelligence" in the form of micro- and opto-electronics.

Since the thermal treatment step is used to induce the phase transformation, changes in the physical properties of most GC materials occur globally. However, there is a subclass of the traditional GC material family, referred to as *photostructurable* or *photosensitive* glass-ceramics (PSGCs), in which the material transformation can be controlled locally by a photoexcitation step, rather than a thermal treatment step. This photochemical process is accomplished by the addition of photoinitiator compounds to the base glass-ceramic matrix. This attribute permits the two- and three-dimensional (2D and 3D) micro-shaping and micro-structuring of PSGC materials via optical lithographic patterning and chemical etching processes. Despite the inclusive heat treatment step, the material transformation is strongly confined to the photoexposed areas. The physical and chemical properties that can be currently controlled by optical excitation include changes in the optical transmission, material strength, and susceptibility to chemical etching. Consequently, the characteristics of the PSGC material, along with the current understanding of the photophysical processes, offer hope that perhaps additional material properties can be similarly controlled.

To realize a material where multiple material properties can be selectively fixed, either local control of the material constituent matter must be achieved or the photo-initiation process must be wavelength-sensitive to allow the site-selective activation of the desired functional attributes. The former approach presents difficulties in the manufacturing of the base glass, while the latter approach requires a detailed understanding of the photophysical processes that affect a significant change in the material property. Recently, considerable effort has been focused on elucidating photo-induced effects in the chalcogenide glasses [8]. For example, in the amorphous chalcogen (i.e., non-oxide S, Se, Te) glasses, there is a photochemical effect that induces metastability in the glass network structure. The result is a change in the band gap and refractive index [9]. Since this process is reversible, the photo-induced effect has been applied to the development of read-write optical memory and data storage. Additional studies in the chalcogenide glasses have revealed that Ag^+ ion mobility can be enhanced as a result of photo-exposure. Silver-rich chalcogenide glasses that contain more than 30 at.% silver can be prepared by photolytic action, and have been evaluated as a solid-state electrolyte for battery applications [10].

In comparison with the chalcogenide materials, there has been relatively little recent progress to understand the active photophysical processes in PSGC materials. Many of the investigations were performed in the 1960s and 1970s with non-intense and incoherent light sources, for applications to mitigate the effects of nuclear and X-ray radiation on glasses [11]. Today, lasers offer the ability to induce photoexcitation by either single or multiple-photon excitation events and the subsequent photophysical processes could be considerably more complex. To our knowledge there have been three commercially prepared PSGCs in the past thirty years: Fotoform manufactured by the Corning Glass Corporation, Corning New

York, USA; PEG3 synthesized by the Hoya Corporation of Tokyo, Japan; and Foturan™ manufactured by the Schott Corporation of Mainz, Germany. Foturan™ is the only PSGC that is commercially available today.

We believe that the commercially produced PSGCs and their variants can be implemented in a diverse array of new applications in which the photostructurable glass-ceramic material acts as the host substrate or support structure and the desired "instrument" is fabricated in the interior (i.e., embedded) or on the surface of the active substrate. Instrument or device fabrication can be accomplished by direct patterning and the subsequent local alteration of the material property. Alternatively, the PSGC substrate can serve as a traditional support platform that is similar to a multi-chip module (MCM) and can be utilized for the direct attachment of microelectronics, photonics, fluidic MEMS, micro-optoelectronic systems (MOEMS), and high-frequency RF communication systems. In this latter application, the PSGC material is used as a multi-purpose substrate that surpasses the capabilities of other current substrate materials (e.g., silicon and low-temperature co-fired ceramics). The consequence is the potential development of complex functional systems such as small mass-producible satellites that are constructed almost entirely out of glass-ceramic materials. An interesting implication is that more fully integrated micro-instruments can be fabricated where intra-system communication is accomplished by photonics, thus obviating the need for electronic vias.

In this paper we present the known photophysical and photochemical processes of a particular photostructurable glass-ceramic material trademarked as Foturan. The data will demonstrate that, by a careful examination and detailed understanding of the relevant photophysical processes, it is possible to locally and precisely alter the PSGC chemical etching rate, the material strength, and the optical and infrared transmission wavelengths. Based on the ability to control these three material properties, we provide examples of several structures and devices that can be fabricated with true 3D control.

This paper is divided into the following sections. Section 11.1 serves as the Introduction, Section 11.2 formally presents the GC and PSGC materials and describes the traditional processing approaches and the respective limitations. Section 11.3 describes the ultraviolet (UV) laser direct-write processing technique, the measured optical spectroscopy results and the relevant photophysics. Section 11.3 also includes information on the chemical etching process. Section 11.4 presents representative 3D structures that can be processed and fabricated using the UV laser techniques. Finally, Section 11.5 includes the conclusion and a discussion of future directions and impacts of laser processing of PSGC materials.

11.2
Photostructurable Glass-ceramic (PSGC) Materials

Glass can be described as a disordered infinite network structure [12] that is commonly based on a 4-bond coordinated element (e.g., Si, Ge) or a 3-bond coordinated element (e.g., B, As). A 4-bond coordinated species network forms a 3D interconnected system and includes the oxide glasses. In contrast, a 3-bond coordinated species network, such as the chalcogens, forms a 2D planar network system. In general, glass can be defined by its three main essential components: (a) the network-*formers* (e.g., Si, Ge, B, P) that comprise the backbone of the glass matrix and have a coordination number of 3 or 4; (b) the network-*modifiers* (e.g., Li, Na, K, Ca, Mg) that preserve coordination numbers greater than 6, and (c) the network-*intermediates* (e.g., Al, Nb, Ti) that can either reinforce the network (coordination number of 4) or weaken the network (coordination numbers of 6–8) [13]. The network-intermediate species cannot form a glass. Minor constituents also appear in concentrations representative of a dopant and can serve as photo-initiators or nucleating agents.

To a first approximation and for the oxide glasses, changes in the optical absorption can be correlated with alterations in the strength of the network oxygen bond. These alterations can occur by the inclusion of additives (e.g., modifiers) and chromophore compounds, and the precipitation of colloids or clusters with characteristic resonances. Weakening a network oxygen bond shifts the UV absorption edge to the red. Ultimately, it is the nature of the chemical bonds that determines the physical and the optical properties (e.g., the coefficient of thermal expansion, CTE; the absorption coefficient, a) of the glass, and these properties then dictate the material response to light and laser radiation. For example, a highly connected network system results in a glass with a high transition temperature (T_g) and a low CTE. Finally, the glass state does not represent the lowest energy state of the system. Rather, it is the crystalline or ceramic state in which the glass achieves minimization of its free energy. If the ceramization process is not well-controlled or understood, especially for complex glasses, it will result in the formation of stress centers and physical defects that will affect the material properties.

The glass-ceramics represent a particular glass formulation whereby the crystallization (ceramization) process is controlled by the addition of nucleating agents that act as a precursor to crystallization. In the absence of nucleating agents, random crystallization will occur at the lower energy surface sites. One consequence of random crystallization is that strong physical distortions can result, when two such crystals meet in a plane of weakness. In the presence of internal nucleating agents, crystallization occurs uniformly and at high viscosities with the result that the transformation from glass to ceramic proceeds with little or no deviation from the original shape. The criteria for enacting a controlled crystallization process include maintaining a consistently high nucleation frequency throughout a given volume and the subsequent growth of very small crystallites of uniform dimension. The unique advantage of the glass-ceramics, over conventional ceramics, is the ability to use high-speed plastic forming processes to create complex shapes free of inhomogeneities [14].

The invention of glass-ceramics by S. D. Stookey at Corning Glass Works in the 1950s was the result of serendipity and the ability to recognize an extraordinary event. As described by W. Höland and G. Beall [15], Stookey was focused on the development of a new photographic medium, a photosensitive glass that operated by UV (~ 300 nm) exposure and the subsequent precipitation of silver particles

[16]. The glass was an alkali silicate that contained small concentrations of metals (e.g., Au, Ag, Cu) and photosensitizers (e.g., Ce). The typical development process required the heating of the exposed sample to just above the glass transition temperature $\sim$ 450 °C. One night the furnace overheated to 850 °C and, rather than finding a pool of glass melt, Stookey found a white material that had not changed shape. History says that Stookey accidentally dropped the sample, and from the manner in which it broke, he recognized that the material retained unusual strength. By exchanging silver with titania as the nucleating agent in the aluminosilicate glass, the development of thermally shock resistant lithium disilicate ($Li_2Si_2O_5$) glass-ceramic materials began.

The photostructurable glass ceramics are a subclass of the glass-ceramics. These materials are formed with other additives that delineate the nucleation and crystallization (ceramization) process into two distinct steps. Similar to GC materials, nucleating agents are added but these agents are triggered into action only after an optical excitation event. The key is to engineer a photosensitive glass that operates similarly to the known photographic process. Two well-investigated approaches are: (a) a photosensitive glass based on the formation of metal clusters and colloids; and (b) a photosensitive glass based on partial crystallization in lithium barium silicate systems. Table 11.1 presents examples of compositions for these two cases. For the first case, the photoexcitation process results in the formation of metal clusters and colloids via oxidation and reduction chemistry. The colloid

Tab. 11.1 Glass melt composition for the formation of photosensitive glass (adapted from W. Vogel [13].

Photosensitive glass based on formation of metal clusters		Mass %	Photosensitive glass based on partial crystallization		Mass %
Base glass	SiO_2	< 75	Base glass	Li_2O	5–25
	Na_2 or K_2O	< 20		And/or BaO	3–45
	CaO, PbO, ZnO,	< 10		SiO_2	70–85
	CdO	< 2		Na_2O, K_2O,	small
	Al_2O_3			BeO, MgO,	small
				CaO, SrO	small
				Al_2O_3, B_2O_3	small
Dopants	Cu, Ag, Au	< 0.3	Nucleation	Cu_2O, Ag_2O,	
	CeO_2	< 0.05	agents	or Au_2O	0.001–0.3
				CeO_2	0.005–0.05
Thermal	SnO_2	< 0.2	Sensitizers	F^-	1–3
sensitizers				Cl^-	0.01–0.2
				Br^-	0.02–0.4
				I^-	0.03–0.6
				SO_4^{2-}	0.05–0.1

serves as a protocrystal for heterogeneous nucleation. For the second case, the photoexcitation process results in the formation of an immiscible microphase that is rich in Li or Ba. In the thermal development process, these regions have higher nucleation rates and preferentially form Li or Ba metasilicate phases.

The present study utilized a PSGC material obtained from the Schott Corporation under the trade name Foturan™. Foturan is an alkali-aluminosilicate glass and consists primarily of silica (SiO_2: 75–85 wt%) along with various stabilizing oxide admixtures, such as Li_2O (7–11 wt%), K_2O and Al_2O_3 (3–6 wt%), Na_2O (1–2 wt%), ZnO (<2 wt%) and Sb_2O_3 (0.2–0.4 wt%). The photoactive component is cerium (0.01–0.04 wt% admixture Ce_2O_3) and the nucleating agent is silver (0.05–0.15 wt% admixture Ag_2O). The photo-initiation process (latent image formation) and subsequent "fixing" of the exposure (permanent image formation) proceed via several generalized steps (Eqs. 1–3). Upon exposure to actinic radiation $\lambda < \sim 350$ nm, the nascent cerium ions are photo-ionized resulting in the formation of trapped electrons with defect electronic state absorptions (Eq. 1). These trapped (defect) states correspond to the latent image and have been associated with impurity hole centers and electron color centers. Thermal treatment is then used to convert the latent image into a fixed permanent image. The nucleation process is initiated by the scavenging of the trapped electrons by impurity silver ions as described by Eq. 2. During thermal processing, the atomic silver clusters agglomerate and nucleate to form nanometer-scale Ag clusters as shown in Eq. (3). The formation of metallic clusters corresponds to "fixing" of the exposure and permanent image formation in the glass matrix. The formation of the silver cluster is dictated by the concentration of trapped electrons and silver ions and the nucleation kinetics of both the neutral and ionic silver (Ag^0, Ag^+) species; these processes have been shown to be highly temperature dependent [17]. The growth of the soluble ceramic crystalline phase (lithium metasilicate, Li_2SiO_3) is initiated when the temperature is further increased to $\sim 600\,°C$.

$$Ce^{3+} + h\nu \rightarrow Ce^{4+} + e^- \tag{1}$$

$$Ag^+ + e^- + \Delta H \rightarrow Ag^0 \tag{2}$$

$$xAg^0 + \Delta H \rightarrow (Ag^0)_x \tag{3}$$

Cerium is the most commonly used species for the photoexcitation because the photoactive Ce^{3+} oxidation state can be readily stabilized in the glass matrix. The Ce^{3+} ion has the $4f^1$ electronic configuration and therefore the ground state is $^2F_{5/2}$. The only allowed f–f transition is to the $J = 7/2$ electronic state and this occurs in the IR region (~ 0.248 eV) [18]. The Ce^{3+} absorption spectrum reveals no discernible absorption features in the visible wavelength region. However, several bands are present in the UV that arise from 4f–5d transitions; the crystal field can split the strong ^2D electronic state up to five levels. Unlike the 4f orbitals, the 5d orbitals are exposed to significant interaction with the surrounding atoms and ions. This interaction leads to covalent interactions and a decrease in the energy of the 5d levels. Consequently, the UV spectra for cerium-containing glasses are influenced by the glass composition [19]. In silicate glasses (SiO_4) containing Na_2O, the Ce^{3+} absorption is an asymmetrical band with a single maximum at 314.5 nm (3.94 eV) and a FWHM of ~0.5 eV [20]. In comparison, the Ce^{3+} absorption maxima in borax glass (BO_4), phosphate glass (PO_4), and water (H_2O) are 4.9 eV, 5.3 eV, and 5.5 eV, respectively. Thus, we conclude that the Ce^{3+}–O bond is more covalent (i.e., the bonding electrons are shifted more toward the Ce^{3+} atom) in the silicate system. Consequently, the Si–O bonds near the Ce^{3+} species must be less covalent.

For the silver concentration found in Foturan, the "fixing" temperature is near 500 °C and is below the temperature required for crystallization. At higher temperatures, a new crystalline ceramic phase "precipitates" on the silver clusters. This precipitation process is initiated when the neutral silver atoms agglomerate to form silver clusters with a critical size of ~ 8 nm. Two major ceramic phases can be grown via thermal treatment. For a 75–85 wt% mixture of SiO_2 and a 7–11 wt% mixture of Li_2O, a lithium metasilicate crystalline phase (LMS; Li_2SiO_3) grows at temperatures near 600 °C. The structure is a chain silicate and the crystallization of this compound proceeds dendritically. By reducing the fraction of the silicate to 60.8 mol% and increasing the fraction of Li_2O to 35.6 mol%, the metasilicate phase can be observed at temperatures as low as ~ 440–500 °C. However, homogeneous metasilicate formation at these temperatures requires induction times of ~20 to 150 minutes [21]. At higher temperatures of ~ 700–800 °C, a lithium disilicate (LDS; $Li_2Si_2O_5$) crystalline phase is observed. The LDS crystalline phase is a layered silicate structure [22], and retains a 75% degree of "polymerization" similar to that of Muscovite (mica; $KAl_2(AlSi_3O_{10})(OH)_2$) [23]. This layered morphology facilitates machining.

S. D. Stookey was the first to recognize a unique attribute of the lithium metasilicate crystalline phase [24]. The LMS phase is soluble in dilute aqueous hydrofluoric (HF) acid, and the solubility ratio between exposed and unexposed material approaches 50 for some PSGC formulations [25]. This large etch rate contrast is attributed to the fact that the metasilicate crystallites retain a lower silica content and are more susceptible to HF attack, compared with the residual and more tenacious amorphous aluminosilicate glass [26]. One indication that the crystalline phase is not identical to the glass composition is the observation that the crystals have a dendritic or skeletal form. When a glass melt crystallizes to a phase identical in composition to the base glass, the crystals are either well faceted (euhedral) or non-faceted (anhedral) but not dendritic or spherule in morphology. Figure 11.1 shows a transmission electron microscope (TEM) image of the dendritic crystallites measured after laser processing and thermal treatment of Foturan. The TEM sample was thinned, polished and slightly chemically etched in 5.0 vol% HF/H_2O.

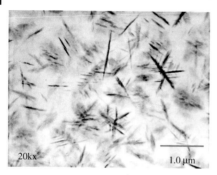

Fig. 11.1 TEM image of UV laser exposed and thermally processed Foturan. The dendritic morphology of the lithium metasilicate crystals is shown after chemical etching.

Utilizing the large etch rate contrast to advantage, Stookey was first to demonstrate that high precision parts could be fabricated from certain formulations of photosensitive glass. The fabrication steps that were developed to process PSGC materials included the following:

1. Exposure using a high power UV lamp that has strong emission from 300 to 350 nm.
2. Use of masks and lithography for patterning.
3. Thermal processing with a specific temperature protocol.
4. Chemical etching in dilute (5–10%) HF.

There are many PSGC compositions, but the most commercially successful PSGCs have a nonstoichiometric composition near the lithium disilicate system (e.g., phyllosilicate crystals, $Li_2Si_2O_5$). Nonstoichiometry implies a SiO_2:Li_2O molar ratio that deviates from 2:1. Table 11.2 provides the general composition of the PSGC Foturan that was obtained from the manufacturer's literature. The actual composition is considered a trade secret.

Tab. 11.2 Constituent compounds in the PSGC, Foturan.

	Constituent compound	Weight %
Base glass	SiO_2	75–85
	Li_2O	7–11
	K_2O and Al_2O_3	3–6
	Na_2O	1–2
	ZnO	<2
Nucleating agents	Ag_2O	0.05–0.15
	Ce_2O_3	0.01–0.04
	Sb_2O_3	0.2–0.4

In the Foturan glass formulation, potash (K_2O) and alumina (Al_2O_3) are added to stabilize the glass by increasing viscosity at the liquid–solid transition temperature during the forming process. Antimony oxide (Sb_2O_3) is a thermal sensitizer added to enhance metal cluster formation and improve nucleation efficiency. The Sb_2O_3 refining agent ultimately improves the resolution of the "developed" exposed image.

This chapter focuses on the active photophysical processes and the role of the photosensitizer in the UV laser processing of PSGC materials. Based on the concentrations listed in Table 11.2, the major constituent that is sensitive to UV light corresponds to the addition of the rare earth compound cerium oxide, Ce_2O_3. Other listed compounds may promote UV absorption, but only at higher concentrations. Both Al_2O_3 and sodium oxide (Na_2O) are known to show absorption bands in SiO_2–R_xO_y systems. For example, absorption features ranging from the UV to the IR have been noted in SiO_2–Na_2O glasses at concentrations of 26–46 mol% Na_2O, and specific bands at 310 nm (4 eV) and 520 nm (2.4 eV) have been identified for concentrations > 35 mol% Na_2O [27]. The addition of Na_2O to silica helps to reduce light scattering (~ 13%) by encouraging relaxation of the structural material [28], and suppresses another UV absorption feature that is associated with the addition of Al_2O_3.

In Na_2O–SiO_2–Al_2O_3 systems, the addition of alumina to silica creates the presence of nonbridging oxygen sites that are characterized by a UV absorption feature at 365 nm (3.4 eV). The Na_2O:Al_2O_3 ratio affects the concentration of nonbridging oxygen centers. When the ratio is unity, all of the nonbridging oxygen atoms are bound into the AlO_4 tetrahedron and a network of SiO_2–AlO_4 is formed [29]. Unfortunately, this ratio cannot be calculated for Foturan without knowledge of the exact concentrations. We presume the ratio to be near unity as a result of the measured absorption spectrum. Figure 11.2 shows the optical absorption spectra acquired for native (unexposed) Foturan samples with cerium (c-Foturan) and without cerium (nc-Foturan). The absorption band of cerium is also shown and was derived from the difference between the c-Foturan and nc-Foturan spectra. The strong absorption in the UV (~ 252 nm, 4.9 eV) cannot be attributed to the optical absorption edge as that occurs at >7 eV (< 177 nm) for amorphous silica [30]. Previous studies have indicated that the defect centers in glassy and crystalline SiO_2 are similar, except under random orientation conditions. Thus, the strong absorption at 250 nm may be attributed to impurities in the sand used for the silica [31], or the result of optically active defects (e.g., oxygen di-vacancy, 4.95 eV; di-coordinated silicon, 3.15 eV [32]; oxygen excess center, 4.8 eV) [33]. However, experimental evidence does exist on a specific lithium metasilicate glass composition (79.29 wt% SiO_2, 11.61 wt% Li_2O, 7.2 wt% Al_2O_3, 2.74 wt% Na_2O, 4.16 wt% K_2O, 0.18 wt% $AgNO_3$, 0.4 wt% Sb_2O_3, 0.07 wt% SnO, 0.065 wt% $CeCl_3$) that attributes the band absorption at 4.8 eV to the material band edge [34].

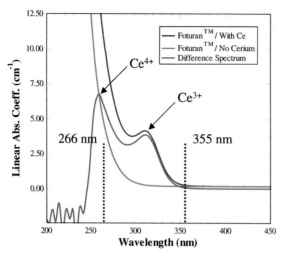

Fig. 11.2 Optical absorption spectra acquired for native Foturan samples with cerium (blue) and without cerium (red). The absorption spectrum of cerium (green) was derived from the difference between the cerium and no-cerium Foturan samples. The two key laser processing wavelengths are denoted by the dashed lines at $\lambda = 266$ nm and $\lambda = 355$ nm.

In Foturan, the intended photosensitizer compound is Ce_2O_3 and is added at concentrations of < 0.04 wt%. Cerium has two valence states (3+ and 4+) and it is the Ce^{3+} oxidation state that is UV photoactive [35] and donates a photoelectron [36]. To ensure that the Ce^{4+} oxidation state is a minority constituent, the compound mannitol ($C_6H_{14}O_6$) is sometimes added to the glass batch [37]. However, there is no mention of $C_6H_{14}O_6$ in the constituent materials list for Foturan.

The results in Fig. 11.2 also indicate two spectral features of the chromophore species that are associated with the oxidation state of cerium. The absorption peak near 315 nm (3.94 eV) has been assigned to Ce^{3+}, while the second peak located at 260 nm is close to the value of 242.5 nm (5.11 eV) reported for Ce^{4+} [38]. If the samples had been exposed to UV light, the 260 nm peak could easily have been assigned to Ce^{3++} (Ce^{3+} + hole). The Ce^{3++} center represents a Ce^{3+} species that has lost an electron, but spectroscopic evidence shows that this species is not identical to Ce^{4+} [39]. The Ce^{3++} center appears after UV radiation and is thermally stable to 400 °C, whereas UV irradiation reduces the Ce^{4+} absorption spectrum peak. Using the derived absorption cross-sections [40] for Ce^{3+} (~2.8 × 10^{-18} cm^2) and Ce^{4+} (~2 × 10^{-17} cm^2) and the measured data in Fig. 11.2, it is possible to estimate the respective cerium concentrations at the peak absorptions. These calculations indicate that the cerium (3+) and cerium (4+) concentrations are approximately 1.3 × 10^{18} Ce^{3+} cm^{-3} and 3.1 × 10^{17} Ce^{4+} cm^{-3}, respectively, and yield a total cerium concentration of 1.6 × 10^{18} Ce cm^{-3}. From this analysis, the Ce^{3+}/Ce^{4+} ion ratio is approximately 4:1 and the percentage of cerium that exists in the Ce^{3+} oxidation state is ~ 81%. For Foturan, the cerium is introduced into the mixture by adding Ce_2O_3. Other formulations utilize CeO_2 as the photosensitzer agent; however, with CeO_2 the cerium ion oxidation state ratio is generally poorer (Ce^{3+}/Ce^{4+} = 7:3) [41]. The calculated percentage of Ce^{3+} can be used along with the data of Paul et al. [42] on cerium in borate glass (near-similar covalency to the silicates) to get an independent verification of the total cerium content in Foturan. A comparison yields that the cerium content in Foturan must be less than 0.029% and this value falls within the range noted in Table 11.2.

The aforementioned spectroscopic measurements are useful when processing PSGC materials with low intensity light, but should only be used as a general guide when pulsed lasers are applied. There are two additional aspects that must be considered when processing PSGCs with pulsed lasers. One important issue is the magnitude of multi-photon excitations or multiple absorption events on the photoexcitation process. In practice, a nonlinear change in the photo-induced carrier density is a manageable problem because the laser intensity (fluence or irradiance) can be modulated while the carrier population is monitored via an *in situ* spectroscopic probe. A second and more difficult issue to address concerns the localized heating of the substrate from the intense laser irradiation. In contrast to laser ablative micromachining of dielectrics where the effects (e.g., internal stresses and fracture) of laser-induced heating can be minimized by reducing the incident laser fluence, the effects of laser induced heating in the PSGC materials are more subtle. The photo-generated carriers in PSGCs are mobile and the localized temperature rise that occurs during laser exposure may alter the initial conditions to adversely affect the processing kinetics for nucleation and ceramization. The problem is compounded by the fact, that during the exposure process, the irradiated zone typically receives more than one laser pulse. The additional heat and subsequent perturbation in the kinetics could result in the loss of exposure fidelity. Multiple laser pulse exposure is commonly utilized to compensate for fluctuations in the spatial distribution of the laser intensity and to accommodate potential nonuniformity in the photosensitizer density. In the case of short pulse lasers (i.e. femtosecond pulses) there is also the potential for accessing additional absorption channels via multiple photon excitations. The consequence may be the accumulation of sufficient thermal energy to initiate the unintended commencement of heterogeneous crystallization.

The laser processing regimes that might be affected due to thermal perturbations can be examined by calculating the time required to attain thermal equilibrium. The characteristic time (τ) needed to achieve a thermal steady state is determined by the thermal diffusivity (D_t) of the material and the area of the irradiated zone (ψ^2) and is defined as: $\tau = \psi^2/(4D_t)$. The thermal diffusivity is defined as $D_t = K/\rho c$, where K ($W\,mK^{-1}$) is the thermal conductivity, c ($J\,gK^{-1}$) is the specific heat, and ρ ($g\,cm^{-3}$) is the density [43]. Using the published values [44] for K (1.35 $W\,mK^{-1}$ at $T = 20\,°C$), c (0.88 $J\,gK^{-1}$ at $T = 25\,°C$), and ρ (2.37 $g\,cm^{-3}$), a thermal diffusivity value of $D_t = 6.47 \times 10^{-7}\ m^2\,s^{-1}$ can be calculated. For an irradiated spot diameter of 2 µm, the time taken to achieve thermal equilibrium in Foturan is approximately 1.2 µs. The actual value will be slightly longer since the thermal diffusivity will decrease with increasing temperature. Assuming that the temperature rise and fall times are equivalent, these calculations suggest that different laser processing "windows" might be employed for lasers with repetition rates approaching 400 kHz compared to lasers with repetition rates ≪ 400 kHz. Although this value is considered a high repetition rate for a laser, there are now commercially available lasers that exceed these rates by a factor of 100. However, note that if the processing zone is increased by a factor of 100 (i.e., a factor of 20 increase in laser spot size diameter) there is a commensurate hundredfold decrease in the laser repetition rate (i.e., 4 kHz). Employing repetition rates above this value would necessitate an evaluation of the thermal load from the prior laser pulses.

11.3
Laser Processing Photophysics

The optical patterning and chemical etching techniques associated with pulsed UV laser PSGC material processing are unique when compared with other traditional ceramic processing approaches. Perhaps the most significant outcome of the UV laser direct-write patterning technique is the capability of fabricating structures with true 3D fidelity as opposed to the extruded prismatic shapes or 2.5D features that are formed using mask lithography processing. Figure 11.3 shows a representative example of a true 3D patterned structure that was easily fabricated in PSGC using laser direct-write processing. The structure is a functional counter-rotating double turbine that has been patterned to appear as if it had been fabricated using "cut-glass" techniques. However, the structure is only 1 cm in diameter with tapered blades that measure 200 micrometers in height and 20 micrometers in thickness at the narrowest point. The floor pattern between the blades has been obliquely patterned to give the cut-glass effect. In addition, the articulated design surrounding the center shaft hole is comprised of a repeating pattern of cone pyramids where the height is sequentially decreased as the center hole is approached. The spacing and nominal peak height for the cone pyramid pattern is 50 micrometers. No masks were used to pattern the double turbine, and the articulate 3D shapes are the direct result of utilizing the dependence of the chemical etching rate on the laser irradiation exposure dose.

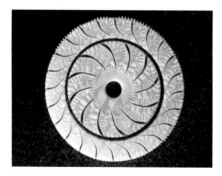

Fig. 11.3 A counter-rotating double turbine structure that was patterned and fabricated in a PSGC material via UV laser direct-write processing.

The structure displayed in Fig. 11.3 cannot easily be fabricated using standard mask lithography processing techniques. Mimicking the realized 3D fidelity would require more than ten masking step operations. The capability for fabricating true 3D structures in glass ceramic material without the need for a milling machine or laser ablation has been the most significant advantage of laser patterning PSGC materials.

Several groups have investigated the photophysical processes and the applications of PSGC materials exposed by pulsed laser irradiation. Nearly all of the investigations have utilized the Schott product Foturan; there is much less data on a similar glass formulation manufactured by Hoya (PEG3). One general conclusion that has been derived from these investigations is that the laser processing wavelength need not fall within the absorption band of the photosensitizer to promote efficient exposure and metasilicate formation. PSGC material containing Ce^{3+} that has a peak absorption at 314.5 nm (3.94 eV, FWHM ~ 0.5 eV) has been successfully exposed using nanosecond pulse lasers at 193 nm (6.42 eV) [45], 248 nm (5.0 eV) [46], 266 nm (4.66 eV) [47], and 355 nm (3.49 eV) [48] and with femtosecond pulse lasers at 775 nm (1.6 eV) [49] and 800 nm (1.55 eV) [50]. These results suggest that, either additional photoelectron donor species are present that can initiate exposure, or multiple photon absorption processes are in effect. The data will show that both processes are operating in the case of Foturan. Using an ultrafast laser, Masuda et al. [51] has shown that the exposure to induce chemical etching in Foturan proceeds by a 6-photon absorption process at 775 nm. In a different experiment that measured the change in the material transmittance, Kim et al. [52] have shown a 3-photon dependence for an ultrafast laser operating at 800 nm. Finally, Aerospace experiments using nanosecond lasers have shown that, for both 266 nm and 355 nm wavelengths, the *threshold* exposure required for the formation of a connected network of etchable crystalline phases has a quadratic dependence on the single shot laser fluence [53]. Figure 11.4 shows the results of a parameterized model that was developed to relate the threshold dose that is required to initiate chemical etching (D_c) to the laser fluence (F) and the number of laser shots (n) at 355 nm. The data could be fit to the equation $D_c = F^m n$ with a quadratic dependence on the fluence. A similar dependence was measured for laser irradiation at 266 nm [54].

In an analogous subsequent experiment that measured the critical dose threshold needed for chemical etching, Sugioka et al. determined a fluence dependence of 1.5 for a nanosecond pulsed laser operating at 308 nm (in-band laser excitation) which is near the peak of the chromophore absorption peak [55]. Previous experiments which were conducted with a cw in-band UV lamp showed a linear photon dependence for the formation of an etchable crystalline phase [56].

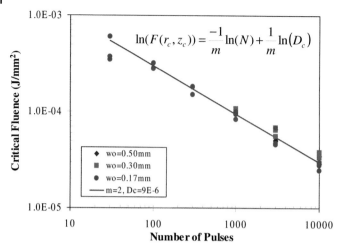

Fig. 11.4 Data representing the critical fluence necessary for initiating chemical etching as a function of laser fluence and laser pulse number. The fluence dependence for $\lambda = 355$ nm is quadratic, $m = 2$. Three data sets are plotted and represent three different spots sizes, w_o.

At first glance, the composite data suggest that the laser processing of Foturan in the UV-VIS regions can be described by a multiple-photon excitation scheme. The measurement of a 6-photon dependence suggests that the nonlinear excitation process is likely a stepwise excitation process as opposed to a simultaneous excitation event, which would yield a very small cross-section. However, upon more careful inspection there appears to be a discrepancy in the fluence or photon dependence results for the ultrafast laser studies. To resolve this discrepancy, one must first ask what is being measured and how does it relate to the fundamental excitation process. For all of the laser experiments, each irradiated spot size receives multiple laser pulses. Previous studies have revealed that the photosensitive glass absorption is dependent on both the incident laser fluence and the acquired number of laser pulses [57]. Due to these dynamic changes in the material absorptivity, it is difficult to describe the multiple-photon results as a measurement of the initial excitation of the virgin sample. To elucidate the fundamental excitation process, experiments must be performed where the absorption properties are measured on a shot-by-shot basis. Cavity ring-down experiments are currently being conducted at The Aerospace Corporation to address these issues.

The apparent conflict between the various photon dependence results can be partly ameliorated by examining the specific laser parameters used in each experiment. The incident laser fluences ranged from 17–71 mJ cm^{-2} for Masuda et al. [58], 500–1000 mJ cm^{-2} for Kim et al. [59], and 3–30 mJ cm^{-2} for the Aerospace studies [60]. Clearly, the photon dependence studies correspond to a wide range of incident laser fluences and total absorbed energy conditions. The photon dependence data correspond to a total energy absorbed of 9.6 eV for Masuda et al.,

4.65 eV for Kim et al., and 7–9.3 eV for the Aerospace experiments. The correlation between the Aerospace and Masuda et al. experiments suggest that the total absorbed energy is related to the excitation fluence. The 3-photon fluence dependence measured by Kim et al. is likely due to a different excitation mechanism that is accessible at the higher incident laser fluences. Although these results do not explicitly reveal the fundamental excitation mechanism for pulsed laser irradiation of PSGC materials, the nonlinear absorption results can be practically applied to fabricate embedded structures [61].

The fluence dependence experiments reveal the photolytic activation of an absorbing species. The results suggest that this species can interact with the incident laser pulse, and is sufficiently long-lived to interact with subsequent laser pulses. Figure 11.5 shows the results of a pulsed UV laser exposure experiment conducted at Aerospace using the PSGC, Foturan [62]. The data represents optical transmission spectra as a function of incident laser irradiance for two laser processing wavelengths ($\lambda = 266$ nm and $\lambda = 355$ nm). The absorption of the unexposed native Foturan glass has been subtracted to reveal the change in absorption due to the applied pulsed UV laser irradiation. The data is plotted in irradiance units rather than in laser fluence units, to reflect the fact that each irradiated spot received an average of 30 laser pulses. Furthermore, the digitized spectra were analyzed and converted to linear absorption coefficient (a) values, where a is defined by $I/I_o = K \exp(-ad)$. K accounts for surface reflections and I/I_o is the fraction of light transmitted by a glass sample of a thickness d. The a can be related to the population density by the relation $a = \sigma n$, where σ is the cross-section for absorption (cm^2) and n is the population density of the absorbing species (species cm^{-3}). Depicting the optical transmission data in these units permits a direct comparison of the data sets for the two laser excitation wavelengths ($\lambda = 266$ nm and $\lambda = 355$ nm).

Figure 11.5 reveals the appearance of absorption features that grow with increasing laser irradiance. For $\lambda = 355$ nm irradiation, the absorption is confined to a narrow band centered at ~ 265 nm and ranges from ~ 250 to 290 nm with a smaller feature located to the red at ~ 350 nm. For $\lambda = 266$ nm excitation, there is a similar absorption feature centered at ~ 280 nm; however, there is also a broad, featureless absorption that extends beyond 460 nm into the visible wavelength region. The peak absorption for $\lambda = 266$ nm laser excitation is nearly twice as large compared with the peak absorption for $\lambda = 355$ nm laser excitation. The absorption features for both excitation wavelengths are associated with trapped electron defect states since the native (unexposed) glass spectra show no such features. The absorption at ~ 265–280 nm is attributed to impurity hole centers $(Ce^{3+})^+$ or electron color centers $(Ce^{4+})e^-$ [63]. Talkenberg et al. has measured similar absorption features and trends [64]. However, in the data shown in Fig. 11.5 there is an increasing positive absorption with increasing laser fluence. Talkenberg et al. also observed this behavior except that they also measured a curious negative induced absorption feature (i.e., optical "bleaching") near ~ 250–260 nm at high laser fluences. The Aerospace experiments do not observe optical bleaching after irradiation at $\lambda = 355$ nm for small number of laser shots; an increase in transmittance near 250 nm was measured for a small (0.007 mW μm^{-2}) laser exposure at $\lambda =$ 266 nm. Aerospace experiments have shown optical bleaching at 355 nm but only after the irradiation of a large number of laser shots. The transmittance values in the wavelength range 240–270 nm are approximately a few percent at these applied irradiances (fluences) and accurate measurement becomes more difficult so that more care is necessary to ensure reliable data. Detailed experiments are now under way at Aerospace to investigate the photobleaching properties of Foturan.

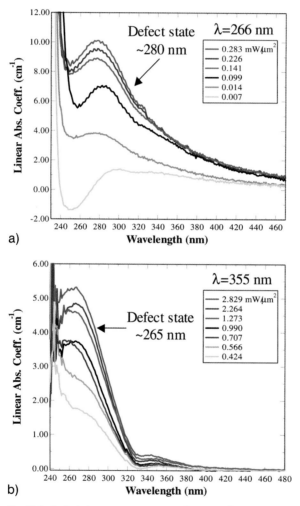

Fig. 11.5 Optical absorption spectra as a function of incident laser irradiance for $\lambda = 266$ nm (a) and $\lambda = 355$ nm (b). The spectra correspond to the Latent Image (exposed) state and are defined as: Latent (exposed) – Native (unexposed).

Regardless of the photobleaching effects, both the Aerospace and Talkenberg et al. data confirm the formation of photo-induced absorbers by pulsed laser exposure. Previous studies on cerium-containing PSGC materials using CW UV lamp sources, have observed the formation of two bands following exposure. Stroud [65] measured an absorption band centered at ~ 245 nm and assigned this feature to the species Ce^{3++} (Ce^{3+}+hole). These studies also measured a spectral band at ~ 270 nm when the Ce^{4+} concentration in the virgin material was sufficiently high. Based on the calculation of a relatively high Ce^{3+}/Ce^{4+} ratio (4:1), our optical spectroscopy results suggest that the formation of trapped (defect) states is not correlated with Ce^{4+} photochemistry. Berezhnoy et al. [66] have also observed a UV-induced absorption band at the 270 nm band and have demonstrated that the absorption is due to trapped electrons in the glass and does not correspond to ion or defect states of cerium or silver. Both Stroud and Berezhnoy et al. observed an additional broad absorption feature centered at ~ 350 nm and trailing to the visible wavelength region. Based on electron spin resonance (ESR) spectroscopy results, Stroud attributed this absorption to trapped photoelectrons and labeled this feature as the f_1 band. The f_1 absorption can be erased optically via exposure to radiation at $\lambda > 350$ nm. Similarly, Berezhnoy et al. demonstrated that the 270 nm absorption band could be erased by thermal treatment at 450 °C.

A recent series of experiments were performed at Aerospace to identify the constituent source of the measured absorption [67]. These studies employed Foturan samples *with* cerium (c-Foturan) and *without* cerium (nc-Foturan). The effectiveness of the Ce^{3+} chromophore in generating photoelectrons was measured at two laser wavelengths ($\lambda = 266$ nm and $\lambda = 355$ nm) that lie above and below the peak absorption band of the photosensitizer (~ 312 nm). Prior to use, the cerium Foturan and noncerium Foturan samples were sent to an independent laboratory (Galbraith Laboratories, Inc., Knoxville, TN) for compositional analysis and cerium content verification. The samples were quantitatively tested using inductively coupled plasma mass spectroscopy (ICP-MS). The cerium content was measured to be ~ 9 ppm and < 64 ppb for the c-Foturan and nc-Foturan samples, respectively.

Figure 11.6 shows the optical absorption spectra for c-Foturan and nc-Foturan that were measured following laser irradiation at $\lambda = 266$ nm and $\lambda = 355$ nm. The incident laser irradiances were 0.283 mW μm^{-2} and 2.829 mW μm^{-2} for $\lambda = 266$ nm and $\lambda = 355$ nm, respectively. The absorption spectra correspond to the latent image state and illustrate the effect of cerium on the laser-induced defect state [68]. Several conclusions can be derived from the spectroscopy data displayed in Fig. 11.6. First, the two spectra measured for the c-Foturan samples show the presence of a 260 – 280 nm absorption feature following laser irradiation at either $\lambda = 266$ nm or $\lambda = 355$ nm. Based on previous studies [69] and the absorption features identified in Fig. 11.5, the 260–280 nm absorption band is correlated with impurity hole centers and electron color centers and is ascribed to photoelectrons generated by the cerium photo-initiator ($Ce^{3+} + h\nu \rightarrow Ce^{4+} + e^-$). It is interesting to note that the absorption peak generated by 266 nm irradiation is red shifted in comparison to that generated by 355 nm. Second, the two spectra measured for the nc-Foturan samples do not show a distinct absorption feature in the range 260–280 nm. Instead, a broad featureless absorption is observed from ~ 240 nm to > 480 nm for $\lambda = 266$ nm and $\lambda = 355$ nm irradiation. This broad absorption feature is associated with trapped photoelectrons that were generated from non-cerium donors (e.g., other admixture compounds, base glass matrix). Finally, laser irradiation at $\lambda = 266$ nm is more efficient at generating this broad absorption compared with laser excitation at $\lambda = 355$ nm. In addition, the integrated absorption of the broad, featureless band (following exposure at $\lambda = 266$ nm) represents a significant fraction of the total laser-induced absorption. This suggests that cerium may play a minor role as a photoelectron donor when processing PSGC materials at $\lambda = 266$ nm.

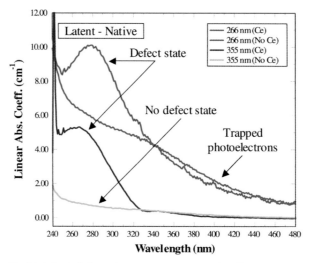

Fig. 11.6 Optical absorption spectra comparing the effect of cerium on the formation of UV laser-induced absorption states.

To examine the specific role and necessity of cerium in the laser processing of PSGCs, an experiment was assembled, which permits the controlled delivery of a precise laser exposure dose to a calibrated PSGC sample. Optical transmission spectroscopy (OTS) and x-ray diffraction (XRD) were used to monitor the sequential changes in the optical absorption of Foturan during the individual processing steps, i.e., to measure the changes in the material from the native glass state, to the exposed state (called the latent image state), to the cluster/colloid formation state, and finally to the growth and formation of the etchable, crystalline lithium metasilicate phase.

Figure 11.7 shows the experimental setup used for the laser exposure studies. The pulsed UV lasers were Q-switched, diode-pumped Nd:YVO$_4$ systems manufactured by Spectra-Physics (OEM Models J40-BL6-266Q and J40-BL6-355Q). Typical laser pulse durations of 6.0 ± 0.5 ns (FWHM) were achieved in the Q-switched

TEM_{00} operation mode. The pulse-to-pulse stability was ± 5.0% at a nominal pulse repetition rate of 10.0 kHz. The irradiation dose was applied to the glass sample using an automated direct-write, laser-patterning tool that included a circular neutral density (CND) filter (CVI Laser Corp., CNDQ-2-2.00) coupled to a micropositioning motor. A raster scan pattern was used to expose a 1.5 mm × 8.0 mm area with a laser spot diameter of 3.0 μm and a XY stage velocity of 1.0 mm s^{-1}. During exposure patterning, the incident laser surface power was controlled in real time by a closed-loop LabVIEW software program that monitored the power at an indicator position in the optical train. This "indicated" power could be related to the incident surface power by applying the appropriate calibration factors [70]. Figure 11.8 shows the CND filter calibration data and the power resolution achievable at the sample surface. A UV grade, achromatic 10× microscope objective (OFR, LMU-10 ×-248) was used as the focusing element.

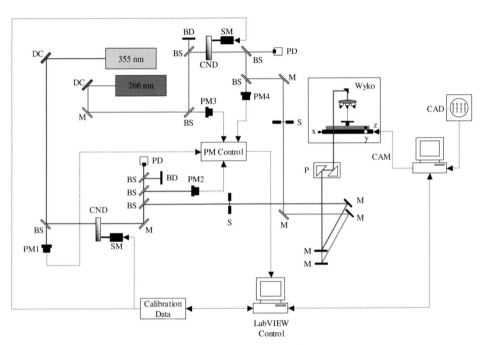

Fig. 11.7 Schematic representation of the experimental layout. BD, beam dump; BS, beamsplitter; CND, circular neutral density filter; DC, dichroic mirror; M, high-reflectance mirror; P, periscope assembly; PD, photodiode; PM, power meter; S, shutter; SM, stepper motor. The XYZ motion system is integrated into a non-contact, white light optical interferometer stage (WYKO/Veeco Corp.) and is controlled by computerized CADCAM software.

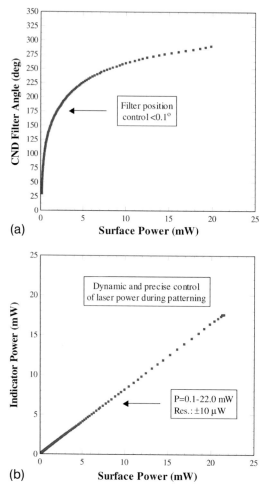

Fig. 11.8 (a): Calibration of the CND filter angle setting versus the incident surface power. (b): Calibration of the indicator (monitor) power versus the incident surface power.

Several Foturan wafers (100 mm diam., 1 mm thick) were cut into 1 cm × 1 cm square coupons and thinned to 200 μm and 500 μm thicknesses. The coupons were then polished to achieve an optically flat finish. The 200 μm and 500 μm coupons were used for the $\lambda = 266$ nm (absorptivity = 10.05 cm^{-1}) and $\lambda = 355$ nm (absorptivity = 0.27 cm^{-1}) studies, respectively. The thicknesses were selected to ensure that the laser penetration depth exceeded the glass sample thickness and to minimize gradients in the exposure volume. The samples were cleaned using a RCA cleaning protocol and were subsequently handled using gloves and clean-room techniques. Six laser irradiance settings were used for the $\lambda = 266$ nm experiments and seven laser irradiance settings were used for the $\lambda = 355$ nm experiments; the total sample size exceeded 150 coupons. The laser irradiances values were chosen to encompass the measured laser processing window [71] which is defined as the range from the minimum threshold irradiance required to initiate chemical etching (etch contrast 1:1) to the irradiance at the saturation limit (etch contrast 30:1).

Optical transmission spectroscopy was performed on the exposed ($\lambda = 266$ nm and $\lambda = 355$ nm) and unexposed (reference) samples using a Perkin Elmer Lambda 900 spectrophotometer. The spectrometer was calibrated prior to each session and a reference (unexposed) spectrum was obtained at the beginning and end of each session. Spectra were acquired between 180 nm and 2.0 μm at a resolution of 2 nm and a scan speed of 240 nm min^{-1}. A 1.0 mm × 6.0 mm mask was used to define the area of analysis. The experiment produced four sets of samples for each exposure level. These samples were analyzed sequentially during the laser and thermal processing steps:

Step 1. The unexposed glass samples (Native state).
Step 2. The exposed *only* glass samples (Latent Image state).
Step 3. The exposed and bake I (500 °C) glass samples (Cluster Formation state).
Step 4. The exposed and bake II (605 °C) glass samples (Etchable Ceramic Phase state).

The baked samples were kept at the temperature plateaus of 500 °C and 605 °C for one hour. The digitized spectra were analyzed and converted to linear absorption coefficient values. In the case of the Etchable Ceramic phase state samples, it was not possible to assign a spectroscopic feature to the lithium metasilicate crystalline phase. Consequently, XRD was utilized to monitor the changes in the PSGC material following laser exposure and bake II thermal treatment.

Figure 11.9 presents the OTS results for Foturan samples that have been processed beyond the Latent Image state and first bake step to the Cluster Formation state. The material has been baked to permit migration of the nascent silver ions to the trapped electron defect site. The results at the two laser exposure wavelengths ($\lambda = 266$ nm and $\lambda = 355$ nm) are shown and the plotted data reflect *the changes* in the spectra from the prior state; i.e., the spectra from the prior state have been subtracted from each displayed spectra.

Comparison of the data shown in Fig. 11.9 with that in Fig. 11.5 (Latent Image state spectra) reveals several noteworthy trends. First, the absorption features associated with the trapped defect state appear as a negative absorption, while a new feature is shown to grow at 420 nm. The feature at 420 nm corresponds to the well-known plasmon absorption band for colloidal silver particles Ag_x [72]. Second, both the depletion of the defect state region and the growth of the cluster band region depend on the incident laser irradiance. Finally, the density of silver clusters formed via $\lambda = 266$ nm laser irradiation is much greater than the density formed with $\lambda = 355$ nm laser irradiation, as determined by the intensity of the 420 nm feature. This difference reflects the formation of more photoelectron carriers for $\lambda = 266$ nm laser excitation than for $\lambda = 355$ nm laser excitation. A comparison of the peak absorption values in Fig. 11.5 supports this conclusion; the absorption coefficients are $a_{266nm} \approx 10$ cm^{-1} and $a_{355nm} \approx 5.5$ cm^{-1} for the highest laser irradiance values. Note that the laser irradiance values related to 355 nm laser processing are nearly 10 times higher than for 266 nm laser processing. However, the absorptivity [73] of the virgin material at 266 nm is nearly 149 times that at 355 nm, which suggests a more efficient conversion of the absorbed laser energy to form photoelectron defects at 355 nm.

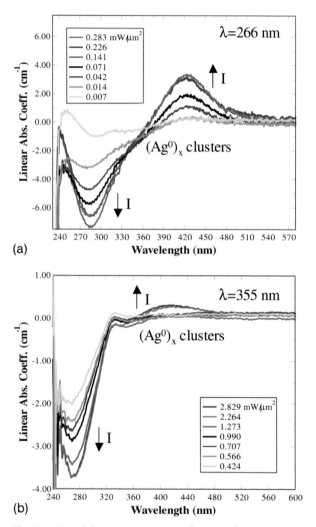

(a)

(b)

Fig. 11.9 Optical absorption spectra as a function of incident laser irradiance for $\lambda = 266$ nm (a) and $\lambda = 355$ nm (b). The spectra correspond to the Cluster Formation state and are defined as: Cluster Formation state (Bake I) – Latent Image state (exposed). The arrows show the change in the spectra with increasing laser irradiance.

The efficiency of the Bake I phase protocol in scavenging labile photoelectrons can be estimated by comparing the data shown in Figs. 11.6 and 11.10. Using the high laser irradiance data as a guide, it is estimated that > 75% of the trapped photoelectron species are neutralized by the nascent silver ions. This value is considered a lower limit since the 260–280 nm band has an absorption feature on the long-wavelength shoulder that grows during the Bake I phase and complicates the analysis for determining the net reduction in the photoelectron population. This absorption feature can be seen as a small bump at ~ 330 nm in the right hand panel of Fig. 11.9, and is attributed to the formation of Ag^0 centers. The energies of the resonance doublet of the free Ag atom are 328 nm (3.78 eV: $5sS_{1/2} \to 5pP_{1/2}$) and 338.3 nm (3.665 eV: $5sS_{1/2} \to 5pP_{3/2}$) and correlate well with the measured absorption.

The general conclusion of the data shown in Figs. 11.6 and 11.10 can be summarized as follows. Photoelectrons are indeed generated at the two laser wavelengths and the population of this defect state appears to "feed" the concurrent growth of the metallic silver clusters. To better elucidate whether the laser photo-induced electrons induce the formation of the metallic clusters, a separate experiment was conducted in which Foturan samples were first exposed at a single constant laser irradiance then individually processed for *progressively longer bake times* at 500 °C. At 500 °C (Bake I phase), there is a strong propensity for cluster formation, but not the precipitation of the chemically soluble metasilicate phase. XRD studies conducted on these samples revealed that the concentration of the metasilicate phase was below the detectable limit. A total of seven bake-time durations were employed and ranged from 10 to 200 minutes.

The optical absorption spectra of the Cluster Formation state following laser exposure at $\lambda = 355$ nm and thermal treatment at 500 °C are shown in Fig. 11.10. The results reveal that the $(Ag^0)_x$ cluster absorption band at ~ 420 nm increases with increasing Bake I time. In contrast, the trapped defect state absorption band (~ 265 nm) is observed initially to decrease with increasing Bake I time. This observation is depicted more clearly in the right panel in Fig. 11.10, which shows the integrated peak areas of the trapped electron state and the $(Ag^0)_x$ cluster species as a function of bake I time. For visual reference, the data corresponding to bake I times that ranged from 10 minutes to 100 minutes have been fit using a linear least-squares regression analysis. The results in Fig. 11.10 (b) indicate that there is a concomitant monotonic decrease in the trapped electron defect state population and an increase in the Ag cluster state population. For Bake I times > 100 minutes, there is a continued increase in the Ag cluster state population. However, the 265 nm absorption band is also observed to increase for Bake I times > 100 minutes. The increase in the 265 nm band is likely associated with the diffusion and coalescence of outlying neutralized Ag atoms to form small Ag clusters (e.g., dimers, trimers, etc.). Preliminary spectroscopic evidence suggests that the increase in the 265 nm absorption band with extended Bake I times corresponds to the growth of $(Ag^0)_2$ species. These results suggest that the 260–280 nm bands and the 420 nm band can be used as spectral monitors to address the relative propensity of the photoelectrons in neutralizing Ag^+ ions to form $(Ag^0)_x$ species.

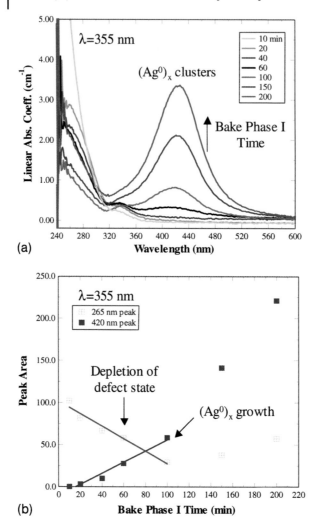

Fig. 11.10 (a) Optical absorption spectra of the Cluster Formation state as a function of Bake I time following laser irradiation at $\lambda = 355$ nm. The incident laser irradiance was 2.829 mW μm^{-2}. (b) The integrated peak areas of the trapped (defect) electron state and the $(Ag^0)_x$ cluster species as a function of Bake I time.

Figure 11.11 shows the optical absorption subtraction spectra that correspond to the silver Cluster Formation state following laser exposure and thermal treatment at 500 °C. The negative absorption values in the 260–280 nm region are indicative of the subtraction process and correspond to the depletion of the trapped photoelectron states during the Bake I phase. The absorption trends that occur at ~ 420 nm characterize the population density of $(Ag^0)_x$ clusters. The results demonstrate that the *exclusion* of the cerium photo-initiator *does not preclude* the formation of $(Ag^0)_x$ clusters for $\lambda = 266$ nm laser irradiation. Clearly, the trapped electrons that are derived from the noncerium photoelectron donors (broad featureless absorption) can reduce the nascent silver ions in the glass matrix. In contrast, a detectable feature at 420 nm is not observed for the nc-Foturan samples following irradiation at $\lambda = 355$ nm and Bake phase I heat treatment; i.e., the density of photoelectrons generated from noncerium donors at $\lambda = 355$ nm is too small to induce measurable silver cluster formation. These results suggest that the cerium photo-initiator is required for $\lambda = 355$ nm laser processing.

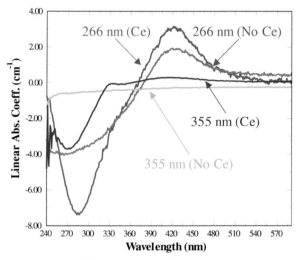

Fig. 11.11 Optical absorption spectra for c-Foturan and nc-Foturan samples following laser irradiation at $\lambda = 266$ nm and $\lambda = 355$ nm and thermal treatment at 500 °C. The spectra correspond to the Cluster Formation state and are defined as: Cluster Formation state (Bake I) – Latent Image state (exposed). The incident laser irradiances were 0.283 mW μm^{-2} and 2.829 mW μm^{-2} for $\lambda = 266$ nm and $\lambda = 355$ nm, respectively.

The cerium photo-initiator efficiency in the pulsed UV laser exposure process becomes readily apparent by summarizing the spectroscopy data in a single graph. Figure 11.12 presents a compilation of the integrated peak areas (i.e., concentration) of the $(Ag^0)_x$ cluster species measured versus laser irradiance at $\lambda = 266$ nm and $\lambda = 355$ nm for c-Foturan and nc-Foturan samples. Several notable trends are apparent. (a) The integrated $(Ag^0)_x$ peak areas show an initial increase with laser irradiance and turn over at higher laser irradiance. This saturation behavior is influenced by the kinetics of Ag diffusion and cluster growth, as well as by our specific thermal treatment protocol. (b) Comparison between the c-Foturan and nc-Foturan results reveals that the saturation behavior begins at higher laser irradiance values for the glass samples without cerium. This observation is likely associated with the generation of fewer photoelectrons in the non-cerium Foturan for a given laser irradiance. (c) The laser irradiance values that are required for $\lambda = 266$ nm processing are a factor of ~10 smaller compared with the laser irradiance values that are necessary for $\lambda = 355$ nm processing. This result is attributed to the large difference in the native glass absorptivity at the two laser excitation wavelengths. (c) Photoelectron generation and silver cluster growth are more efficient with $\lambda = 266$ nm laser excitation than with $\lambda = 355$ nm laser excitation. (d) For $\lambda = 266$ nm laser exposure, $\sim 33\%$ of the $(Ag^0)_x$ clusters are formed via *non-Ce* photoelectron donors. In contrast, for $\lambda = 355$ nm laser exposure nearly all of the $(Ag^0)_x$ clusters are formed via *Ce* photoelectron donors; i.e., the cerium photoinitiator is required for pulsed UV laser processing at $\lambda = 355$ nm.

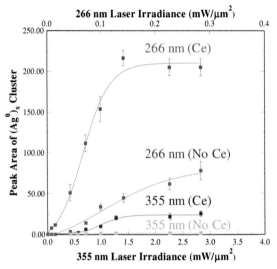

Fig. 11.12 Integrated peak areas of the $(Ag^0)_x$ cluster species versus laser irradiance at $\lambda = 266$ nm and $\lambda = 355$ nm for c-Foturan and nc-Foturan.

Figure 11.13 shows a comparison of the XRD results for c-Foturan and nc-Foturan samples following laser exposure at $\lambda = 355$ nm and thermal treatment through Bake phase II (Ceramic Phase state). The cerium-containing sample exhibits distinct features that are associated with crystalline lithium metasilicate, while the noncerium sample shows no crystalline feature. A distinct difference in the saturation behavior of the silver cluster density is also apparent from the results displayed in Fig. 11.12. The c-Foturan data measured at $\lambda = 266$ nm and $\lambda = 355$ nm show clear saturation of the silver cluster population at high laser irradiances. However, the nc-Foturan samples retained a markedly different rate of Ag cluster formation and saturation ("turn-over") of the Ag cluster density was not observed; this is most readily apparent for the nc-Foturan data corresponding to $\lambda = 266$ nm laser exposure. The formation of the silver clusters is initially dictated by the Ag^+ mobility and the proximity of the Ag^+ ion to a labile electron. After neutralization of the Ag^+ species, the formation of silver clusters is then controlled by the mobility of the Ag^0 species to form clusters. Given that the samples undergo the same thermal processing protocol, the mobility of the Ag^+ and the Ag^0 species should not be affected by the absence of a minor constituent such as cerium (~ 0.01–0.04 wt%). The difference in the saturation behavior of the cerium and noncerium results shown in Fig. 11.12 is intriguing but cannot yet be explained.

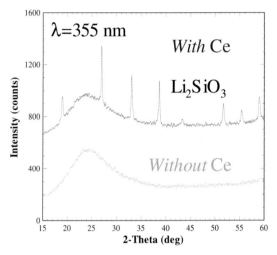

Fig. 11.13 XRD data showing the relative extent of crystallinity of c-Foturan and nc-Foturan samples. The glass samples were exposed to $\lambda = 355$ nm laser irradiation (2.829 mW μm^{-2}) and thermally processed to the metasilicate crystalline phase.

An additional set of Foturan samples were irradiated using calibrated exposures at $\lambda = 266$ nm and $\lambda = 355$ nm, and thermally processed to the Etchable Ceramic Phase state (Bake II). The initial intent was to identify a spectroscopic signature that could be attributed to the metasilicate phase and could be used as an *in situ* monitor during laser material processing. Figure 11.14 shows the subtraction spectra that were measured following Bake step II and crystallization. The data show a significant increase in the absorption value (80–150 cm^{-1}) in the Ag cluster band region, along with the growth of an additional band at ~ 280 nm that has been tentatively assigned to the silver dimer (Ag_2) species. Further analysis is necessary to identify a spectroscopic feature that can be uniquely associated with the growth of the lithium metasilicate phase.

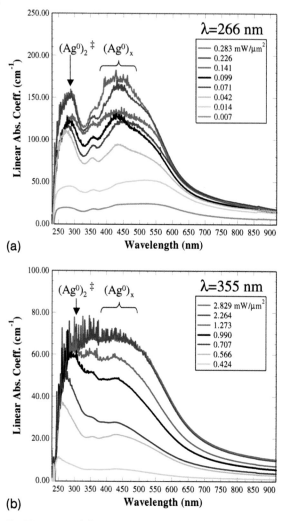

Fig. 11.14 Optical absorption spectra as a function of incident laser irradiance for λ = 266 nm (a) and λ = 355 nm (b). The spectra correspond to the Etchable Ceramic Phase state and are defined as: Etchable Ceramic Phase (Bake II) – Cluster Formation state (Bake I).

XRD analysis was conducted on the Bake phase II (Etchable Ceramic Phase state) samples and the results are displayed in Fig. 11.15. The XRD data correspond to 2-theta scans that ranged from 15 degrees to 60 degrees for λ = 266 nm and λ = 355 nm samples and a lithium metasilicate reference sample. The XRD data show the presence of a broad background feature that is associated with the native amorphous glass, and indicate that the exposed PSGC material is not fully crystalline following the Bake II thermal treatment step. Stookey et al. [74] have estimated that the maximum extent of crystallinity in the exposed volume of Foturan is approximately 40%. This extent of crystallinity is qualitatively consistent with the TEM results measured after UV laser exposure and thermal processing (cf. Fig. 11.1).

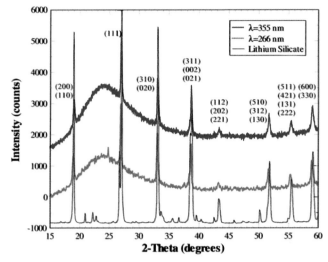

Fig. 11.15 XRD data measured following UV laser exposure at $\lambda = 266$ nm (red) and $\lambda = 355$ nm (blue) and thermal treatment through bake phase II. XRD data obtained from a lithium metasilicate reference sample (black) is also shown. The XRD results have been offset for visual clarity.

Fig. 11.16(a) shows the XRD data corresponding to the <111> diffraction plane as a function of laser irradiance at $\lambda = 355$ nm. Figure 11.16(b) shows the calculated lithium metasilicate crystal diameters as a function of laser irradiance at $\lambda = 266$ nm and $\lambda = 355$ nm. Several conclusions can be derived from the data presented in Fig. 11.16. First, the lithium metasilicate crystallite concentration increases with laser irradiance and saturates at higher laser irradiances. Second, the lithium metasilicate crystallite diameters remain constant as a function of incident laser irradiance. Third, the metasilicate crystallites formed using $\lambda = 266$ nm laser irradiation are larger than the crystallites formed via $\lambda = 355$ nm laser excitation. The average lithium metasilicate crystallites diameters derived from the XRD studies were 117.0 ± 10.0 nm and 91.2 ± 5.8 nm for $\lambda = 266$ nm and $\lambda = 355$ nm, respectively. This 28% increase in the crystallite diameter at $\lambda = 266$ nm is noticeably outside the experimental error of the XRD data. This appreciable difference in crystallite size may be correlated with the concentration dependence of the silver cluster formation kinetics. The number of silver clusters that are formed per unit diffusion volume is ~8 times larger for $\lambda = 266$ nm compared with the silver cluster number density for $\lambda = 355$ nm; this increased density of metal colloids could enhance the size of the metasilicate crystallites under the current thermal treatment protocol.

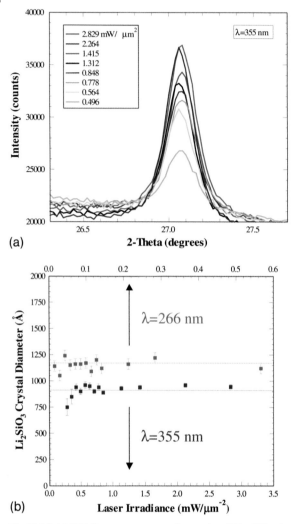

Fig. 11.16 (a) XRD features corresponding to the <111> diffraction plane (2-theta = 27.1°) as a function of incident laser irradiance at $\lambda = 355$ nm. (b) Lithium metasilicate (Li_2SiO_3) crystal size as a function of laser irradiance at $\lambda = 266$ nm and $\lambda = 355$ nm, as determined by XRD peak widths.

During the laser processing of PSGC materials, the exposed surface area typically receives multiple laser pulses. To address the photophysics of the sequential delivery of laser energy into the photosensitive glass substrate, we have conducted experiments to compare the effect of a single large exposure (i.e., single irradiance) versus multiple smaller exposures (i.e., several sequential irradiances). The experiments explore the use of a single exposure at 0.848 mW μm^{-2} versus two multiple (sequential) exposures that each have an irradiance of 0.424 mW μm^{-2}. The optical spectroscopy and XRD data measured for Foturan samples that were thermally processed to the Bake phase II stage are shown in Fig. 11.17. Figure 11.17(a) shows the absorption spectra which correspond to defect state formation following laser exposure at $\lambda = 355$ nm. The single small exposure spectrum is plotted along with the double exposure and the single large exposure. The results show that the *single large irradiance* exposure and the *double exposure smaller irradiance* produce comparable photoelectron defect state densities as defined by the absorption profiles. However, differences appear following thermal processing and lithium metasilicate formation. Figure 11.17(b) shows the XRD peak corre-

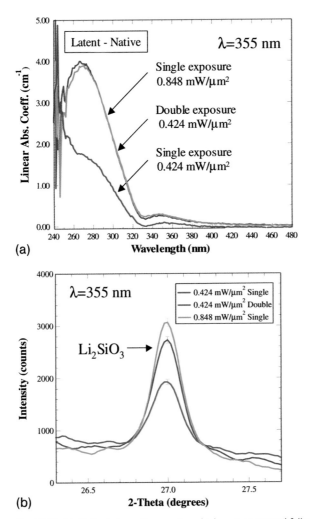

(a)

(b)

Fig. 11.17 (a) Optical absorption spectra which were measured following single and multiple laser exposure conditions at $\lambda = 355$ nm. The spectra correspond to the Latent Image (exposed) state and are defined as: Latent (exposed) – Native (unexposed). (b) XRD data obtained following single and multiple-laser exposures and thermal treatment.

sponding to the <111> diffraction for the single small exposure, the double small exposure, and the single large exposure. A comparison of the XRD data reveals that the metasilicate density that is formed from the double small exposure is approximately twice as high when compared with the density formed from the single small exposure. However, the metasilicate crystallite density that is generated from the double small exposure is approximately 15–20% lower than the density created from the single large exposure. Clearly, if the aim were to maximize the lithium metasilicate crystal density, the optimum processing protocol would require exposure at a high irradiance, rather than multiple exposures at a lower irradiance. Given that the total photoelectron defect densities are comparable for a single large exposure and serial multiple small exposures, the issue that remains unanswered relates to the specific nature of the defects that are generated at the higher exposures.

11.4
Laser Direct-write Microfabrication

A comprehensive list of the material properties of the PSGC Foturan is provided in Table 11.3. Two common states of the material exist and are denoted as the vitreous glass state and the ceramic state. There is also an intermediate state of the material that is very similar to the vitreous state; this state has undergone slight conversion to the ceramic form via homogenous nucleation. The intermediate material state corresponds to a region of a sample that has been thermally treated without exposure. Material property data for this intermediate state is virtually nonexistent. The data presented in Table 11.3 for the vitreous and ceramic glass states has been extracted from the product literature [75], except for the dielectric constant and loss tangent at 10 GHz [76]. The loss factor ($\tan(\delta)$) characterizes the material absorption in the RF region. For Foturan, the loss factor is nearly equivalent to that of FR4 fiberglass, which is a material that is utilized in special RF applications. The Foturan loss factor is ~ 100 times larger when compared with the loss factors associated with typical RF materials such as alumina (Al_2O_3, 0.0001) and sapphire (0.0001). Foturan has a large loss factor due to the mobility of the lithium atoms in the vitreous and intermediate states. In the fully ceramic state, the lithium atom is more confined in the lithium disilicate crystal ($Li_2Si_2O_5$) and the loss factor should improve.

Table 11.3 Material properties of the PSGC Foturan.

Property (glass – ceramic) Foturan Schott Glass	Vitreous state	Ceramic state
Young' Modulus (Gpa)	78	88
Poissons's Ratio	0.22	0.19
Hardness Knoop (Mpa)	4600	5200
Modulus – rupture (Mpa)	60	150
Density (g/cm^3)	2.37	2.41
Thermal expansion @ 20–300 °C (10^{-6} K^{-1})	8.6	10.5
Thermal conductivity @ 20 °C (Wm^{-1} K^{-1})	1.35	2.73
Transformation temp. (°C)	465	
Maximum safe operating temp (°C)		750
Water durability DIN/ISO 719 ((μg)Na$_2$O/g)	468	1300
Acid durability DIN 12116 (mg dm^{-2})	0.4	0.9
Porosity (gas–water)	0	0
Electrical conductivity		
@ 25 °C (Ohms – cm)	8.1×10^{12}	5.6×10^{16}
@ 200 °C (Ohms – cm)	1.3×10^{7}	4.3×10^{7}
Dielectric constant @ 1 MHz. 20 °C	6.5	5.7
Loss factor tan(δ) @ 1 MHz. 20 °C	65	25
Dielectric constant @ 10 GHz. 20 °C	6.2	
Loss factor tan (δ) @ 10 GHz. 20 °C	0.011	

Material processing via a direct-write (DW) technique is particularly appealing since it provides a means for variegated control in the process. These attributes become valuable in materials processing where a high degree of customization is desired, e.g., during rapid prototyping operations. Direct-write processing is commonly distinguished by its serial processing approach that does not require the use of masking layers. Processing in serial fashion is often considered a detriment because of the inherent limit to fabrication throughput. However, advantages are gained since the maskless processing removes complexity and reduces the processing cost. In the development of direct-write or laser direct-write (LDW) processing tools, one must balance the benefits of serial processing (maskless, variegated control, customization) along with the related cost. The scale is often tipped in favor of DW or LDW when the goal is to attain true 3D and free-form fabrication. Moreover, when the list of available direct-write tools is compared, the laser stands out because it provides the ability to control a variety of experimental parameters (e.g., laser wavelength, laser energy, pulse duration, pattern velocity). These experimental parameters can be implemented to alter many critical properties of a material [77]. If the advantages of serial processing and batch processing could be combined and integrated, a far superior material processing approach could be realized. The achievement of this advanced material processing method is possible if the material (such as PSGCs) is an "active" participant in the process.

The Aerospace Corporation has developed such a processing technique for PSGC materials that merges the advantages of the serial processing and the batch processing approaches. The approach utilizes the laser to set the initial conditions in the substrate, which subsequently induce a change in a specific material property. Consequently, the processing speed in the serial patterning stage can be substantially increased. The laser is *not* used to remove the material. Instead, material removal occurs via a batch-process mode. The Aerospace technique exploits the unique properties of the PSGC materials by employing the knowledge of the fundamental photophysical interactions and the related changes which can be induced in the material. The pulsed laser direct-write patterning of true 3D shapes was first demonstrated by Aerospace in the PSGC, Foturan [78]. The processing technique utilized a computer controlled XYZ motion stage for patterning and pulsed UV lasers with high precision power control (cf. Figs. 11.8 and 11.9) [79].

In a series of experiments, it was determined that the aggregate laser photon dose affects the etch depth at a fixed laser fluence (J cm^{-2}). The measured etch depths versus the number of laser pulses (total laser dose) at $\lambda = 266$ nm are displayed in Fig. 11.18. The results suggest that the depth which can be achieved following the chemical etching step can be controlled by monitoring the number of incident laser pulses at a volume element (i.e., voxel) location. Thus, the microstructure aspect ratio can be precisely defined and regulated during the laser patterning stage. Although this feature is useful for 3D material processing, it suffers from the practical inconvenience that deep patterned structures will require longer etch times than the shallow structures. Consequently, the shallow depth structures would have to be protected while the deep patterned structures are chemically excavated. This process is similar to that applied in MEMS fabrication technology when there is a wide variation in aspect ratios. Fortunately, subsequent experiments established that the chemical etching rate is strongly dependent on the laser irradiance [80]. This conclusion can also be derived from the photophysical data presented earlier. The ability to vary the chemical etching rate by altering the photon flux (laser exposure dose) has profound implications for the laser processing of Foturan.

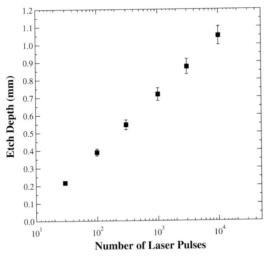

Fig. 11.18 Etch depth versus number of laser pulses (λ = 266 nm).

Figure 11.19 presents the etch depth results measured for exposed and native (unexposed) Foturan. The Foturan samples were chemically etched at room temperature using dilute aqueous hydrofluoric acid (5.0 vol% HF/H$_2$O). Figure 11.19(a) shows the measured etch depths as a function of etch time for Foturan which has been exposed to laser irradiation at λ = 355 nm and thermally processed. The results clearly show that the chemical etch rate increases with a concurrent increase in the laser irradiance. Figure 11.9(b) displays the etch depth as a function of etch time for the native unexposed glass. The native glass etch rate was determined to be 0.62 + 0.06 μm min^{-1}. Whereas the data shown in Fig. 11.18 correlates the increase of achievable etch depth and the aggregate laser exposure dose, the results shown in Fig. 11.19 (a) indicate that the *etch rate itself is dependent on the laser irradiance.*

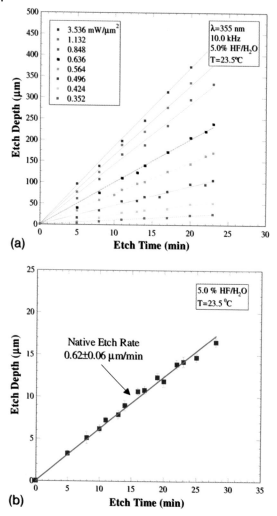

Fig. 11.19 Measured etch depths as a function of etch time for laser-irradiated and thermally-processed Foturan at λ = 355 nm (a) and native unexposed Foturan (b).

The data presented in Fig. 11.19 can be recast in the form of an etch-rate ratio which represents the etch contrast between the exposed and unexposed material. Figure 11.20 shows the etch-rate ratios as a function of laser irradiance at λ = 266 nm (a) and λ = 355 nm (b). The results shown in Fig. 11.20 reveal two distinct laser-processing regimes. For low laser irradiances, the etch-rate ratios increased nearly linearly versus laser irradiance. The initial increase in the etch-rate contrasts were fit using linear least-squares regression analysis to yield the following values: slope = 435.9 ± 46.7 μm^2 $(mW)^{-1}$ for λ = 266 nm and slope = 46.2 ± 2.3 μm^2 $(mW)^{-1}$ for λ = 355 nm. For high laser irradiances, the measured etch-rate ratios reached a plateau region in which the contrast remained constant at ~ 30:1. The average maximum etch rate was determined to be $18.62 \pm 0.30 \, \mu m \, min^{-1}$ and was independent of the exposure wavelength. However, the value of the maximum etch contrast is dependent on the thermal treatment protocol employed in this work. UV lamp exposure studies performed on a PSGC, similar to Foturan, demonstrated that increasing the concentration or the temperature of the acid solution could decrease the solubility differential [81]. For example, these studies measured etch-rate ratios of 50:1, 30:1, and 13.5:1 using HF solutions with 1 wt% HF, 10 wt% HF, and 20 wt% HF, respectively [82].

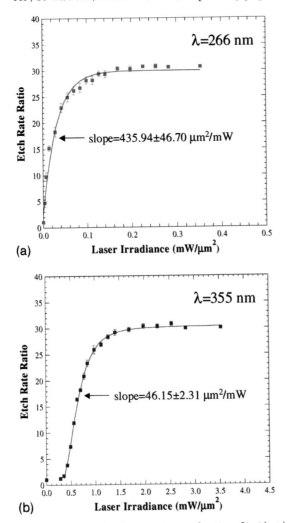

(a)

(b)

Fig. 11.20 Measured etch-rate ratios as a function of incident laser irradiance at λ = 266 nm (a) and λ = 355 nm (b). The solid squares correspond to the measured etch-rate results and the solid lines represent optimized Hill equation fits to the experimental data.

In our opinion, the key result presented in Figs. 11.19 and 11.20 is not the absolute value of the maximum etch contrast, but the implication of the monotonically increasing segment of the data. These results suggest that if the exposure is kept within the linear portion of the measured data, it is possible to precisely vary the chemical etch rate on a local scale by merely varying the laser irradiance during patterning. A change in laser irradiance by as little as 0.01 mW μm^{-2} corresponds to changes in the etch contrast of 4.4 and 0.46 for $\lambda = 266$ nm and $\lambda = 355$ nm, respectively. For a laser spot area of 3 μm^2, the required corresponding change in the input power is 30 μW. The primary advantage that is attained by varying the laser irradiance during patterning is the ability to locally alter the etch rate to produce the desired aspect ratio. The outcome of this type of patterned wafer is that *only one chemical etch time is necessary to release all of the components regardless of aspect ratio.*

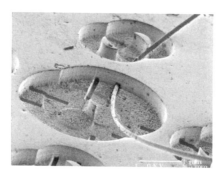

Fig. 11.21 A scanning electron microscope (SEM) image showing several fluidic reservoirs that are interconnected by an embedded channel. The channels were fabricated by selectively exposing embedded layers in the glass. A human hair has been threaded through one channel and the scale bar is 1 mm.

To implement this capability, Aerospace designed and constructed a semi-automated laser direct-write exposure tool that consists of a computer controlled three-axis motion system that can process 100 mm diameter wafers. A desired pattern is initially drawn in 3D using a computer-assisted design (CAD) solid modeling (SolidWorks™, Dassault Systemes of America Corp.) software tool. The design is then converted to a sequence of tool path motions by a computer-assisted manufacturing (CAM) software tool (MasterCAM, CNC Software Corp.). The tool path output is a three-dimensional CNC (computer numeric control) code that is interpreted by the three-axis motion system for synchronous motion. In addition, a second program script is generated that contains the required laser irradiance values for a particular tool path sequence [83].

The results presented in Fig. 11.20 also reveal that a threshold irradiance value (laser exposure dose) is required to yield an interconnected network of etchable crystals. Figure 11.21 shows the application of this knowledge for the fabrication of embedded fluidic channels that serve as reservoir interconnects [84]. The channel widths are approximately 100 μm and a human hair has been threaded through one of the channels. The fluid channels were fabricated by applying the appropriate laser irradiance that induced exposure and etchable crystal formation only in the laser focal volume. The XYZ motion control system was then directed to pattern an embedded channel. The developed processing technique enables two unique capabilities:

1. The formation of undercut structures, which are particularly important in the development of MEMS in glass ceramic materials.
2. The fabrication of embedded channels, which could reduce packaging steps in the development of microfluidic devices.

The quality, length and precision of undercut structures and embedded channels are strongly dependent on the ability to efficiently transport the reagent to the etching interface and remove the reaction products. Several agitation approaches have been utilized to enhance this process, including slosher baths and ultrasonic agitators. Ultrasonic agitators are often used, but these systems induce an increase in the solution temperature. The increase in the temperature reduces the etch contrast between the exposed and unexposed regions. Aerospace employs a high-pressure sprayer assembly to deliver the chemical etchant to the glass substrate surface. Figure 11.22 shows a schematic layout of the high-pressure sprayer

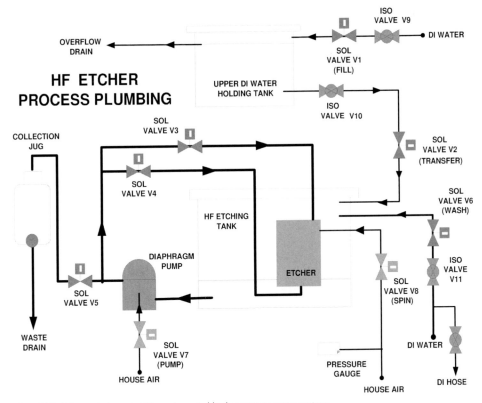

Fig. 11.22 Schematic layout of the automated high-pressure spray system used for chemical etching.

system. The high-pressure sprayer is timer-controlled and contains two eight-nozzle arrays. The nozzle arrays are located above and below the glass sample and can be operated independently to deliver the etchant to the desired surface of the glass substrate. This automated system is capable of processing a single 100 mm diameter wafer at room temperature and the etchant concentration can be varied as needed. During the etching process, the glass sample is rotated at ~ 20 rev sec^{-1} to facilitate uniform HF delivery and efficient removal of the etched products. The implementation of the high-pressure sprayer system reduces the etching times by nearly a factor of two compared with the etch times obtained using a slow agitating slosher bath system.

Figure 11.23 presents a sequence of images which characterize the Aerospace variable laser exposure process. Figure 11.23(a) represents an optical micrograph of a 100 mm diameter wafer which contains numerous 3D laser-patterned struc-

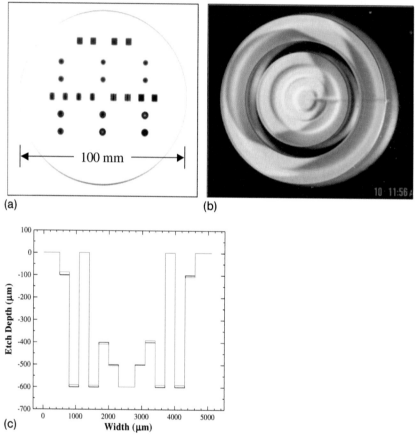

(a)　　　　　　　　　　　　　　(b)

(c)

Fig. 11.23 (a) Optical microscope photograph of a 100 mm diameter wafer that contains numerous laser-exposed and thermally processed structures with varying aspect ratios. (b) An example of a 3D etched "cityscape" structure that was fabricated and released from the wafer. (c) Crosscut comparison between the calculated CAD specifications (black) and experimentally measured (grey) etch depths and feature dimensions.

tures. Although it is not possible to show it in clear visual detail, these structures correspond to a series of test diffractive optical elements for operation in the far-IR. These structures contain features with highly variegated high and low aspect ratios. Figure 11.23(b) corresponds to a low magnification optical microscope image of a realized etched 3D structure that we have denoted as the "cityscape". The structure is 4.1 mm in diameter and contains a variety of topographic and aspect ratio features. Figure 11.23(c) displays the CAD file that corresponds to the cityscape pattern. The optical profilometry measurements that were performed on the realized part are also co-plotted. Measurement of the etch depth by optical profilometry shows a typical relative error of < 10% compared with the dimensions specified in the CAD file. The entire composite set of structures on the wafer were subjected to a *single* timed chemical etch process for structure release. The realized structure in Fig. 11.23 demonstrates that features at depths of 600 micrometers could be co-fabricated alongside features that are 100 micrometers in depth. This processing technique is found to be very useful for fabricating complex 3D structures that would be very difficult to achieve by other methods. Given the development of stable and higher repetition rate pulsed lasers, the variable exposure process is limited primarily by the uniformity of the photosensitizer and nucleating agent densities that can be achieved in the manufacturing of PSGC materials.

The ability to control the laser irradiance during patterning also allows the control of several other attributes of the processed Foturan structure. For example, the unexposed glass is transparent in the visible spectrum, but by appropriately controlling the laser exposure, it is possible to impart local changes in color and absorptivity in the visible and IR regions. Figure 11.24 shows an optical microscope photograph of an optical element that corresponds to the realized structure shown in Fig. 11.23, but prior to chemical etching. Several color bands are observed and are related to the laser exposure irradiance. For applications where the color must survive the chemical etch treatment (e.g., during the release of a component), the image could be embedded below the glass ceramic surface during the LDW patterning phase.

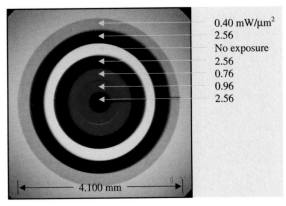

0.40 mW/μm^2
2.56
No exposure
2.56
0.76
0.96
2.56

4.100 mm

Fig. 11.24 Optical microscope photograph of the "cityscape" pattern following variable laser exposure and heat treatment, but prior to chemical etching.

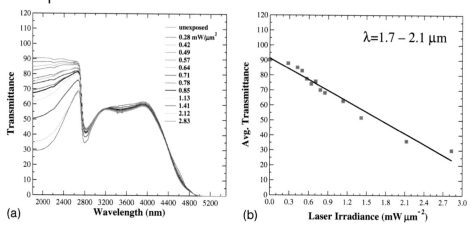

Fig. 11.25 (a) Optical transmission spectra in the infrared region as a function of incident laser irradiance at $\lambda = 355$ nm. (b) Measured average transmission in the 1.7–2.1 μm band as a function of laser irradiance.

Figure 11.25 shows the optical transmission spectra which were measured in the IR region and correspond to Foturan that has received laser exposure and thermal treatment to the partially ceramic lithium metasilicate crystalline phase. (a) shows the optical transmission spectra as a function of incident laser irradiance at $\lambda = 355$ nm, and (b) shows the average transmission within a narrow ($\lambda = 1.7$–2.1 μm) IR band region as a function of laser irradiance at $\lambda = 355$ nm. The results suggest that calibrated IR filters can easily be patterned and fabricated with precision on the surface and inside PSGC materials.

During the past 20 years, MEMS technology has routinely presented evidence for the fabrication of useful compliance into microstructures. However, these materials and structures would have an impractical compliance figure if they were similarly fabricated on the macroscale. The PSGC materials are no different from silicon if it were feasible to fabricate the structures with microscale resolution. The LDW processing approach has been utilized to fabricate structures in Foturan which exhibit useful compliance. Several examples of compliant structures which were fabricated in Foturan are shown in Fig. 11.26. The structures have overall dimensions on the mesoscale (~1 cm), but have integrated feature dimensions on the microscale (< 100 μm) that retain useful compliance. Figure 11.26(a) represents a wishbone spring which is loading a piston where the bow dimension tapers down to 80 μm. Figure 11.26(b) corresponds to a close-up view of a coil spring in the shape of a hotplate where the coil width is ~ 15 μm. For relative comparison, a human hair < 60 μm thick has been overlapped on the hotplate structure. Finally, (c) shows an eight-blade miniature turbine fan which is 1 mm in height and retains a blade width (at the thickest portion) of 100 μm. The microturbine has been operated at speeds of up to 180 000 RPM using an air jet without mechanical failure. These structures show compliant behavior in a material which is not normally associated with retaining such properties.

The examples displayed in Fig. 11.26 are all designed with near-uniform material strength. The strength of the fabricated part can also be locally altered by subjecting the component to either a second laser patterning treatment and a subsequent ceramization step or directly embedding "stiffness" in the form of lithium metasilicate crystals. Under these conditions, sections which require more stiffness receive additional exposure which results in a larger crystalline fraction. The bake protocol for the second ceramization step is altered to permit growth of the nonsoluble lithium disilicate crystalline phase and to increase the glass stiffness. By applying this form of process control, the modulus of rupture of the glass material can be locally varied from 90–150 MPa.

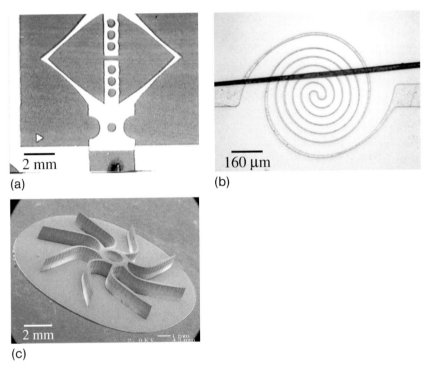

(a)

(b)

(c)

Fig. 11.26 Optical microscope photographs of several compliant structures which were fabricated using the PSGC Foturan. A wishbone spring (a), a hotplate coil spring with a human hair overlaid for visual reference (b), and a microscopic rotary turbine blade (c).

Finally, the examples in Fig. 11.26 all show surface topology that is the remnant of a chemically etched surface. The surface roughness has been measured and is found to be a function of the exposure dose. For $\lambda = 266$ nm, the measured surface roughness varies from 400 to 1200 nm for laser irradiances of 0.283–0.042 mW μm^{-2}. For $\lambda = 355$ nm, the measured surface roughness is larger and varies from 500 to 3000 nm for laser irradiances of 2.829 – 0.495 mW μm^{-2}. For any particular location, the surface roughness increases with increasing depth.

This change is a result of the decreasing light intensity with depth and its effect on the dynamics of the nucleation and crystallization process [85]. These changes in the nucleation and crystallization processes as a function of depth also affect the chemical etching rate. In general, the etching rate decreases with depth, but this effect is surmounted by the larger effect of mass transport limitations in the chemical etching process. Changing the thermal treatment protocol can markedly alter the surface roughness. If the Bake II phase duration is increased, the metasilicate crystals continue to grow and the etched regions are rough. Aerospace has employed this technique to develop high-aspect ratio structures which have very rough topology ($R_a > 3$ μm) with sharp angular features. The devices were to be used as field emission ion sources for miniature satellite propulsion systems. Ions could be generated not only at the single tip of a pyramid, but also along the pyramid face and walls. An alternative approach to smoothing the surface roughness is to anneal the processed part after chemical etching. Experiments conducted by Sugioka et al., show that a post-etch anneal step can reduce the surface roughness from $R_{max} = 53.4$ nm to $R_a = 0.8$ nm [86]. In addition, a photolytic approach can be employed to reduce the surface roughness. By increasing the uniformity and spatial homogeneity of the light pattern, the absorption uniformity and the surface smoothness also increase. The vertical sidewalls of a LDW exposure are especially straight and smooth when the exposed area is within the depth of focus (i.e., confocal parameter) of the writing laser beam.

11.5
Conclusions

This chapter has provided a comprehensive overview of the experimental results which correspond to the fundamental photochemistry and photophysics of PSGC laser processing. The review specifically included experimental data that relate to the UV laser exposure process, the thermal treatment protocol, and the chemical etching behavior of the native and irradiated glass material. The results have been measured and presented as a function of laser irradiance at two common laser processing wavelengths ($\lambda = 266$ nm and $\lambda = 355$ nm). These spectroscopic and kinetic studies were initiated to improve our overall understanding of laser-glass ceramic interactions and to elucidate and mitigate the problems associated with experimental parameter variations (e.g., laser power, processing speed, bake temperatures, and etchant concentration). Finally, using a series of microfabricated structures as examples, we have shown that, by carefully controlling the laser exposure, it is possible to locally vary the material strength, chemical solubility, color and transmission in the IR wavelength region.

Although our results clearly demonstrate that we can locally control these three PSGC material properties, we have yet to compare the photophysical results (optical spectroscopy and XRD) and the chemical etching results. The etch-rate ratio and the integrated peak areas for the trapped (defect) state and silver clusters are co-plotted versus laser irradiance in Fig. 11.27. Several trends are noteworthy:

(a) For both laser irradiation wavelengths, the concentration of defect species and silver clusters turn over at a laser irradiance which is commensurate with the turn-over region for the etch-rate ratio. (b) The Ag cluster concentration saturates with increasing laser irradiance, while the defect concentration does not saturate. (c) The Ag cluster concentration is larger for $\lambda = 266$ nm irradiation (~ 200 units) than for $\lambda = 355$ nm irradiation (~ 20 units). Despite the significant difference in Ag cluster concentration for $\lambda = 266$ nm and $\lambda = 355$ nm irradiation, the maximum etch-rate ratios and absolute etch-rates measured for both wavelengths were equivalent. These results suggest that a threshold concentration of Ag clusters is necessary to induce nucleation and growth of the etchable Li_2SiO_3 crystalline

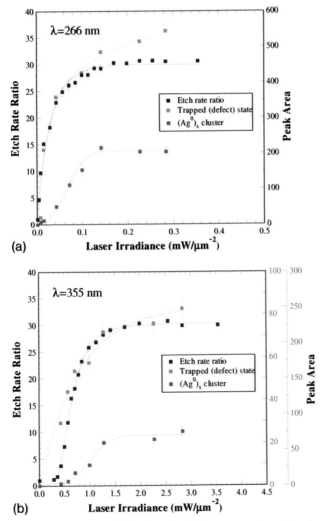

Fig. 11.27 Comparison of the photophysical measurements and the chemical etching data for laser exposure processing at $\lambda = 266$ nm (a) and $\lambda = 355$ nm (b).

phase. A silver cluster concentration that exceeds a critical level does not enhance the etching rate. The results presented in Fig. 11.27 may help to refine the protocol for PSGC processing. The Ag cluster state may be a beneficial monitor to identify the saturation and limits of the etchable phase. The defect-state concentration, on the other hand, appears to be a valuable monitor of the etch-rate ratio prior to saturation. These results could facilitate a detailed understanding and refinement of the variable-dose laser-processing technique. Clearly, the measured dependence of the etch-rate ratio and saturation behavior are complex processes. The results are related to the material doping density and the specific laser-processing protocol, as well as the crystalline fraction which can be accommodated in an exposed volume. To help clarify these critical issues, current experiments are examining the influence of baking protocol and photo-initiator doping.

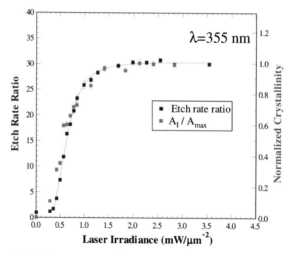

Fig. 11.28 Comparison of the measured etch-rate ratios and normalized XRD results which correspond to the density of lithium metasilicate crystallites.

A comparison of the XRD data and chemical etching results is presented in Fig. 11.28. The etch-rate ratios and normalized integrated peak areas which correspond to the <111> diffraction plane (i.e., lithium metasilicate crystal density) are co-plotted in Fig. 11.28. The results are shown as a function of laser irradiance at $\lambda = 355$ nm. The results displayed in Fig. 11.28 indicate that the lithium metasilicate crystal density saturates and closely follows the etch-rate ratio curve. These results argue that the saturation behavior of the etch-rate ratio is controlled by the saturation of the silver cluster density. However, the absolute etch rate does not show a strong dependence on the silver cluster concentration. It is possible that only a select cluster size affects the growth of lithium metasilicate crystals and the concentration of this species is the same for both $\lambda = 266$ nm and $\lambda = 355$ nm laser irradiation. The species in question would have a spectroscopic signature within the 420 nm absorption which cannot easily be deconvoluted.

We have presented experimental evidence that allows photolytic control of several material properties. Material removal was accomplished without the need for ablation; and material strength could be tailored without the need to bake the entire part. We believe this approach to material processing represents the proper approach for the development of next-generation integrated devices. The synthesis and development of new materials which can be similarly processed are required. As a material class, the PSGCs have the potential for further "tuning" that will enable the localized engineering of other unique properties. Finally, by taking steps to better understand the underlying laser processing mechanisms, we have significantly enhanced our ability to process with higher precision and fidelity.

Acknowledgments

The authors are indebted to the Aerospace Independent Research and Development (IR&D) Program and the Air Force Office of Scientific Research (Dr. Howard Schlossberg, Program Manager) for financial support and for trusting our ability to develop a process which would structure glass-ceramic materials without the need for laser ablation. The authors also acknowledge the valuable contributions and helpful discussions of Dr. Peter Fuqua (critical fluence measurements), Dr. Paul M. Adams (optical spectroscopy and XRD), Mark E. Ostrander (cutting and polishing of glass samples), and William W. Hansen.

References

1 Horace, the Odes, Book II, odes iv, line 65, 23 B.C.
2 MRS Bulletin, "Photonic Materials for Optical Communications" **28** (2003).
3 B. Derby and N. Reis, "Inkjet Printing of Highly Loaded Particulate Suspensions," MRS Bulletin **28** (2003) pg. 815.
4 J. Wne, G.L. Wilkes, "Organic/Inorganic Hybrid Network Materials by the Sol-Gel Approach," Chem. Mater. **8**, (1996) pg. 1667.
5 K-H Haas and H. Wolter, "Synthesis, Properties and Applications of Inorganic-Organic Co-polymers (ORMO-CERs)," Current Opinion in Solid State & Material Science **4** (1999) pg. 571.
6 R. Buestrich, F. Kahlenberg, M. Popall, P. Dannberg, R. Müller-Fiedler and O. Rösch, "ORMOCERs for Optical Interconnection Technology," J. of Sol-Gel Science and Tech. Vol. **20**, (2001) pg. 181.
7 A. Berezhnoi, *Glass-ceramics and Photo-sitalls* English translation of Russian text, (Plenum Press, NY, 1970).
8 *Photo-Induced Metastability in Amorphous Semiconductors*, A. V. Kolobov, Ed., (Wiley-VCH Press, Weinheim, 2003).
9 K. Shimakawa, "Dynamics of Photo-Induced Metastability in Amorphous Chalcogenides" in *Photo-Induced Metastability in Amorphous Semiconductors*, A. V. Kolobov Ed. (Wiley-VCH Verlag & Co. Press, Weinheim, Germany, 2003), pg. 58.
10 T. Kawaguchi, "Photo-Induced Deposition of Silver Particles on Amorphous Semiconductors" in *Photo-Induced Metastability in Amorphous Semiconductors*, A. V. Kolobov Ed. (Wiley-VCH Verlag & Co. Press, Weinheim, Germany, 2003), pg. 58.
11 S.M. Breknovskikh, Yu, N. Viktorova, Yu. L. Grinshteyn, and L. M. Landa,

"Principles of the Radiation Study of Glass and Ceramic Materials" Translation of Osnovy Radiatsionnogo Materialovedeniya Stekla I Keramiki, (Foreign Science and Technology Center Press, US Army Materiel Command, 1973); J. S. Stroud, "Color Centers in a Cerium-Containing Silicate Glass," J. Chem. Phys. **37**, No. 4, (1962) pg. 836.

12 W.J. Zachariasen, J. Am. Ceram. Soc. **54** (1932) pg. 3841; B. E. Warren Z. Kristallogr. Mineralog. Petrogr. **86** (1933) pg. 349.

13 W. Vogel, *Chemistry of Glass* (The Am. Ceram. Soc. Press, Columbus OH, 1985) pg. 35.

14 W. Hölland and G. Beall, *Glass-Ceramic Technology* (Am. Ceram. Soc. Press, Westerville OH, 2002).

15 ibid, pg. XVI.

16 S.D. Stookey, "Photosensitive Glass," Ind. and Eng. Chem. **41** (1949) 856.

17 U. Kreibig, "Small Silver Particles in Photosensitive Glass: Their Nucleation and Growth," Appl. Phys. **10** (1976) pg. 255.

18 R. Reisfeld, " Spectra and Energy Transfer of Rare Earths in Inorganic Glasses" in *Structure and Bonding* Vol. 13. J. D. Dunitz, P. Hemmerich, J. A. Ibers, C. K. Jorgensen, J. B. Neilands, R. S. Nyholm, D. Reinen and J. P. Williams, Eds. (Spring-Verlag, New York, 1973) pg. 53.

19 J.A Duffy and G.O. Kyd, "Ultraviolet Absorption and Fluorescence Spectra of Cerium and the Effect of Glass Composition," Phys. Chem. Glasses **37**, No. 2, (1996) pg. 45.

20 J.S. Stroud, "Photoionization of Ce^{3+} in Glass", J. Chem. Phys. **35**, No. 3 (1961) pg. 844.

21 G.A. Sycheva, "Nucleation Kinetics of Lithium Metasilicate in Photosensitive Lithium Aluminosilicate Glass," Glass Phys. and Chem. **25** (1999) pg. 501.

22 F. Liebau, "Untersuchungen an Schichtsilikaten des Formeltyps $A_m(Si_2O_5)_n$. I. Die Kristallstruktur der Zimmertemperaturform des $Li_2Si_2O_5$," Acta Crystallogr. **14**, (1961) pg. 389; R. Dupree, D. Holland and M. G. Mortuza, "A MAS-NMR Investigation of Lithium

Silicate Glasses and Glass-Ceramics," J. Non-Cryst. Solids **116** (1990) pg. 148.

23 W. Hölland and G. Beall, *Glass-Ceramic Technology* (Am. Ceram. Soc. Press, Westerville OH, 2002), pg. 7.

24 S.D. Stookey, "Chemical Machining of Photosensitive Glass," Ind. Eng. Chem., Vol. 45, (1953) 115.

25 A. Berezhnoi, *Glass-Ceramics and Photo-Sitalls*, (Plenum, New York, 1970).

26 G.H. Beall, "Design and Properties of Glass-Ceramics," Annu. Rev. Mater. Sci. (1992) pg. 119.

27 S.M. Breknovskikh, Yu, N. Viktorova, Yu. L. Grinshteyn, and L. M. Landa, "Principles of the Radiation Study of Glass and Ceramic Materials" Translation of Osnovy Radiatsionnogo Materialovedeniya Stekla I Keramiki, (Foreign Science and Technology Center Press, US Army Materiel Command, 1973) pg. 28.

28 A.J. Ikushima, T. Fujiwara, K. Saito, "Silica Glass: A Material for Photonics," Appl. Phys. Rev. **88**, No. 3, (2000) pg. 1201.

29 ibid, pg. 30.

30 N. Itoh, and A. M Stoneham, *Materials Modification by Electronic Excitation*, (Cambridge University Press, Cambridge, UK, 2001) pg. 287.

31 J.S. Stroud, "Photoionization of Ce^{3+} in Glass," J. Chem. Phys. **35** (1961) pg. 844.

32 L. Skuja, "Optically Active Oxygen-deficiency-related Centers in Amorphous Silicon Dioxide," J. of Non-Crystalline Solids **239** (1998) pg. 16.

33 D.L. Griscom, "Optical Properties and Structure of Defects in Silica Glass," The Ceramic Soc. Japan **99** (1991) pg. 923.

34 M. Talkenberg, E. W. Kreutz, A. Horn, M. Jacquorie and R. Poprawe, "UV Laser Radiation-Induced Modifications and Microstructuring," Proc. SPIE **4637** (2002) pg. 258.

35 J.S. Stroud, "Photoionization of Ce^{3+} in Glass," J. Chem. Phys. **35** (1961) pg. 844.

36 M. Tashiro, N. Soga, and S. Sakka, "Behavior of Cerium Ions in Glasses Exposed to X-rays," J. Ceram. Assoc. Japan **87** (1960) pg. 169.

37 J.S. Stroud, "Color Centers in a Cerium-containing Silicate Glass," J. Chem. Phys. **37** (1962) pg. 836.

38 J.S. Stroud, "Photoionization of Ce^{3+} in Glass," J. Chem. Phys. **35** (1961) pg. 844.

39 ibid.

40 ibid.

41 H-J Kim and S-C Choi, "Effect of Sb_2O_3 and Raw Materials on the Crystallization of Silver Containing Glasses," Phys. Chem. Glasses **4** (2000) pg. 55.

42 A. Paul and R. W. Douglas, "Cerous-ceric Equilibrium in Binary Alkali Borate and Alkali Silicate Glasses," Physics and Chemistry of Glasses **6** (1965) pg. 212.

43 *LIA Handbook of Laser Materials Processing*, J. F. Ready Ed. (Laser Institute of America Press, Orlando, FL, 2001) pg. 173.

44 *Foturan – A Material for Microtechnolgy*, Schott Glaswerke Publication, Optics Division, Mainz, Germany.

45 M. Talkenberg, E. W. Kreutz, A. Horn, M. Jacquorie and R. Poprawe, " UV Laser Radiation-Induced Modifications and Microstructuring," Proc. SPIE Vol. 4637, (2002) pg. 258.

46 W.W. Hansen, S.W. Janson, and H. Helvajian, "Direct-write UV Laser Microfabrication of 3D Structures in Lithium Aluminosilicate Glass", Proc. SPIE **2991** (1997) pg. 104; M. Talkenberg, E. W. Kreutz, A. Horn, M. Jacquorie and R. Poprawe, "UV Laser Radiation-Induced Modifications and Microstructuring", Proc. SPIE **4637** (2002) pg. 258.

47 ibid.

48 W.W. Hansen, S.W. Janson, and H. Helvajian, "Direct-write UV laser Microfabrication of 3D Structures in Lithium Aluminosilicate Glass", Proc. SPIE **2991** (1997) pg. 104; M. Talkenberg, E. W. Kreutz, A. Horn, M. Jacquorie and R. Poprawe, "UV Laser Radiation-Induced Modifications and Microstructuring," Proc. SPIE **4637** (2002) pg. 258.

49 M. Masuda, K. Sugioka, Y. Cheng, N. Aoki, M. Kawachi, K. Shihoyama, K. Toyoda, H. Helvajian and K. Midorikawa, "3-D Microstructuring Inside Photosensitive Glass by Femtosecond Laser Excitation," Appl. Phys. A. **76** (2003) pg. 857.

50 J. Kim, H. Berberoglu, and X. Xu, "Fabrication of Microstructures in FOTURAN using Excimer and Femtosecond Lasers," in Photon Processing in Microelectronics and Photonics II, Proc. SPIE **4977** (2003) pg. 324.

51 M. Masuda, K. Sugioka, Y. Cheng, N. Aoki, M. Kawachi, K. Shihoyama, K. Toyoda, H. Helvajian and K. Midorikawa, "3-D Microstructuring Inside Photosensitive Glass by Femtosecond Laser Excitation," Appl. Phys. A. **76** (2003) pg. 857.

52 J. Kim, H. Berberoglu, and X. Xu, "Fabrication of Microstructures in FOTURAN using Excimer and Femtosecond Lasers," in Photon Processing in Microelectronics and Photonics II, Proc. SPIE **4977** (2003) pg. 324.

53 P.D. Fuqua, D.P. Taylor, H. Helvajian, W.W. Hansen, and M.H. Abraham, "A UV Direct-Write Approach for Formation of Embedded Structures in Photostructurable Glass-Ceramics," Mater. Res. Soc. Proc. **624** (2000) pg. 79.

54 ibid.

55 K. Sugioka, Y. Cheng, M. Masuda, and K. Midorikawa, "Fabrication of Microreactors in Photostructurable Glass by 3D Femtosecond Laser Direct-write," Proc. SPIE **5339** (2004) pg. 205.

56 L.Y. Lui, P.D. Fuqua, and H. Helvajian, "Measurement of Critical UV Dose in Lamp Exposure of a Photostructurable Glass-Ceramic," Aerospace Report No. ATR-2001(8260)-1.

57 P.D. Fuqua, D.P. Taylor, H. Helvajian, W.W. Hansen, and M.H. Abraham, "A UV Direct-Write Approach for Formation of Embedded Structures in Photostructurable Glass-Ceramics," Mater. Res. Soc. Proc. **624** (2000) pg. 79; J. Kim, H. Berberoglu, and X. Xu, "Fabrication of Microstructures in FOTURAN using Excimer and Femtosecond Lasers," in Photon Processing in Microelectronics and Photonics II, Proc. SPIE **4977** (2003) pg. 324.

58 M. Masuda, K. Sugioka, Y. Cheng, N. Aoki, M. Kawachi, K. Shihoyama, K. Toyoda, H. Helvajian and K. Midorikawa, "3-D Microstructuring Inside Photosensitive Glass by Femtosecond

Laser Excitation," Appl. Phys. A **76** (2003) pg. 857.

59 J. Kim, H. Berberoglu, and X. Xu, "Fabrication of Microstructures in FOTURAN using Excimer and Femtosecond Lasers," in Photon Processing in Microelectronics and Photonics II, Proc. SPIE **4977** (2003) pg. 324.

60 P.D. Fuqua, D.P. Taylor, H. Helvajian, W.W. Hansen, and M.H. Abraham, "A UV Direct-Write Approach for Formation of Embedded Structures in Photostructurable Glass-Ceramics," Mater. Res. Soc. Proc. **624**, (2000) pg. 79.

61 P. D. Fuqua, D. P. Taylor, H. Helvajian, W. W. Hansen, and M. H. Abraham, "A UV Direct-Write Approach for Formation of Embedded Structures in Photostructurable Glass-Ceramics," Mater. Res. Soc. Proc. **624**, (2000) pg. 79; J. Kim, H. Berberoglu, and X. Xu, "Fabrication of Microstructures in FOTURAN using Excimer and Femtosecond Lasers," in Photon Processing in Microelectronics and Photonics II, Proc. SPIE **4977** (2003) pg. 324.

62 F.E. Livingston, P.M. Adams, and H. Helvajian, "Active Photo-Physical Processes in the Pulsed UV Nanosecond Laser Exposure of Photostructurable Glass Ceramic Materials", SPIE Proc. Laser Precision Microfabrication **5662** (2004) pg. 44.

63 J.S. Stroud, "Photoionization of Ce^{3+} in Glass," J. Chem. Phys. **35** (1961) pg. 844; J.S. Stroud, "Color centers in a cerium-containing silicate glass," J. Chem. Phys. **37** (1962) pg. 836.

64 M. Talkenberg, E.W. Kreutz, A. Horn, M. Jacquorie and R. Poprawe, "UV Laser Radiation-Induced Modifications and Microstructuring," Proc. SPIE **4637** (2002) pg. 258.

65 J.S. Stroud, "Photoionization of Ce^{3+} in Glass," J. Chem. Phys. **35** (1961) pg. 844.

66 A.I. Berezhnoy, A.M. Gel'berger, A.A. Gorbachev, S.E. Piterskikh, Yu. M. Polukhin and L.M. Yusim, "Light Sensitive Properties of Lithium Alumosilicate Glasses as a Function of Silver and Cerium Concentration," Soviet J. Opt. Technol. **36** (1969) pg. 616.

67 F.E. Livingston, P.M. Adams, and H. Helvajian, "Active Photo-Physical Processes in the Pulsed UV Nanosecond Laser Exposure of Photostructurable Glass Ceramic Materials," SPIE Proc. Laser Precision Microfabrication **5662** (2004) pg. 44; F.E. Livingston, P. M. Adams, and H. Helvajian, "Influence of Cerium on the Pulsed UV Nanosecond Laser Processing of Photostructurable Glass Ceramic Materials," Appl. Surf. Sci.-Laser Interactions in Materials: Nanoscale to Mesoscale **247** (2005) pg. 526.

68 F.E. Livingston, P.M. Adams, and H. Helvajian, "Influence of Cerium on the Pulsed UV Nanosecond Laser Processing of Photostructurable Glass Ceramic Materials," Appl. Surf. Sci.-Laser Interactions in Materials: Nanoscale to Mesoscale **247** (2005) pg. 526.

69 J.S. Stroud, "Photoionization of Ce^{3+} in Glass", J. Chem. Phys. **35** (1961) pg. 844; J. S. Stroud, "Color Centers in a Cerium-containing Silicate Glass", J. Chem. Phys. **37** (1962) pg. 836; M. Talkenberg, E.W. Kreutz, A. Horn, M. Jacquorie and R. Poprawe, "UV Laser Radiation-Induced Modifications and Microstructuring", Proc. SPIE **4637** (2002) pg. 258.

70 F.E. Livingston and H. Helvajian, "True 3D Volumetric Patterning of Photostructurable Glass Using UV Laser Irradiation and Variable Exposure Processing: Fabrication of Meso-Scale Devices," Proc. SPIE **4830** (2003) pg. 189.

71 F.E. Livingston, W.W. Hansen, A. Huang and H. Helvajian, "Effect of Laser Parameters on the Exposure and Selective Etch Rate of Photostructurable Glass," Proc. SPIE **4637** (2002) pg. 404; F. E. Livingston and H. Helvajian, "True 3D Volumetric Patterning of Photostructurable Glass Using UV Laser Irradiation and Variable Exposure Processing: Fabrication of Meso-Scale Devices," Proc. SPIE **4830** (2003) pg. 189.

72 U. Kreibig, "Small Silver Particles in Photosensitive Glass: Their Nucleation and Growth," Appl. Phys. **10** (1976) pg. 255.

73 *LIA Handbook of Laser Materials Processing*, J. F. Ready Ed. (Published by Laser Institute America, Orlando, FL, 2001) pg. 176.

74 S.D. Stookey, "Chemical Machining of Photosensitive Glass," Ind. and Chem. Eng. **45** (1953) pg. 115.

75 Schott Corporation, Technical Glass Division, Yonkers, NY, Foturan product literature F10/1999.

76 S.S. Osofsky, W.W. Hansen and H. Helvajian, "Measurement of the RF Dielectric Properties of the Lithium Alumino-Silicate Glass-Ceramic Foturan in X-Band (8-12.4 GHz) and R-Band (26.5-40 GHz), in preparation.

77 H. Helvajian, "3D Microengineering via Laser Direct-write Processing Approaches," in *Direct-Write Technologies for Rapid Prototyping Applications*, (Academic Press, NY, 2002), pg. 415.

78 W.W. Hansen, S.W. Janson and H. Helvajian, "Direct-Write UV Laser Microfabrication of 3D Structures in Lithium Aluminosilicate Glass," Proc. SPIE **2991** (1997) pg. 104.

79 F.E. Livingston, W.W. Hansen, A. Huang and H. Helvajian, "Effect of Laser Parameters on the Exposure and Selective Etch Rate of Photostructurable Glass," Proc. SPIE **4637** (2002) pg. 404; F. E. Livingston and H. Helvajian, "True 3D Volumetric Patterning of Photostructurable Glass Using UV Laser Irradiation and Variable Exposure Processing: Fabrication of Meso-Scale Devices," Proc. SPIE **4830** (2003) pg. 189.

80 F.E. Livingston and H. Helvajian, "True 3D Volumetric Patterning of Photostructurable Glass Using UV Laser Irradiation and Variable Exposure Processing: Fabrication of Meso-Scale Devices," Proc. SPIE **4830** (2003) pg. 189.

81 S.D. Stookey, US Patent 2628160 (1953).

82 A. Berezhnoi, *Glass-Ceramics and Photo-Sitalls* (Plenum Press, NY 1970) pg. 153.

83 F.E. Livingston, W. W. Hansen, A. Huang and H. Helvajian, "Effect of Laser Parameters on the Exposure and Selective Etch Rate of Photostructurable Glass," Proc. SPIE **4637** (2002) pg. 404.

84 P.D. Fuqua, D.P. Taylor, H. Helvajian, W.W. Hansen and M.H. Abraham, "A UV Direct-Write Approach for Formation of Embedded Structures in Photostructurable Glass-Ceramics" in *Materials Development for Direct-Write Technologies*, edited by D.B. Chrisey, D.R. Gamota, H. Helvajian, and D.P. Taylor, (Mater. Res. Soc. Proc. **624** Pittsburgh, PA, 2000) pg. 79; H. Helvajian, P. D. Fuqua, W. W. Hansen and S. Janson, "Laser Microprocessing for Nanosatellite Microthruster Applications" RIKEN Review **32** (2001) pg. 57.

85 G.A. Sycheva, "Nucleation Kinetics of Lithium Metasilicate in Photosensitive Lithium Aluminosilicate Glass," Glass Phys. and Chem. **25** (1999) pg. 501.

86 Y. Cheng, K. Sugioka, M. Masuda, K. Shihoyama, K. Toyoda, and K. Midorikawa, "Three-dimensional Micro-optical Components Embedded in Foturan Glass by a Femtosecond Laser," Proc. SPIE **5063** (2003) pg. 103.

12
Applications of Femtosecond Lasers in 3D Machining

Andreas Ostendorf, Frank Korte, Guenther Kamlage, Ulrich Klug, Juergen Koch, Jesper Serbin, Niko Baersch, Thorsten Bauer, Boris N. Chichkov

Abstract

Femtosecond lasers and related machining systems have been developed which allow efficient material processing on the micro and nanoscale. In order to generate feature sizes with higher quality and smaller dimensions compared to conventional long-pulse laser machining, it is important to understand the basic process phenomena initiated by the interaction of ultrafast laser pulses with a different kind of matter. The following chapter describes the latest state-of-the art of laser source development and analyzes the general machining process characteristics. Based on the achieved models, the potential applications for 3D femtosecond laser micromachining, ranging from microelectronics to MEMS, are described. If nonlinear absorption effects are explored, even nanostrucures far below the conventional diffraction limit, can be achieved. Potential future applications in the biomedical and nanophotonics area are discussed.

12.1
Machining System

12.1.1
Ultrafast Laser Sources

The recent development of lasers with short output pulses in the femtosecond ($1 \text{ fs} = 1 \times 10^{-15}$ s) range, has opened up a wide variety of new applications. Current investigations in scientific laboratories have led to significant progress in various disciplines and have given new insight in light–matter interaction with almost any kind of material (Stuart et al. 1996, Chichkov et al. 1996, Her et al. 1998, Momma et al. 1996). Driven by the broad field of potential applications and advanced requirements, these lasers are continuously being optimized (Aus der Au et al. 2000, Druon et al. 2000).

3D Laser Microfabrication. Principles and Applications.
Edited by H. Misawa and S. Juodkazis
Copyright © 2006 WILEY-VCH Verlag GmbH & Co. KGaA, Weinheim
ISBN: 3-527-31055-X

Since the discovery of the mode-locking effect of laser pulses and thus the realisation of picosecond lasers, pulse durations have consequently been reduced. Especially with the introduction of dye and Ti:Sapphire laser media, powerful laser systems have been developed with extremely short pulses of less than 5 fs (Baltuska et al. 1979). Today, the availability of laser media with superior performance has made dye lasers obsolete in most applications. Therefore, dye lasers, which were the first fs-laser sources, have been widely replaced by Ti:Sapphire systems. Currently, new laser materials such as Yb:KGW as well as new laser concepts are competing with Titanium-based systems with regard to stability, compactness and performance (Keller et al. 1992).

Different concepts aim at the improvement of the stability and availability of amplified, ultrafast laser systems. The driving force behind this is the need for highly stable power output, regardless of the changing environmental conditions. As femtosecond laser systems are increasingly being used as tools for experiments, "hands free" operation of the systems is therefore needed. This demands highly reliable and stable oscillators and amplifiers.

While first oscillators are mainly conventional designs using crystals and discrete optics, modern oscillators often rely on fiber-based concepts. Fiber lasers offer advantageous characteristics for highly stable seed lasers. One main advantage, compared to conventional solid state crystals, is the waveguiding structure of a fiber and the resulting propagation of light within the fiber. When light is focused into a crystal, only a small area shows high intensity. However, within a fiber, the light propagates through the full length without changing its diameter. This results in a large interaction length with high intensity, which is not limited by focusing. Changes of intensity in a fiber are only caused by diffraction, absorption or amplification. Due to the large interaction length at high intensities, the threshold for optical pumping decreases, compared with other concepts. Furthermore, the high intensity causes nonlinear effects, which sum up over the length of the fiber. Dispersion effects also have a significant influence on the propagating signal. Furthermore, the advantageous surface-to-volume ratio of fibers, as well as the small distance between the surface and fiber core, are advantageous for heat removal. For this reason, fiber lasers typically do not require active cooling.

Due to the significant advances in fiber technology, especially with the introduction of photonic crystal fibers, one large drawback of the fiber has been overcome, namely, dispersion effects, which hinder the delivery of ultrashort laser pulses with significant pulse energy within fiber-based, ultrashort pulse lasers. Recent developments of fiber lasers and oscillators have made higher pulse energies of several microjoules accessible.

There are two different types of femtosecond lasers in use: high repetitive systems with "low" pulse energy, and low repetitive systems delivering "high" pulse energy. Examples of the first laser type are fiber lasers and oscillators, typically delivering several MHz repetition rate, but significantly less than microjoule pulse energy. For micromachining processes, laser developments also show tendencies towards high pulse energy at about millijoule level, and increased repetition rates. Currently, such systems are limited to 20–50 kHz repetition rates. Another focus of development is the improvement in the average power, which is presently limited to approximately 5 W for laboratory systems.

Especially developments in new resonator concepts such thin disk lasers as well as the availability of new crystal materials, like KGW, are currently improving average power, repetition rate and pulse energy (Sorokin et al. 2001, Brunner et al. 2000, Krainer 1999).

12.1.2
Automation, Part-handling and Positioning

Precise positioning and part-handling is the key issue for successful, high-quality performance in micromachining and micromanipulation. Regarding the use of micropositioning systems, there is principally no difference between ultrashort pulsed lasers or any other micromachining tool. However, due to the smaller dimensions for direct micromanipulation via ultrashort pulsed lasers, the requirements for accuracy in fast beam and part positioning, have dramatically risen. Miniscule processing spots achieved with focal diameters down to several microns means that handling of the ultrashort pulsed laser process is highly susceptible to minor beam displacements that would generally be tolerable in conventional laser micromachining processes, but which cannot be accepted in micromachining with ultrashort pulsed lasers. At the same time, the processing area has to be increased up to several centimeters. Thus, (a) the choice of the translation stage type (air bearing or stepping motor systems), (b) the design of the relevant axes/optics (degrees of freedom), and (c) the interaction of the modules involved, have to be carefully considered. On the one hand, *relative* 2D positioning is less critical and relatively easy to establish with a precision three-axis translation stage plus rotational stage, or with a scanner plus a precision lift table, e.g., for the generation of contiguous structures and cuts. On the other hand, *absolute* 2D and 3D positioning in a reliable manner is still not only a costly undertaking, but also a tricky challenge for the process designer, for two reasons. First, the focus position of short pulsed lasers is subject to fluctuations depending, e.g., on the pulse energy. However, the remote processing spot must be in the correct position with the first opening of the laser shutter. Consequently, sophisticated monitoring and control of the focus position has to be implemented individually for short pulsed laser applications. Second, usually more than three axes and moving optics equipped with linear measuring systems, respectively, are utilized for full functionality in 3D machining, tremendously increasing the effort for accurate beam delivery. Regarding the achievable complexity of positioning systems, beam-path deviations that typically accompany inevitable readjustments of the laser optics, and tuning of the amplifier's pulse picker are unfortunately often underestimated. Figure 12.1 illustrates schematically a six-axis setup for the finishing of cutting tools by chamfering the cutting edges of drillers and milling tools with picosecond laser pulses (Tönshoff et al. 2003).

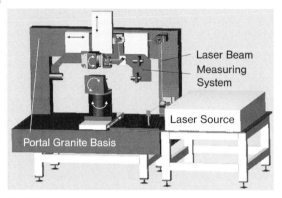

Fig. 12.1 Design of six-axis positioning system.

In the present discussion concerning the integration of short-pulsed lasers into automated processes, the limited long-term stability is certainly the dominating issue. The design of stable laser modules – in terms of being applicable for industrial manufacturing purposes – is still a challenge in the development of amplified femto- *and* picosecond systems. Although intensive efforts have been undertaken to minimize this effect, it will certainly not vanish in future generations of lasers. Therefore, a reliable stability monitoring system, based on the automatic observation of the amplified pulse train and of the single pulse energy is the basis for the implementation of ultrashort pulsed lasers. If feasible solutions for laser stability monitoring, or stability control, and compensation of the inevitable beam-path fluctuations are provided, automated features, e.g., for variable beam shaping and pulse picking, will be applicable.

12.2
Beam Delivery

12.2.1
Transmission Optics

In most laser machining processes, focusing of the laser beam is necessary. Only by decreasing the beam diameter can the laser peak intensity increase to values where material modification becomes possible. Furthermore, a smaller laser–matter interaction area allows a much more precise machining process.

Focusing can be achieved using refractive, diffractive and reflecting optics. This section gives a comparison of these techniques in terms of focusing ultrashort laser pulses.

Because it is relatively easy and cheap to realize, the spherical lens is by far the most common and readily available focusing optic. However, a spherical lens only works perfectly in the paraxial limit. For beam optics, this is the limit where $\sin \varphi$ can be approximated by φ, where φ is the angle of the refracted beam relative to the optical axis. The nonideal focusing of off-axis (or marginal) beams, results in what are called aberrations, which are grouped by order in the Taylor expansion of sin φ at which they are taken into account. There are five primary aberrations of lowest order, termed spherical aberration, coma, astigmatism, field curvature, and distortion (Hecht 2002). All these aberrations have to be considered when designing an imaging system and, in fact, all can be minimized with the proper choice of a multiple-lens system. However, when simply focusing a circular beam which is guided normal to the focusing lens, spherical aberration is the only aberration which deteriorates the focal spot size, because all the other aberrations involve the different focal lengths of oblique beams, either relative to the paraxial rays, or to marginal rays parallel to the optical axis.

In the case of a nonmonochrome beam (e.g., an ultrashort laser pulse) chromatic aberrations also have to be considered. These too can be minimized with the proper choice of a multiple-lens system. Here, an achromatic lens system is required. In general, optimization of the lens system in terms of aberrations, gains in importance with tighter focusing. This explains the complex setup of high-NA optics for nearly diffraction-limited spot sizes (e.g., high magnifying microscope objectives).

Besides refractive optics, like lenses or lens systems, it is possible to use diffractive optics for focusing. Diffractive optical elements (DOEs) are computer-generated holographic devices which can transform an illuminating laser beam into a specified intensity distribution. The diffractive surfaces of these beam-shaping elements are split into arrays of cells, each designed to transform the phase of the coherent illuminating beam by a specified amount. The required intensity profile is produced by interference of the diffracted wave front in the reconstruction plane. Nearly any shape like rectangles, squares, lines, symmetric/nonsymmetric spot arrays and even character sets, are possible. The computer-generated algorithms used to calculate the diffraction patterns incorporate the profile of the illumination laser beam. Therefore, DOEs are not all-purpose. Furthermore, the shape in the reconstruction plane does not contain the entire laser pulse energy.

Another focusing technique, apart from transmission optics, is to use reflecting optics. A single concave mirror is the simplest optic of this type. To achieve a focus position outside the incident beam, the optic has to be turned with respect to the incident beam. This results in astigmatism. The beam shape becomes elliptic above and below the focus position. To avoid this, a Schwarzschild objective can be used. It consists of a small convex and a large concave mirror with common optical axis (Barnes et al. 1990). A Schwarzschild objective allows tight focusing with the advantage of a much wider working distance, compared to a focusing lens objective with the same focus spot size.

When focusing ultrashort laser pulses with transmission optics, there are two points which must be taken into consideration. Firstly, when a short pulse travels through a dispersive medium, the component frequencies are separated in time. Normal dispersive media, like glass, impose a positive frequency chirp on the pulse, meaning that the blue components are delayed with respect to the red. This is a pulse-broadening effect, if it is not compensated by a suitable negative frequency chirp before the medium transmission. Secondly, the high intensities of ultrashort laser pulses induce self-focusing. The refractive index becomes intensity dependent. For a refractive optic, this results in a shift of the focus position towards the optic, with higher laser intensity. With tighter focusing this effect becomes stronger.

The choice of a suitable focusing optic is a question of focus diameter, laser intensity and pulse length. When using transmission optics, a good understanding of dispersion is essential in order to deliver a short pulse to the sample. To avoid problems with nonlinear effects and chromatic aberrations, reflecting optics like Schwarzschild objectives can be used. For short laser wavelengths, scattering should be taken into account. The intensity of the scattered light is proportional to the fourth order of the light frequency.

12.2.2
Scanning Systems

For material processing, it is necessary to control the laser focus flexibly on the sample. Two possibilities can be listed. First, the sample can be moved while keeping the laser focus at a fixed position. Second, the laser beam can be scanned without moving the sample. When moving the sample, sometimes large masses have to be accelerated and decelerated, which limits the speed and/or the precision of the material processing. This drawback can be overcome if galvo scanners are used to move the laser focus on the surface of the sample, since then only the scanning mirrors have to be rotated.

Two different setups are possible for a galvo scanner. First, a moving coil can be placed in the centre of a permanent magnet. This kind of setup was used in the 19th century when electrical currents were measured by means of moving coil instruments. There are two major drawbacks when using this setup: the thermal limit defined by the heat transfer capabilities of the coils, and coil creep and deformation under centrifugal acceleration. The second possible setup is a moving permanent magnet placed in the centre of two pole pieces, which are supplied with two coils thermally connected to the housing. This setup, which is illustrated in Fig. 12.2(a), is generally applied in modern galvo scanners. Compared to moving coil systems, this arrangement can accommodate three times the power dissipation of equivalent inertia and torque constants. A spindle is attached to the magnet, which is fixed in its position by means of two ball bearings. Instead of using a mechanical spring, a servo loop with a position detector (PD) and controller is used. To control the laser focus within the x-y plane, two galvo scanners have to be arranged perpendicular to each other (see Fig. 12.2(b)). The beam is reflected 90° from the mirror of the first scanner to the mirror of the second scanner.

When using standard optics to focus the scanned laser beam, the focal spot moves on a curved surface, as shown in Fig. 12.3(a). This effect limits the scanning area, since the laser focus will strike the surface of the sample only within a small range of angles. Therefore, most scanners are equipped with so called f–Θ lenses (see Fig. 12.3(b)). These lens systems are optimized in such a way that, for

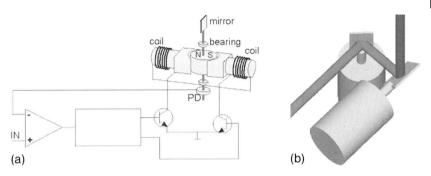

Fig. 12.2 Schematic layout of a galvo scanner (a). Arrangement of two galvo scanners for the positioning of a laser beam in the x-y plane (b).

each scan angle Θ, the foci are located in a plane. However, the so called f–Θ condition must be fulfilled. A second drawback, when scanning the beam rather than moving the sample, is the fact that the angle of incidence on the sample will not be perpendicular for every scanning angle. This especially limits the aspect ratio when using a scanning system to drill deep holes (see Fig. 12.3(c)). This problem is solved by applying telecentric lens systems which are corrected to provide perpendicular incidence for any point within the scanning field (see Fig. 12.3(d)).

The area that can be scanned depends on the focusing optics applied. Standard galvo scanners provide a maximum scanning angle of ± 12.5°. Using a lens system with a focal distance of 100 mm, an area of about (45 × 45) mm^2 can be scanned. The achievable scanning speeds can be in the order of several meters per second, again depending on the focal length of the focusing optics.

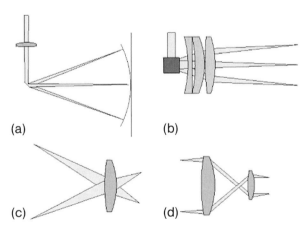

Fig. 12.3 (a) Due to the rotational symmetry of the scanner, the foci of the deflected beams will form a circular or spherical surface when using standard lenses. (b) F–Θ lens: The lens system is optimized in such a way that for each scan angle the foci are located in a plane. (c) Using standard optics, the scanned beam is incident under different angles on the surface of the sample. (d) Telecentric lenses are corrected to provide perpendicular incidence for any point within the scanning field.

12.2.3
Fiber Delivery

Fiber propagation of high-power laser pulses in a single spatial mode is important for many scientific and technological applications, such as for fiber lasers and amplifiers. In a conventional fiber, the light is guided by internal reflection, which is strongly influenced by the influences of optical nonlinearities. Therefore, high intensities lead to undesired side-effects, such as self-focusing effects which can lead to the destruction of the optical devices. Conventional fibers are limited in dispersion, attenuation, nonlinearity and damage threshold, because the most intensive parts of the beam are transmitted through the core of the fiber, resulting in high intensity and strong response on nonlinearities. Ultrashort laser pulses with peak power exceeding the megawatt level cannot be delivered by this type of fiber. At such intensities, self-phase modulation and Raman scattering, combined with dispersion effects result in strong pulse distortion travelling a few centimeters in a fiber.

In order to maintain the beam and pulse characteristics for ultrashort pulsed laser radiation in conventional fibers, the energy per pulse is restricted to values not higher than the nanojoule level.

Recent developments in photonic bandgap fibers (PBGF) show a different mechanism of guiding light. Here the light is guided more by diffraction than by total internal reflection, which means fiber designs can be used, where the light field is confined to a hollow fiber core. Such a fiber design is shown in Fig. 12.4 (Ouzounov et al. 2003). PBGFs are fibers with a two-dimensional array of fine channels running down the full fiber length. One of the main effects is light guidance with low losses in a large central air hole, due to the photonic band gap of the surrounding air/silica matrix (Luam et al. 2004).

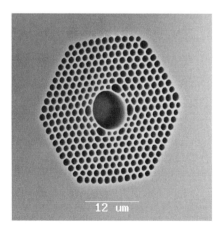

Fig. 12.4 SEM-image of a photonic bandgap fiber.

Such bandgap-guided modes are confined for a limited range of wavelengths, typically 10–15% of the central wavelength for silica-based structures. The performance of such fibers is substantially free from limitations imposed by the core material. Therefore, the limitations are relieved in hollow-core PBGF, sometimes by several orders of magnitude, making a number of previously impossible applications possible (Luam et al 2004). Using such fibers, a beam delivery of megawatt femtosecond radiation of 1550 nm has been demonstrated for fiber lengths of 3 m. In such fibers, the delivery of microjoule femtosecond laser pulses is possible (Ouzounov et al. 2003).

The total dispersion of these fibers is dominated by waveguide dispersion, which is usually anomalous. The need for anomalous dispersion restricts the wavelength beyond the zero-dispersion wavelength of bulk silica, which is 1.27 μm. Recently developed air–silica microstructured fibers provided anomalous dispersion at wavelengths down to 700 nm (Ranka et al. 2000, Price et al. 2002).

The large dispersion slope within the transmission gap and extremely low nonlinearity, allow for high power solitons. Due to the guidance of the light within the hollow, typically air-filled, central channel of the fiber, the refractive index and the nonlinearities of the air contribute to the characteristics of the fiber. Typically, the nonlinearity of air is about 1000 times less than a conventional glass fiber. Due to the low nonlinearity and the large anomalous dispersion of PBGFs, the peak power of the transmitted solitons can be increased up to ~1000 times, compared with solid, singlemode fibers.

The transmitted pulse splits into a soliton part and a dispersive wave part, which experience a Raman shift. The Raman frequency shift limits the maximum distance over which the soliton can propagate within the medium, because the pulse might be shifted outside the transmission band of the fiber. However, by introducing a gas such as Xenon into the core, which does not have a Raman active component, it is possible to avoid the soliton Raman shift. Using this method, the transmission of 75 fs laser pulses of Gaussian shape exceeding a peak power of 5.5 MW has been demonstrated for fiber lengths of more than 5 m, showing potential for fibers longer than 200 m (Ouzounov et al. 2003).

12.3
Material Processing

12.3.1
Ablation of Metals and Dielectrics

The principal interaction phenomen of "long" (nanosecond) laser pulses and ultrashort (femtosecond) laser pulses with solid-state materials, are illustrated in Fig. 12.5. Long pulses applied with sufficient intensities ($I > 10^{10}$ W cm^{-2}) lead to the formation of laser-induced plasma, significantly reducing the amount of radiation that contributes to interaction with the solid-state material. In contrast, ultrashort laser pulses are not shielded by the plasma, and interact directly with the material surface, due to the negligible spatial expansion of the plasma during the extremely short time interval.

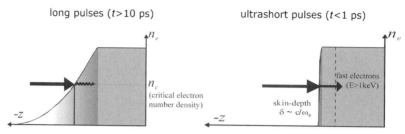

Fig. 12.5 Illustration of the interaction of long and ultrashort laser pulses with solids. The laser radiation can only propagate in the plasma if the electron number density n_e is below the critical value n_c (Source: (Momma et al. 1997)).

In this time regime, the absorbed pulse energy is in a thin, superficial layer that corresponds in its thickness to the optical penetration depth (skin depth, ~10 nm). Thermal diffusion effects into the lattice of the solid-state material are almost irrelevant. Moreover, the classical heat conduction theory, based on the assumption that a material can be characterized by one temperature only, is no longer valid. Instead, electron–lattice interactions have to be taken into account, and the temperatures of both the electrons and the lattice have to be treated separately, as demonstrated below. At very high intensities ($I > 10^{16}$ W cm^{-2}), electrons located within the skin depth are heated up to extremely high temperatures, and additionally, overheated electrons are generated with energies up to the MeV range. Subsequent diffusion of the hot electrons transmits the major part of the pulse energy into subjacent areas, which is the reason for higher ablation rates per pulse, compared with ablation mechanisms of long pulses. Part of the pulse energy is emitted, due to the Bremsstrahlung mechanism, via a broad spectrum of hard x-ray radiation (keV to MeV range).

A detailed, analytical survey on the ablation of metals by Momma et al. (1997) will be discussed here briefly. Ablation of metals with femtosecond lasers is characterized by rapid overheating and thermalization of the electrons within the optical penetration depth. Due to the low thermal capacity of electrons in comparison with the lattice, the electrons are rapidly heated beyond the Fermi level to very high, transient temperatures, forcing an extreme nonequilibrium state between the electron and lattice system. Assuming: (i) the electron thermal capacity to be described by $C_e = 3N_e k_B/2$, with N_e being the electron number density and k_B being the Boltzmann constant; and (ii) the electron thermal conductivity to be given by $k_e = C_e v_F^2 \tau/3$, with τ_F being the Fermi velocity and τ being the electron relaxation constant described in a first approximation by $\tau = a/v_F$, then the electron diffusion coefficient D determined by $D = k_e/C_e$ remains temperature-independent, in order to be applied using the differential equation system below, modeled for the temperature prediction of the electrons (T_e) and of the lattice (T_l) along the solid plane normal (z) and the time scale (t)

$$\frac{\partial T_e}{\partial t} = D\frac{\partial^2}{\partial z^2}T_e - \frac{T_e - T_l}{\tau_e} + \frac{IA\alpha}{C_e}e^{-\alpha z}$$

$$\frac{\partial T_l}{\partial t} = \frac{T_e - T_l}{\tau_l} \tag{1}$$

with the initial and boundary values

$$T_e(z, t = -\infty) = T_i(z, t = -\infty) = T_0 \approx 0$$

$$\left.\frac{\partial T_e}{\partial z}\right|_{z=0} = \left.\frac{\partial T_e}{\partial z}\right|_{z=\infty} = 0$$

The expression for the analytical solution of T_l for a pulse with intensity $I(t)$, and absorption A is then approximately given by

$$T_l \approx \frac{F_a}{C_l}\frac{1}{l^2 - \delta^2}\left[le^{-z/l} - \delta e^{-z/\delta}\right] \tag{2}$$

with F_a being the absorbed fluence, C_l the lattice thermal capacity, the characteristic electron diffusion length $l = \sqrt{D\tau_a}$ with the ablation interval τ_a which is independent of the pulse length in this time regime, and the optical penetration length $\delta = 1/\alpha$. Based on this analytical solution, two border cases can be derived for the lattice temperature distribution:

$$T_l \approx \frac{F_a}{C_l\delta}e^{-z/\delta} \quad (l \ll \delta) \tag{3}$$

and

$$T_l \approx \frac{F_a}{C_l l}e^{-z/l} \quad (l \gg \delta) \tag{4}$$

Equation (3) is valid for negligible electron diffusion ($l \ll \delta$), whereas Eq. (4) is valid in the case when the electron diffusion length is considerably greater than the optical penetration depth ($l \gg \delta$).

Assuming that the conditions for significant ablation are fulfiled if the absorbed energy density is larger than the solid enthalpy of evaporation $C_l T_l \geq \rho\,\Omega$, with the density ρ, and specific enthalpy of evaporation Ω, the following expressions for the ablation depth can be derived from equations (3) and (4):

$$L \approx \delta\ln\left(\frac{F_a}{F_{th}^{\delta}}\right) \quad (l \ll \delta) \tag{5}$$

and

$$L \approx l\ln\left(\frac{F_a}{F_{th}^{l}}\right) \quad (l \gg \delta) \tag{6}$$

with the respective ablation thresholds determined by $F_{th}^{\delta} \approx \rho\,\Omega\delta$ and $F_{th}^{l} \approx \rho\,\Omega l$.

According to the aforementioned simplified, theoretical model, Eqs. (3) and (4) yield two expressions for the ablation depth per pulse as logarithmic functions of the laser fluence, i.e., the pulse energy. Equation (5) is the characteristic expression for the ablation depth per pulse for a mechanism without a significant heat transfer beyond the optical penetration depth. Equation (6) describes the ablation mechanism that is characterized by a significant heat transfer beyond the optical penetration depth due to hot electron diffusion. Both effects are experimentally verifiable for a variety of materials, including metals with excellent thermal conductivities like copper (Momma et al. (1997)).

Ablation of dielectrics with femtosecond lasers is also characterized by nonthermal breakdown of the solid material. However, the free conduction band electrons (CBE) essential for the high-intensity energy diffusion first need to be generated. CBE generation is considered to be described by two competing processes: (i) collisional (avalanche) ionization, and (ii) multiphoton ionization. The dominance of either is determined by the form of energy gain, which is highly dependent on the experimental environment. A detailed study of this topic is presented in Jia et al. (2000). If sufficient CBE densities are achieved, the ablation proceeds similarly to the mechanism observed in the short-pulse ablation of metals. Due to the additional expense in pulse energy for CBE generation, though, the available pulse energy for hot electron diffusion, i.e., the effective ablative power, is reduced, in return decreasing the hot CBE maximal diffusion length. Thus, for predictions of the potential ablation depth of dielectrics, Eq. (5) is more favorable. The mechanisms of CBE generation are manifold and also influence the ablation thresholds. Typically, a significant dependency of the pulse length on the breakdown threshold in dielectrics can be observed (as reported, e.g., in Soileau et al. (1984) and Stuart et al. (1996)), and in fact must be considered for correct determination.

12.3.2
fs-laser-induced Processes

In addition to direct ablation, material processing can also be done using fs-laser-induced processes, e.g., for structural and optical changes within the bulk of transparent materials, and for subsequent etching processes. This section presents such effects that can be employed in different application fields.

In the case of large-band-gap materials, where laser machining is based on non-linear absorption of high-intensity pulses for energy deposition, femtosecond lasers can be used for local photon absorption within the bulk material, allowing three-dimensional micromachining. The extent of the structural change produced by femtosecond laser pulses can be as small as or even smaller than the focal volume (Ferman et al. (2003)).

In principle, laser pulses above a certain material-dependent energy threshold, produce permanent structural changes in the material. Several different mechanisms can lead to these changes, each producing a different morphology. This can lead to density and refractive index variations due to nonuniform resolidification from the molten phase, but also to color centers, or at higher pulse energies, to voids. Such voids are surrounded by a halo of higher density, caused by an explosive expansion of the focal volume.

In order to confine structural changes to the focal volume, experiments in various transparent materials have suggested that pulse durations of a few hundred femtoseconds or less are necessary. With such pulses and the respective process parameters, the produced structures can be tailored to a particular machining application. The small density and refractive index change produced near the threshold are suitable for direct writing on optical waveguides and other photonic devices, as indicated in Fig. 12.6 (Korte et al. (2000)), and the voids formed at higher energy are ideal for binary data storage, because of their high optical contrast.

Fig. 12.6 Principle of the setup used for optical waveguide microfabrication in transparent materials.

A related fs-laser-induced process is the etching of quartz glass at significantly increased rates, which can be achieved when etching the substrate after irradiation with low-fluence fs laser pulses. Using this technique, locally increased etching depths have been achieved, which can be of interest in MEMS applications. As an example, the etched depth of Pyrex glass treated with a 5% HF solution for four hours has been reported to increase by a factor of ten, if low-fluence femtosecond laser pulses are applied before the treatment (Chang et al. (2003)). In other investigations, irradiating silica glass samples near the bottom surface and applying a selective chemical etch to the bottom surface has produced clean, circular, submicrometer diameter holes, that were spaced as close as 1.4 μm to one another. Such techniques could be used for the fabrication of periodic microstructures in glass, and could be applicable to the production of two- and three-dimensional photonic band-gap crystals (Taylor et al. (2003)).

Another fs-laser-induced effect connected with etching is the formation of microscopic spikes, by applying short pulses to a silicon surface in an SF_6 atmosphere. This application is also demonstrated in Section 12.5.3, about surface structuring. In this spike-formation process, in which silicon surfaces develop arrays of sharp conical spikes, both laser ablation and laser-induced chemical etching of silicon are involved. The spikes have the same crystallographic orientation as the bulk silicon, and always point along the incident direction of the laser pulses. The base of the spikes has an asymmetric shape and its orientation is determined by the laser polarization. The height of the spikes decreases with increasing pulse duration or decreasing laser fluence.

Figure 12.7 shows sharp conical spikes produced on Si(100) using 500 laser pulses of 100 fs with a fluence of 10 kJ m^{-2} in an SF_6 atmosphere, viewed from the top of the surface (a) and from the side of the surface (b) (Her et al. (2000)). The distribution of the spikes reflects the intensity variation over the spatial profile of the laser beam, the spike separation increasing sharply with increasing laser fluence.

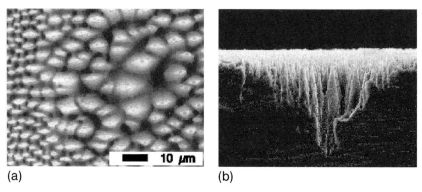

Fig. 12.7 Conical spikes generated with 100 fs pulses at a fluence of 10 kJ m^{-2}, shown from the top (a) and side (b).

In investigations with a fixed fluence, spike separation has been shown to decrease sharply as the pulse duration increases from 0.1 to 1 ps, and to remain approximately constant between 1 and 10 ps. At 250 ps, the surface is corrugated, but no sharp spikes are observed.

Cones with a morphology similar to those on silicon have been observed on metals, dielectrics, and oxides after ion or nanosecond laser sputtering. Their formation has been attributed to shielding of the underlying substrate by sputtering-resistant impurities on the surface, resulting in cones as the surrounding material is removed by the sputtering particles. A similar argument can be used to explain the formation of the spikes on silicon. Initial fluctuations give rise to preferential removal of material at certain locations, explaining the crystallographic orientation and pointing of the spikes (Her et al. (2000)).

Silicon surfaces that have been structured this way are highly light-absorbing and turn deep black. In addition to near-unity absorption in the visible range, the irradiated surface absorbs over 80% of infrared light for wavelengths as long as 2500 nm. This makes the microstructuring process interesting for the production of photodiodes with high responsivity in both the visible and infrared range, and to make use of the extended absorption range for silicon solar cells, which convert more of the spectrum of sunlight into electricity.

The examples presented in this section demonstrate that, in the field of material processing, there is a wide range of possible femtosecond laser applications besides direct ablation, most of which must still be explored.

12.4
Nonlinear Effects for Nano-machining

12.4.1
Multiphoton Ablation

In laser-processing technologies, the minimum achievable structure size is determined by the diffraction limit of the optical system, and is of the order of the radiation wavelength. However, this is different for ultrashort laser pulses. By taking advantage of the well-defined ablation (in general, modification) threshold, the diffraction limit can be overcome by choosing the peak laser fluence slightly above the threshold value (Pronko et al. (1995)). In this case, only the central part of the beam can modify the material, and it is possible to produce subwavelength structures. On transparent materials, there is a further possibility to overcome the diffraction limit. This will be discussed in this section.

Due to the nonlinear nature of the interaction of femtosecond laser pulses with transparent materials, the simultaneous absorption of several photons is required to initiate ablation. Multiphoton absorption initially produces free electrons that are further accelerated by the femtosecond laser beam electric field. These electrons induce avalanche ionization and optical breakdown, and generate a microplasma. The subsequent expansion of the microplasma results in the fabrication of a small structure at the target surface. The diameter of such a femtosecond laser-drilled hole not only depends on the energy distribution in the laser-matter interaction area and the ablation threshold. There is also a dependence on the energy band gap of the material. To overcome a wider band gap, more photons are needed. This results in a higher limitation of the ionization process to the peak intensity region of the beam profile. The smaller absorption volume leads to a reduced hole diameter. This, combined with the technique described above, is a further way of producing subdiffraction-limited structures.

The minimum structure size (for constant fluence ratio F/F_{th}) that can be produced using femtosecond laser pulses in materials with a certain energy band gap can be calculated by the equation

$$d = \frac{k\lambda}{\sqrt{q}NA}$$

where λ is the radiation wavelength, q is the number of photons required to overcome the energy band gap, NA is the numerical aperture of the focusing optics, and k is a proportionality constant ($k = 0.5\dots1$). To illustrate this equation, Fig. 12.8 shows the effective beam profile and the ablation hole diameter for materials where one, two and four photons are needed for each ionization process.

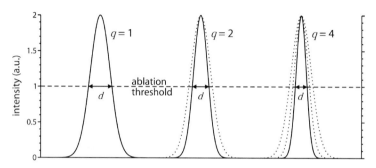

Fig. 12.8 Effective beam profile and ablation hole diameter *d* for materials where *q* = 1, 2 and 4 photons are needed for each ionization process.

The reproducibility of this technique depends on two factors. First, material defects can locally change the band gap energy (and the ablation threshold), resulting in a change of structure size. Second, avalanche ionization increases in importance with longer laser pulses. This results in an increasing influence on the number of free electrons existing in the laser–matter interaction area before the onset of ionization. For tight focusing, one cannot speak of a homogeneous distribution of these electrons (coming from material defects or doping). Therefore, high, reproducible laser structuring of transparent materials with small structure sizes requires short pulse lengths (e.g., for SiO_2: < 100 fs, better: < 50 fs (Kaiser et al. 2000)) so that avalanche ionization is negligible.

12.4.2
Two-photon Polymerization

Several groups have recently demonstrated that nonlinear optical lithography based on two-photon polymerization (2PP) of photosensitive resins allows the fabrication of true 3D nanostructures (Kawata et al. 2001, Cumpston et al. 1999, Sun et al. 1999, 2000, 2001/a,b,c, Galajda and Ormos 2001, Maruo and Kawata 1997, 1998, Kuebler et al. 2001, Tonaka et al. 2002 and Serbin et al. 2003), and therefore of 3D photonic crystals.

When tightly focused into the volume of a photosensitive resin, the polymerization process can be initiated by nonlinear absorption of femtosecond laser pulses within the focal volume. By moving the laser focus three-dimensionally through the resin, any 3D nanostructure with a resolution down to 100 nm can be fabricated. There are many well known applications for single-photon polymerization (1-PP) like UV-photolithography or stereolithography, where a single UV-photon is needed to initiate the polymerization process on the surface of a photosensitive resin, as shown in Fig. 12.9 (a). Depending on the concentration of photo-initiators and of added absorber molecules, UV light is absorbed within the first few μm. Thus, single-photon polymerization is a planar process restricted to the surface of the resin. On the other hand, the photosensitive resins used are transparent to near-infrared light, i.e., near-IR laser pulses can be focused into the volume of the resin (Fig. 12.9 (b)). If the photon density exceeds a certain threshold value, two-photon absorption occurs within the focal volume, initiating the polymerization process. If the laser focus is moved three-dimensionally through the volume of the resin, the polymerization process is initiated along the track of the focus, allowing the fabrication of any 3D microstructure.

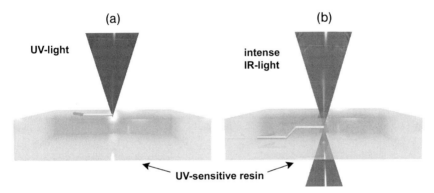

(a) (b)

UV-light intense IR-light

UV-sensitive resin

Fig. 12.9 The principle of single-photon polymerization (a) and two-photon polymerization (b).

The 3D movement of the laser focus can either be realized by scanning the laser in the *x-y* plane, using a galvo-scanner while moving the sample in the *z*-direction, or by moving the sample three-dimensionally, using a 3D piezo stage. Obviously, there are several advantages of 2-PP compared to 1-PP: First, since polymerization can be initiated within the volume of the resin, 2-PP is a true 3D process, whereas 1-PP is a planar process. Applying 1-PP, 3D structures can only be fabricated by means of working 2.5-dimensionally, i.e., working layer by layer. Second, when photopolymerization occurs in an atmosphere with oxygen, quenching of the radicalized molecules on the surface of the resin, and hence a suppression of the polymerization process takes place. This drawback can be overcome by working in the volume rather than at the surface, as is done in 2-PP. Third, the two-photon excited spot is smaller then a single-photon excited spot, allowing the fabrication of smaller structures.

Due to the threshold behaviour of the 2-PP process, a resolution beyond the diffraction limit can be realized by controlling the laser-pulse energy and the number of applied pulses. To predict the size of the polymerized volume (or voxel), it is necessary to define a polymerization threshold. We assume that the resin is polymerized as soon as the particle density of radicals $\rho = \rho(r, z, t)$ exceeds a certain minimum concentration (threshold value) ρ_{th}. For the same initiator, this value is independent of the particular initiation process that leads to the generation of radicals, and should be the same for single and two-photon absorption. The density of radicals ρ produced by femtosecond laser pulses can be calculated by using a simple rate equation:

$$\frac{\partial \rho}{\partial t} = (\rho_0 - \rho)\sigma_2 N^2$$

where $\sigma_2 = \sigma_2^a/\eta$ is the effective two-photon cross-section for the generation of radicals [cm⁴ s], which is defined by the product of the ordinary two-photon absorption cross-section σ_2^a, and the efficiency of the initiation process $\eta < 1$. $N=N(r,z,t)$ is the photon flux, and ρ_0 is the primary initiator particle density. Inserting a Gaussian light distribution within the focal volume, leads to a voxel diameter of

$$d(N_0, t) = r_0 \left[\ln\left(\sigma_2 N_0^2 n\tau_L/C\right)\right]^{1/2}$$

where $n = vt$ is the number of applied pulses, v is the repetition rate of the laser system, t is the exposure time, τ_L is the duration of the laser pulses, and $C = \ln[\rho_0/(\rho_0 - \rho_{th})]$. Figure 12.10 shows the measured voxel diameter for varying laser power and irradiation time.

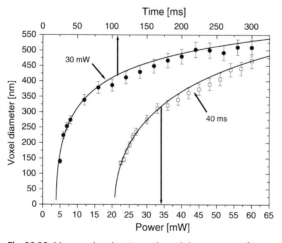

Fig. 12.10 Measured and estimated voxel diameter as a function of time and laser power.

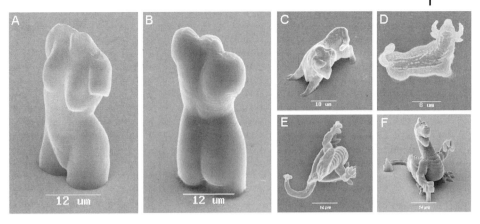

Fig. 12.11 Images of different computer generated micro-models fabricated by means of 2-PP.

By using appropriate algorithms for the processing of 3D CAD files that are similar to those applied in stereolithography, a 3D model can be fabricated on a μm scale as shown in Fig. 12.11.

12.5
Machining Technology

12.5.1
Drilling

Conventional mechanical drilling tools are effective for drilling holes in metals down to approximately 200 μm in diameter at depths of approximately 1 mm. For finer structure sizes or irregular shapes, electron beam, electro discharge (EDM) or conventional lasers are used. Conventional lasers such as CO_2, Nd:YAG, or copper vapor laser machine materials based on localized heating. All of these micro-fabrication techniques heat the material to the melting or boiling point, resulting in thermal stress to the remaining material and often a heat-affected zone. Higher precision is difficult to achieve using these techniques. Only ultrashort laser pulses (picosecond and femtosecond pulses) offer the advantage of removing material without significant transfer of energy to the surrounding areas (Chichkov et al. 1996, Stuart et al. 1996, Simon and Ihlemann 1996, Kautek et al. 1996, Nolte et al. 1997, Horwitz et al. 1999). The important shape characteristics of a laser-drilled hole are the entrance and exit diameter, roundness, wall roughness and hole profile. Control of the hole profile is important because some applications require tapered holes, whereas others require cylindrical ones.

Generally, there are three laser drilling techniques used in industrial applications, which are also relevant to femtosecond laser drilling. These techniques are: single pulse drilling, percussion drilling and trepanning (see Fig. 12.12). Single pulse drilling is used when a high processing speed is necessary and quality is secondary. In percussion drilling, more pulses are necessary, either to increase the volume of material removed (greater depth) or to increase the accuracy of the process, by removing a smaller volume per pulse. During trepanning, the focused laser beam moves along a circular path relative to the workpiece. Trepanning is the standard technique for drilling larger hole diameters up to several millimeters (Dausinger, 2001).

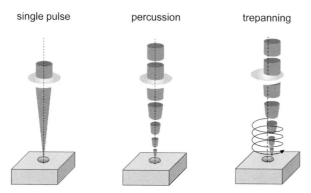

single pulse   percussion   trepanning

Fig. 12.12 Laser drilling techniques.

Applying the trepanning technique leads to excellent results concerning the roundness of the micro holes. Due to the fact that the focused beam moves in circles, the roundness of the hole is less dependent on the beam profile than when simply focusing the beam without rotation. Figure 12.13 demonstrates the difference in hole quality between percussion and trepanning drilling results.

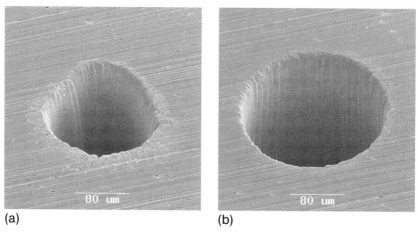

(a)   (b)

Fig. 12.13 Femtosecond laser drilling in stainless steel using percussion (a) and trepanning technique (b).

The main advantages of material processing using femtosecond laser pulses are: (i) efficient, fast and localized energy deposition; (ii) well-defined deformation and ablation thresholds; (iii) minimal thermal and mechanical damage of the substrate material. These advantages make high-quality drilling of solids at relatively low laser fluences, close to the ablation threshold, possible.

However, for many practical applications, when a high processing speed is required, much higher laser fluences well above the ablation threshold are necessary. At these laser fluences, the energy coupled into a workpiece, and the corresponding thermal load, are high. The ablation depth per pulse is determined by the energy transfer into the target, due to the electron thermal conduction and/or the generated shock wave. This is similar to ablation using nanosecond, and longer, laser pulses. Therefore, in the high-fluence regime, considerable advantages for material processing with femtosecond lasers are not usually expected. They can probably drill more efficiently and produce deeper holes, due to the lower thermal losses and negligible hydrodynamic expansion of the ablated material during the femtosecond laser pulse. With the latter, "plasma shielding"-induced radiation losses can be avoided.

For deep drilling of metals, which is necessary in many industrial applications, the most important criteria are the hole geometry and quality. The use of more expensive femtosecond lasers could be justified if they could fabricate holes with a special geometry, superior quality, and high reproducibility. It has been demonstrated that femtosecond lasers are close to this goal. They have the potential to provide a simple processing technique which does not require any additional post-processing or special gas environment. This can be considered as a real advantage of using femtosecond lasers.

The question that needs to be answered at this point is how, even at high laser fluences, can high-quality holes be drilled? The answer to this question is quite simple. High-femtosecond-pulse laser intensities are required to rapidly drill a through-hole. After the through-hole is drilled, the high intensity part of the laser pulse propagates directly through the hole without absorption. Only at the edge of the laser pulse, at laser intensities close to the ablation threshold, is a small amount of material removed. Starting from this point, the interaction of femtosecond laser pulses with the workpiece occurs in the low-fluence regime, where all advantages of femtosecond lasers can be realized. This can be considered to be "integrated" low-fluence femtosecond laser post-processing, which is responsible for the excellent hole quality, or as a low-fluence finishing (Kamlage et al. 2003). The high quality of the single holes and their high reproducibility is demonstrated in Fig. 12.14.

In conclusion, femtosecond laser material processing at high laser fluences has been shown to be a practical tool for high-quality, deep drilling of metals. So far, however, industrial users have been reluctant to integrate femtosecond laser drilling in mass production, as the laser sources are known to be quite unreliable and expensive. But the market perception is changing, thanks to recent advances in lasers concerning ease of use, reliability and overall lifetime. New femtosecond laser sources are entering the market, so in the not-too-distant future, mass produced femtosecond lasers can be expected.

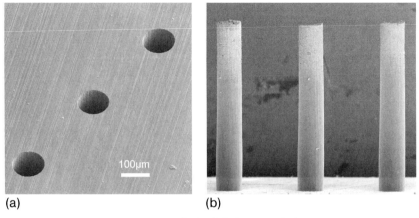

(a)   (b)

Fig. 12.14 Scanning microscope image of holes drilled in 1 mm thick stainless steel plate using femtosecond laser pulses, and their replicas.

12.5.2
Cutting

Today, lasers are well established and very flexible tools for producing various micro-structures. Beside marking, drilling, and welding applications, cutting is the most frequent laser machining process.

For metals, the laser-cutting process can be performed by melting, by oxygen reaction, and by sublimation (Beyer et al. 1985). The advantage of sublimation cutting is that melting and heat-affected zones can be minimized or even avoided. The disadvantage is, however, that the cutting speed is relatively low due to the high evaporation enthalpy of metals. This is the major reason why current industrial laser cutting is predominantly based in melting and oxygen-cutting processes. Oxygen cutting has the advantage that high machining speeds can be realized, and the disadvantage that oxidation occurs. Both processes lead to high burr formation and depositions on the material surface. The burr and the depositions must be removed in further production steps.

Cutting with conventional laser sources (e.g. Nd:YAG), which produce pulses with durations in the range of nanoseconds to milliseconds, has several limitations. First of all, the thermal load in the material, due to heat conduction, is relatively high, which results in large heat-affected zones and melting. The melting again leads to a significant burr formation at the cutting edges and to a deposition of solidified droplets sticking on the surface. In many applications, both have to be removed via several subsequent treatments. Due to the thermal load during laser cutting, the minimum producible structure size is limited. Moreover, a variety of materials cannot be structured using this conventional technique. Thus, there is a demand for an alternative, less invasive, laser machining technique.

Such a technique is femtosecond laser machining. With femtosecond lasers almost all materials (e.g., metals, ceramics, glass, polymers, organic tissue, etc.)

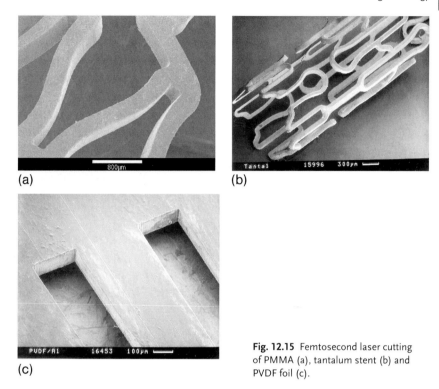

(a)

(b)

(c)

Fig. 12.15 Femtosecond laser cutting of PMMA (a), tantalum stent (b) and PVDF foil (c).

can be structured with minimum or no thermal damage. Materials can be processed which otherwise could not be structured with conventional (laser) techniques. The cuts are very smooth and completely burr-free, and the surface is free of depositions. Figure 12.15 shows three examples. Except for a simple ultrasonic treatment in alcohol, no post-processing was applied here. The stent in Fig. 12.15(a) has very smooth struts which cannot be machined using conventional laser techniques. The potential of femtosecond machininig technique is even more pronounced in the case of extremely sensitive and delicate materials. The SEM images in the middle and the bottom of Fig. 12.15 show cuts in a PVDF foil and PMMA, respectively. Both polymers are optically transparent and very heat-sensitive materials. At present, no other suitable machining process for cutting and structuring these polymers achieves the quality of femtosecond laser machining. For many applications, the nonthermal nature of the femtosecond ablation process is necessary to ensure that the material properties are maintained, even adjacent to the cut area.

Where conventional (laser) techniques are not suitable for cutting heat-sensitive materials, or for very high precision, the use of lasers producing ultrashort pulses with pulse durations in the femtosecond regime is a promising possibility. With this kind of laser, nearly all kinds of material can be treated with minimum mechanical and thermal damage.

12.5.3
Ablation of 3D Structures

For drilling and cutting applications, a longer processing time and higher total applied energy can be used to achieve higher wall-surface qualities. Superficial ablation, however, also requires a smooth ablation ground, and finishing steps with additional laser pulses to polish the walls are not possible. Such ablation is mostly realized by using especially low laser fluences for the whole process, involving comparatively low processing speeds. Achievable structure sizes vary in the range from several tens of microns down to submicron dimensions, depending particularly on the pulse energy and focusing strategy.

For example, by scanning a surface using respective laser parameters leading to low energy densities, precise square planes can be ablated, as shown in Fig. 12.16 for silicon and NiTi.

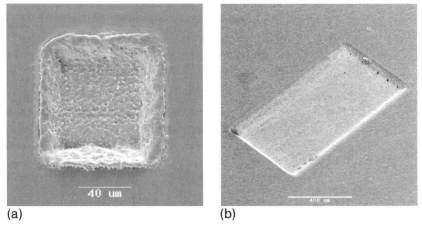

(a)   (b)

Fig. 12.16 Precise planar ablation using fs laser scanning.
(a) Superficial ablation of silicon using 600 fs laser pulses.
(b) Ablation of NiTi with 150 fs laser pulses.

To achieve high-precision ablation of small shapes, mask projection methods can be used to process a workpiece with laser fluences slightly above the ablation threshold. Fig. 12.17 (a) shows ablation of a chromium layer from a glass substrate in a defined area using 150 fs laser pulses (Korte et al. 2001). The picture on the right shows single-pulse ablation of sapphire with submicrometer resolution. The blind holes are processed with very short pulse durations, but also very low pulse energies, in order to achieve extremely small structure sizes, using a fluence that is again slightly above the ablation threshold (Korte et al. 2003).

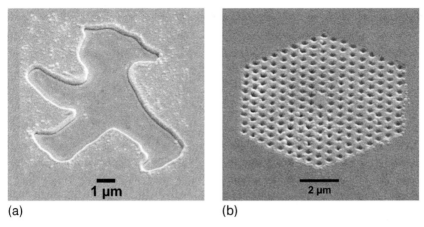

(a)   (b)

Fig. 12.17 Sub-micron structures from static exposure. (a) Mask exposure of chromium to 150 fs laser pulses. (b) Holes in sapphire from single focused 30 fs pulses.

To create areas of periodic microstructures, femtosecond laser pulses are often applied using a direct writing process, i.e., using a scanner system or periodic workpiece movements to subsequently process single tracks. Such structures mainly consist of grooves that can be designed to different dimensions and angles by adapting the process parameters. Figure 12.18 (a) shows a groove in steel with straight vertical walls and an aspect ratio in the range of 1, produced with multiple overlapping femtosecond pulses (Ostendorf et al. 2004). The second picture shows an array of cone structures in fused silica, which was made by superposition of grooves shifted by 90°. The pictures demonstrate machining qualities directly after the process without subsequent cleaning. If optical qualities are needed, such surfaces can be smoothed by subsequent processes like thermal annealing and etching (Tönshoff et al. 2001).

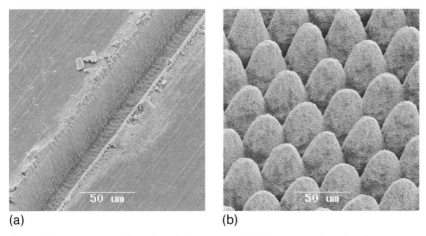

(a)   (b)

Fig. 12.18 Microstructures from direct fs laser writing. (a) Groove in steel produced with 150 fs pulses of 100 μJ. (b) Fused silica, structured using 120 fs pulses of 70 μJ.

Besides periodic microstructures, superficial patterns can also have a stochastic character. With the help of femtosecond laser pulses applied in an SF_6 atmosphere, silicon surfaces have been randomly patterned with microscopic spikes. By adjusting parameters like intensity and gas pressure, the shape can be changed from round and short to sharp and tall spikes, with tip sizes down to several hundred nanometers and spike lengths up to several tens of micrometers. Figure 12.19 shows samples of such structures that are often referred to as "black silicon" due to high light absorption for nearly all wavelengths (Carey et al. 2003). As chemical reactions are involved in this ablation method, further information is provided in Section 12.3.2 about fs-laser-induced processes.

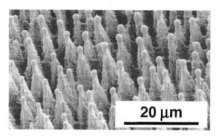

Fig. 12.19 Microspikes in silicon, generated with 100 fs pulses of 200 µJ in an SF_6 atmosphere.

When using ablation strategies such as those presented in this section, femtosecond lasers are a good tool for precise structuring of different materials that can fulfil a wide range of specifications.

12.6
Applications

12.6.1
Fluidics

In the manufacturing industries, such as the aerospace, automotive, biomedical and microelectronics industries, the production of fluidic components are carried out routinely using high-power Nd:YAG or CO_2 lasers, which are used mainly for metals. Excimer lasers and other UV lasers, such as frequency converted (diode-pumped) Nd:YAG and CV lasers, are integrated into production lines as well, mainly for the processing of plastics, ceramics and semiconductors. For the majority of fluidic applications, high-precision microdrilling is the predominant machining process. Even though industrial laser drilling is the oldest production technique using lasers, the number of industrial applications lags far behind those of laser marking, laser cutting and welding. This is due to the fact that conventional laser machining is accompanied by large thermal influences on the bulk material. This leads to problems like burr formation and the creation of microcracks and recast layers.

In some applications, like laser drilled combustion chamber components, or filter components, the desired function is achieved by drilling a great number of holes where a certain degree of inaccuracy in diameter and shape as well as a thin recast layer, can be tolerated. In many other applications, the requirements can only be met by reducing the laser pulse duration down to picoseconds or femtoseconds. This offers great opportunities for the new ultrafast laser technology (Knowles et al. 2001, Dausinger 2001 a, Kamlage et al. 2003).

High-precision hole drilling for fluidic components can be found, e.g., in ink jet printers, fuel injector systems, flow/dosage regulators or analytical microfluidic sensors (Walsh and Berdahl 1983, Weigl et al. 2000, Martin and Friedman 1975, Drzewiecki and Macia 2003, Mehalso 1999). As an example, the highly competitive automotive industry, where the manufacturing of diesel injector nozzles involves microdrilling of a tremendous number of microholes with diameters in the range of 100–200 µm, the development of processes that allow the drilling of smaller, more precise holes at a higher production rates, would be a vital step forward. For drilling nozzles, usually electro-discharge machining (EDM) is the only method that can achieve the necessary precision with acceptable production costs, but EDM is equipment intensive and has limitations. Important quality characteristics of injection holes are burr-free edges, smooth wall surfaces and a round and cylindrical hole profile. Control of the hole profile is particularly important because some applications require cylindrical holes, whereas others require tapered ones. Figure 12.20 shows an injector nozzle which was drilled using a femtosecond laser.

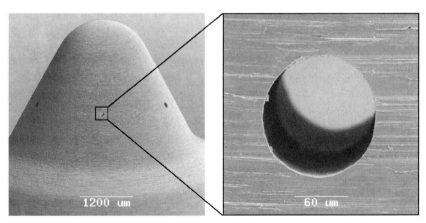

Fig. 12.20 Microdrilling of fuel injector nozzles using a femtosecond laser.

Ink-jet printing is a dot-matrix printing technology in which droplets of ink are directly jetted from a small aperture to a specified position on a media to create an image. The mechanism by which a liquid stream breaks up into droplets was first described by Lord Rayleigh in 1878. However, the first practical printing device was not patented before 1951. Since then, many drop-on-demand ink-jet and bubble ink-jet ideas and systems were invented, developed and commercially produced. Today, the ink-jet technologies most used in laboratories and on the market are the thermal and piezoelectric drop-on-demand ink-jet methods. The trends in industry are in jetting smaller droplets for higher imaging quality, faster drop frequency, and a higher number of nozzles for print speed, while the cost of manufacturing is reduced. These trends call for further miniaturization of the ink-jet design. One of the most critical components in a print head design is its nozzle. Nozzle geometry such as diameter and thickness directly influences drop volume, velocity and trajectory angle. Variations in the manufacturing process of a nozzle plate can significantly reduce the resulting print quality. The two most widely used methods for making the orifice plates are electro-formed nickel and laser ablation on the polyimide. Other known methods for making ink-jet nozzles are electro-discharge machining, micropunching, and micropressing. Because a smaller ink-drop volume is required to achieve higher resolution printing, the nozzle of the print heads has become increasingly small. To date, for jetting an ink droplet of 10 pl, print heads have a nozzle diameter of around 20 μm (see Fig. 12.21). With the trends towards smaller diameters and lower costs, the laser ablation method has become more and more popular for making ink-jet nozzles (Le 1998).

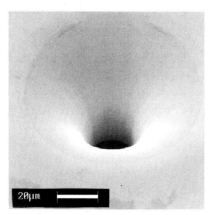 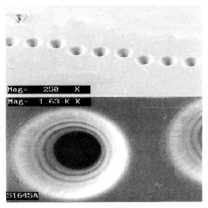

    Electro-formed Ni nozzle   Laser-drilled polyimide nozzle

Fig. 12.21 SEM photographs from the entrance side of Ni and polyimide nozzle plates.

In general, the production of fluidic components like fuel injection and ink-jet nozzles or other dosing devices (hydraulic and pneumatic components) offers excellent prospects for innovative laser machining techniques, such as femtosecond laser machining.

12.6.2
Medicine

As discussed before, materials can be ablated using ultrashort laser pulses at a low energy threshold. Therefore, thermal and mechanical side effects are limited to very small dimensions, typically significantly less than micrometer dimensions. The abundance of side effects opens up the use of femtosecond lasers for a broad field of applications in medicine.

Applications have been demonstrated in dental treatment surgery (Momma et al. 1998) and ophthalmology (Kurz et al. 1998) as well in treatment of cardiovascular diseases. These applications will be demonstrated in the following section:

12.6.2.1 fs- LASIK (Laser in Situ Keratomileusis)

Femtosecond photodisruption offers the possibility of cutting out an intrastromal lens, in order to correct the refraction error of the eye. First attempts for refractive corneal surgery were made to reshape the cornea curvature by means of photodisruption, using picosecond pulses (Remmel et al. 1992). This approach failed, due to higher mechanical side effects by bubble formation inside the corneal tissue. As the size of the laser-induced bubble and the extent of the shock wave depends on the laser pulse energy applied, these side effects have been overcome by reducing pulse duration and with it a reduction of the necessary pulse energy.

As a first step, a lamellar intrastromal cut is performed by scanning the laser in a spiral pattern. This procedure is analogous to the mechanical lamellar cut of a microceratome in the conventional LASIK procedure. In a second step, another cut prepares an intrastromal lenticule with a precalculated shape, dependent on the refractive error of the treated eye. After opening the anterior corneal flap, the lenticule is extracted and the flap can be repositioned on the cornea (Lubatschowski et al. 2002). Histological analyses demonstrate the smooth and precise character of fs tissue-processing by ultrashort laser pulses (Heisterkamp et al. 2000) (Fig. 12.22).

Fig. 12.22 Survey of corneal flap of a rabbit's eye, cut using the fs LASIK procedure.

12.6.2.2 **Dental Treatment**

Currently-used methods in dental surgery, especially turbine drills and Er:YAG-lasers produce thermal and mechanical stress within the dental tissue, which lead to microcracks in the enamel. Those cracks cannot be sealed, due to their small dimensions and can therefore act as starting points for new carious attacks (Xu et al. 1997, Frenzen 2000). The use of femtosecond laser sources for the ablation of dental tissue does not cause shattering of the enamel by mechanical or thermal stress (see Fig. 12.23).

Furthermore, the laser plasma has been used to distinguish carious from healthy tissue. Since the concentration of mineral traces is linked to the state of carious infection, the spectrum and the intensity distribution of the spectrum reveal information about the condition of the tissue.

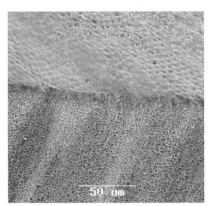

Fig. 12.23 Cavity created by femtosecond laser pulses.

12.6.2.3 **Cardiovascular Implants**

Apart from the direct use of fs lasers in medicine for surgery, structuring of medical implants, e.g. coronary stents, is a promising application with growing industrial interest (Momma et al. 1999).

Coronary stents are used as a minimally invasive treatment of arteriosclerosis, as an alternative for bypass operations. Since the requirements for medical implants (e.g., freedom from burrs, x-ray opacity) are very strict, only a few materials are commonly used. Today, materials typically used for stents are stainless steel or shape memory alloys. For these materials, chemical post-processing techniques have been developed to achieve the required properties. However, these materials are not optimal, concerning several medical aspects (e.g. risk of restenosis, limited biocompatibility, etc.).

New approaches favour stents for temporary use only, which call for bioresorbable materials like Mg-base alloys or special biopolymers (Fig. 12.24).

For these materials, no established post-processing technique is available. Furthermore, most of them react strongly to thermal load. Therefore, it is essential to avoid processing influences on the remaining material in order to retain the specific material properties. While CO_2 laser and excimer lasers show significant material modifications, femtosecond pulse laser material processing meets the requirements of these sophisticated materials (Ostendorf et al. 2002).

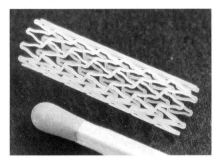

Fig. 12.24 Femtosecond laser-machined cardiovascular implant.

Due to their unique characteristics of minimal invasive ablation of material and tissue, femtosecond lasers offer new possibilities for medical treatment. Besides new methods in surgery, fs-laser material processing also enables access to new materials which it has not yet been possible to machine. It remains yet to be seen to what extent femtosecond lasers will penetrate the market of medical laser systems, since these laser systems are still rather expensive.

12.6.3
Microelectronics

In the field of microelectronics, femtosecond lasers are of particular interest to microelectromechanical systems (MEMS). Today's amazing state-of-the-art in microfabrication has accelerated the development of highly complex microcomponents for more intelligent sensors and more powerful actuators. However, the limited accuracy of microlithographical fabrication, still limits precise mass configuration of mechanical components over an entire wafer, making post-processing of very sensitive, dynamic subsystems inevitable. Femtosecond lasers offer efficient and precise ablation mechanisms with minimal thermal effects on the immediate vicinity of the ablating zone. Remelting, uncontrolled resolidifying, microcracking, and undesired texture alterations can be eliminated almost completely. Consequently, femtosecond lasers are the ideal tool for removal of microvolumes from delicate MEMS structures. They are especially useful for the adjustment of oscillating microcomponents, if active oscillation control of the oscillator is not applicable, and if the correct function depends on the exact fine-tuning of all relevant resonance frequencies. There are two strategies for the desired fine-tuning (trimming) of the oscillator eigenfrequencies. Positive adjustment of the resonance frequencies is achieved by removal of the oscillator mass. Negative adjustment is achieved by decreasing the spring constant by correcting the spring element that has influence on the local geometrical moment of inertia. Depending on the chosen strategy, the design of the components and the process control of the microfabrication have to be designed carefully to assure the one-way adjustability of microcomponents.

Ablation volumes in present applications are in the range of several thousand cubic microns, using low-fluence ($\sim$ 30 J cm^{-2}) femtosecond pulses for a planar superficial ablation step to a depth in the range of 1 µm. Laser beam delivery for the ablation process should use a galvo-scanner to obtain high flexibility, in combination with acceptable processing speeds, considering the trimming process applied on whole sensor arrays of wafers. However, accurate absolute positioning of the laser focus via scanning technologies is still a challenge, and limits the final quality of the post-processed object.

In addition to its use in trimming silicon, femtosecond laser pulses can also be used for superficial ablation of other materials used in microelectronics. An important field is the creation of lithography masks, which can be supported by femtosecond lasers, to repair microscopic defects in existing masks. This process has originally been demonstrated on a 100 nm chromium layer coated onto a quartz substrate, using a Ti:Sapphire laser system with pulses of 100 fs and energies reduced to 5 µJ. These pulses were applied in combination with a near-field optical microscope (SNOM), using a long focal distance to couple the radiation into a hollow fused-silica micropipette with a chrome coating. This micropipette was tapered to an aperture of less than 1 µm, placed slightly above the target surface. Using such a setup, nanostructures can be generated that are not limited either by the laser wavelength or by thermal effects.

Figure 12.25 shows a three-dimensional AFM topographical image of a programmed defect on a lithography mask, demonstrating how this technique can be applied for mask repair and production. The defect was removed using a 690 nm diameter tip held at a constant height of 150 nm above the chrome surface. Using second-harmonic radiation (at 390 nm) and a scanning speed of 5 µm s^{-1}, the defect area was ablated by producing parallel grooves separated by 100 nm (Nolte et al. 1999).

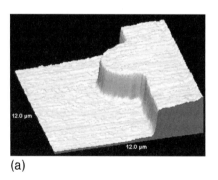 

(a) (b)

Fig. 12.25 AFM images of a lithography mask with a deliberately-produced defect (a), removed with near-field femtosecond laser pulses (b).

In the field of microelectronics, femtosecond laser employment is also an approach for cutting applications. In particular, the cutting of thin silicon in the range of 50 µm and below is a field in which fs lasers have the potential to become a competitive tool.

The best cutting qualities are achieved using low laser fluences in the regime of nonthermal ablation. However, investigations on cutting of thin silicon with amplified short-pulse systems must also focus on the highest available pulse energies, e.g., up to 1 mJ, in order to achieve competitive processing speeds. For this reason, when cutting using the current amplified fs laser systems, linear focus shapes created by a combination of cylindrical lenses can be used to enhance linear cuts through thin substrates. The fluence of the laser pulse is reduced, which leads to high precision and high quality, while on the other hand, the total energy input along the processed line stays the same. Further, due to the low fluence and the more precise focus, the kerf widths are smaller, and lead to a higher total energy input per area for the processed lines. The smoothing effect of the extremely large pulse overlap is another aspect contributing to the quality. Figure 12.26 demonstrates the efficiency of this strategy for silicon ablation (Ostendorf et al. 2004).

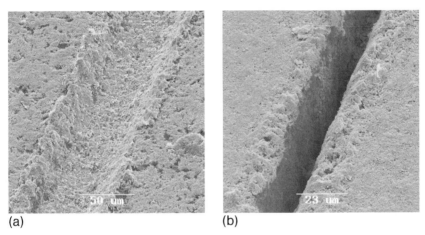

(a) (b)

Fig. 12.26 Silicon, processed four times at velocities of 500 mm min^{-1} using 150 fs pulses of 900 µJ, focusing to a spot (a) and a line (b).

Besides the focusing strategy and basic parameters like material thickness and pulse energy, the substrate temperature is also important. Cutting speeds have been shown to increase by up to 50% compared to room temperature, at temperatures of 300–400° (Wagner 2002). However, the processing of thin wafers at temperatures higher than 100° is usually impossible, due to preconditions in connection with the wafer handling.

One effect which has a major impact on the rear-cut surface is the so-called chipping effect. When the material has not yet been completely cut through by the laser pulses, the remaining layer is burst away by the plasma influence and oscillation effects, especially when working with high fluences in a normal atmosphere, as demonstrated in Fig. 12.27. This effect can be suppressed to a great extent by adapting the process parameters, resulting in reduced processing speeds. For low fluences and low processing speeds, the complete suppression of chipping is simple, as demonstrated in (b). For separation speed optimizations at high pulse energies, there is a distinct feed-rate limit, below which cutting with significant chipping can be avoided (Bärsch et al. 2003).

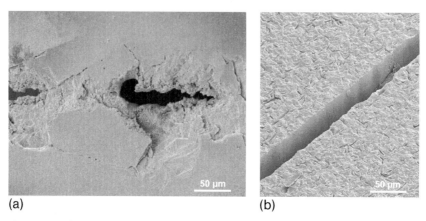

(a)   (b)

Fig. 12.27 Back-side views of silicon cut with 150 fs laser pulses: strong chipping due to incomplete separation at high energies (a), good exit kerf quality using parameters at low process speeds (b).

As indicated, the microelectronic sector offers many possible applications for femtosecond lasers. In particular, the advantage of avoiding thermal influence on the surrounding material makes this micromachining tool most interesting for this field. Lithography mask repair has become one of the first applications in which femtosecond lasers have been established in an industrial process. For many other applications, good results have been demonstrated, and new laser systems with higher average powers and high reliability are expected to increase the industrial establishment of the processes over the next few years.

References

Aus der Au, J., G. J. Spühler, T. Südmeyer, R. Paschotta, R. Hövel, M. Moser, S. Erhard, M. Karszwski, A. Giessen, U. Keller, Optics Lett. 25, 859 – 61 (2000).

Bärsch, N., Körber, K., Ostendorf, A., Tönshoff, H. K., Ablation and cutting of planar silicon devices using femtosecond laser pulses; Appl. Phys. A 77, pp. 237 – 41 (2003).

Baltuska, A., Z. Wie, M. S. Pshenichnikov, D. A. Wiersma, R. Szipöcs, Appl. Phys. B 65, 175 (1997).

Barnes, K., G. Lewis, Beam delivery via the reflecting objective, *Photonics Spectra*, p. 105, 1990.

Beyer, E., O. Märten, K. Behler et al., Laser cutting, *Laser und Optoelektronik*, 3, p. 282 – 90, 1985.

Brunner, F., G. J. Spühler, J. Aus der Au, L. Krainer, F. Morier-Genoud, R. Paschotta, N. Lichtenstein, S. Weiss, C. Harder, A. A. Lagatsky, A. Abdolvand, N. V. Kulsehov, U. Keller, Optics Lett, 25, 1119 – 21 (2000).

Carey, J. E., C. H. Crouch, E. Mazur, Femtosecond-laser-assisted microstructuring of silicon surfaces for new optoelectronics applications; Optics & Photonics News February 2003, pp. 32 – 36 (2003).

Chang, C., T. Abe, M. Esashi.; Glass Etching Assisted by Femtosecond Pulse Modification; Sensors and Materials 15, No. 3, pp. 137 – 45 (2003).

Chichkov, B. N., C. Momma, S. Nolte, F. von Alvensleben, and A. Tünnermann, Appl. Phys. A 63, 109 (1996).

Cumpston, B. H., S. P. Anathavel, S. Barlow, D. L. Dyer, J. E. Ehrlich, L. L. Erskine, A. A. Heikal, S. M. Kuebler, I.-Y. S. Lee, D. McCord-Maughon, J. Qin, H. Röckel, M. Rumi, X.-L. Wu, S. R. Marder, and J. W. Perry, Nature 398, 51 (1999).

Dausinger F.: Drilling of high quality microholes, Laser Manufacturing 22, 25 – 27 (2001).

Druon, F., F. Balembois, P. Georges, A. Brun, A. Courjaud, C. Hönninger, F. Salin, A. Aron, F. Mougel, G. Aka, and D. Vivien, Opt. Lett., 25, 423 – 25 (2000).

Drzewiecki, T. M., N. F. Macia, Fluidic technology: adding control, computation, and sensing capability to microfluidics, Smart Sensors, Actuators, and MEMS, Proceedings of the SPIE, Volume 5116, pp. 334 – 347, 2003.

Ferman, M. E., A. Galvanauskas, G. Sucha, *Ultrafast Lasers – Technology and Applications*; Marcel Dekker, New York (2003).

Frenzen, M., D. Hamrol, Kavitätenpräparation mit dem Er:YAG-Laser, eine histologische Studie, Dtsch. Zahnärztl. Z 55, p (2000).

Galajda, P. and P. Ormos, Appl. Phys. Lett. 78, 249 (2001).

Hecht E., Optics, Addison-Wesley, San Francisco, 2002.

Heisterkamp, A., T. Ripgen, T. Mammon, W. Drommer, H. Lubatschowski, W. Welling, W. Ertmer; Intrastromal cutting effects in rabbit cornea using femtosecond laser pulses, Proc. of SPIE Vol. 4161, pp. 52 (2000).

Her, T.-H., R. Finlay, C. Wu, E. Mazur, Femtosecond laser-induced formation of spikes on silicon, Applied Physics A 70, pp. 383 – 5 (2000).

Her, T. H., R. J. Finlay, C. Wu, S. Deliwala, E. Mazur, Appl. Phys. Lett. 73, 1673 (1998).

see Laser Ablation, Proceedings of the 5th International Conference, Eds. J. S. Horwitz, H.-U. Krebs, K. Murakami, and M. Stuke, in Appl. Phys. A 69, (1999).

Jia, T.-Q., Chen, H., Zhang, Y.-M., Photon absorption of conduction-band electrons and their effects on laser-induced damage to optica materials, Phys. Rev. 61, No. 24, pp. 16522 – 9 (2000).

Jia, T. Q., Xu, Z. Z., Zhao, F. L., Microscopic mechanisms of ablation and micromachinig of dielectrics by using femtosecond lasers, Appl. Phys. Let. 82, No. 24, pp. 4382 – 4 (2003).

Kaiser, A., B. Rethfeld, M. Vicanek, G. Simon, Microscopic processes in dielectrics under irradiation by subpicosecond laser pulses, Phys. Rev. B 61, 17, p. 11437, 2000.

Kamlage, G., T. Bauer, A. Ostendorf,
B. N. Chichkov, Appl. Phys. A 77, 307 –
10 (2003).

Kautek, W., J. Krüger, M. Lenzner, S. Sartania, C. Spielmann, and F. Krausz, Appl.
Phys. Lett. 69, 3146 (1996).

Kawata, S., H.-B. Sun, T. Tanaka, and
K. Takada, Nature 412, 697 (2001).

Keller, U., D. A. B. Miller, G. D. Boyd,
T. H. Chiu, J. F. Ferguson, M. T. Asom;
Optics Lett. 17, 505 (1992).

Knowles, M., A. Bell; G. Rutterford,
T. Andrews, A. Kearsley, CVL laser dril-
ling of fuel injection components, Laser
Manufacturing, 22, 28 – 29 (2001).

Korte, F., S. Adams, A. Egbert, C. Fallnich,
A. Ostendorf, S. Nolte, M. Will,
J.-P. Ruske, B.N. Chichkov, A. Tünnermann, Sub-diffraction limited structuring of solid targets with femtosecond
laser pulses, Optics Express 7, No. 2,
pp. 41 – 49 (2000).

Korte, F., S. Nolte, B.N. Chichkov, C. Fallnich, A. Tünnermann, H. Welling; Sub-
micron structuring of solid targets with
femtosecond laser pulses, Proceedings
of SPIE 4274, pp. 110 – 115 (2001).

Korte, F., J. Serbin, J. Koch, A. Egbert,
C. Fallnich, A. Ostendorf, B. N. Chichkov, Towards nanostructuring with femtosecond laser pulses, Applied Physics
A 77, pp. 229 – 35 (2003).

Krainer, L. , R. Paschotta, J. Aus der Au,
C. Hönninger, U. Keller, M. Moser,
D. Kopf, K. J. Weingarten, Appl. Phys B.
69, 245 – 7 (1999).

Kuebler, S. M., M. Rumi, T. Watanabe,
K. Braun, B. H. Cumpston, A. A. Hei-
kal, L. L. Erskine, S. Thayumanavan,
S. Barlow, S. R. Marder, and J. W. Perry,
J. Photopolym. Sci. Tech. 14, 657 (2001).

Kurz, R. M., C. Horvath, H. H. Liu,
T. Juhasz, Optimal Laser Parameters for
intrastromal Corneal Surgery, SPIE,
VOL.3255, p. 55 – 66 (1998).

Le, H. P., Progress and Trends in ink-jet
printing technology, Journal of Imaging
Science and Technology, Vol. 42, No 1
(1998).

Luam F. et al., Femtosecond soliton pulse
delivery at 800 nm wavelength in hollow-core photonic bandgap fibers,
Optics express, Vol 12, No.5 pp. 835 –
40 (2004).

Lubatschowski, H., A. Heisterkamp, F. Will,
J. Serbin, T. Bauer, C. Fallnich, H. Well-
ing, W. Müller, B. Schwab, A. I. Singh,
W. Ertmer, Ultrafast Laser Pulses for
Medical Applications; Proc. of SPIE,
Vol. 4633, pp. 38 (2002).

Martin, H.R., S. B. Friedman, Some applications of the free-jet liquid fluidic device,
ISA Transactions, vol. 14, no. 1, p. 61 –
7, 1975.

Maruo, S., O. Nakamura, and S. Kawata,
Opt. Lett. 22, 132 (1997).

Maruo S. and S. Kawata, IEEE J. Microelec-
tromech. Syst. 7, 411 (1998).

Mehalso, R., MEMS packaging and micro-
assembly challenges, Proc. SPIE Vol.
3891, p. 22 – 25, Electronics and Structures for MEMS, 09/1999.

Momma, C., B. N. Chichkov, S. Nolte, F. von
Alvensleben, A. Tünnermann, H. Welling, B. Wellegehausen, Optics Communications 129, 134 – 42 (1996).

Momma, C., S. Nolte, B.N. Chichkov, Precise
Micromachining with Femtosecond
Laser Pulses, Laser und Optoelektronik
29, No. 3, pp. 82 – 9 (1997).

Momma, C. , S. Nolte, A. Kasenbacher,
M. H. Niemz, H. Welling, Ablation von
Zahnhartsubstanz mit ps- und fs- Laser-
pulsen, in: *Laser in der Medizin*, Ed.
W. Waidelich, R. Waidelich, J. Wald-
schmidt, Proc. 13th Int. Congr.
LASER97, Springer, Berlin (1998).

Momma, C., U. Knoop, S. Nolte, Laser cutting of slotted tube Coronary stents,
state-of-the-art and future developments, Prog. In Biomed. Res 2. pp. 39
(1999).

Nolte, S., B.N. Chichkov, H. Welling,
Y. Shani, K. Lieberman, H. Terkel,
Nanostructuring with spatially localized
femtosecond laser pulses, Optics Letters, Vol. 24, No. 13, pp. 914 – 6 (1999).

Nolte, S., C. Momma, H. Jacobs, A. Tünnermann, B. N. Chichkov, B. Wellegehausen, and H. Welling, J. Opt. Soc. Am. B
14, 2716 (1997).

Ostendorf, A., T. Bauer, F. Korte, J. Howorth,
C. Momma, N. Rizvi, F. Saviot, F. Salin,
Development of an industrial femtosec-
ond laser micro-machining system,
Proc. of SPIE VOL: 4633, pp. 128
(2002).

Ostendorf, A., C. Kulik, T. Bauer, N. Bärsch, Ablation of metals and semiconductors with ultrashort-pulsed lasers: improving surface qualities of microcuts and grooves; Proceedings of SPIE, Vol. 5340, 153–163 (2004).

Ouzounov, D. G. et al., Generation of Megawatt optical solitons in hollow-core Photonic band-gap fibers; science 301, pp. 1702 – 4 (2003).

Price, J. H. V., K. Furusawa, T.M. Monro, L. Lefort, D.J. Richardson, J. Opt. Soc. AM. B.19, 1286 (2002).

Pronko, P. P., S. K. Dutta, J. Squier, J. V. Rudd, D. Du, and G. Mourou, Machining of sub-micron holes using a femtosecond laser at 800 nm, Opt. Commun. 114, p. 106, 1995.

Ranka, J.K., R. S. Windeler, A. J. Stentz, Opt.Lett. 25, 796 (2000).

Remmel, M., C. M. Dardenne, J. F. Bille, Intrastromal tissue removal using an infrared picosecond Nd:YLF opthalmic laser operating at 1053 nm, Lasers Opthalmol 4 (3/4), pp. 169 (1992).

Serbin, J., A. Egbert, A. Ostendorf, B. N. Chichkov, R. Houbertz, G. Domann, J. Schulz, C. Cronauer, L. Fröhlich, and M. Popall, Opt. Lett., 28(5): 301 – 3, 2003.

Simon, P. and J. Ihlemann, Appl. Phys. A 63, 505 (1996).

Soileau, M.J. , W. E. Williams, T.F. van Stryland et al., Temporal dependence of laser induced breakdown in NaCl and SiO$_2$, in: *Laser-Induced Damage in Optical Materials*: 1982, Benett, H. E., Guenther, A. H., Milam, D.; et al. (eds), Natl. Bur. Stand. (U.S.) Spec. Publ. 669, pp. 387–405 (1984).

Sorokin, E., I. T.Sorokina, E. Wintner, Appl. Phys. B 72, 3–14 (2001).

Stuart, B. C., M. D. Feit, S. Herman, A. M. Rubenchik, B. W. Shore, and M. D. Perry, Optical ablation by high-power short-pulse lasers, J. Opt. Soc. Am. B 13, 459 (1996).

Sun, H.-B., T. Kawakami, Y. Xu, J.-Y. Ye, S. Matuso, H. Misawa, M. Miwa, and R. Kaneko, Opt. Lett. 25, 1110 (2000).

Sun, H.-B., V. Mizeikis, Y. Xu, S. Juodkazis, J.-Y. Ye, S. Matsuo, and H. Misawa, Appl. Phys. Lett. 79, 3173 (2001).

Sun, H.-B., K. Takada, and S. Kawata, Appl. Phys. Lett. 79, 3173 (2001).

Sun, H.-B., T. Tanaka, K. Takada, and S. Kawata, Appl. Phys. Lett. 79, 1411 (2001).

Sun, H.-B., S. Matsuo, and H. Misawa, Appl. Phys. Lett. 74, 786 (1999).

Tanaka, T., H.-B. Sun, and S. Kawata, Appl. Phys. Lett. 80, 312 (2002).

Taylor, R.S., C. Hnatovsky, E. Simova, D.M. Rayner, V.R. Bhardwaj, P.B. Corkum, Femtosecond laser fabrication of nanostructures in silica glass, Optics Letters 28, Issue 12, pp. 1043 – 5 (2003).

Tönshoff, H. K., A. Ostendorf, K. Körber, T. Wagner: Micromachining of Semiconductors with Femtosecond Lasers, in: Proceedings of ICALEO, Dearborn, USA. (2000).

Tönshoff, H. K., A. Ostendorf, F. Korte, J. Serbin, T. Bauer; Generation of periodic microstructures with femtosecond laser pulses; Proceedings of SPIE, Vol. 4426 pp. 177 (2001).

Tönshoff, H. K., A. Ostendorf, C. Kulik, F. Siegel, Finishing of Cutting Tools Using Selective Material Ablation, In: Proceedings of 1 st International CIRP Seminar on Micro and Nano Technology, Copenhagen, Denmark, November 13 – 14, 2003.

Wagner, T., Abtragen von kristallinem Silizium mit ultrakurzen Laserpulsen, Thesis, Hannover (2002).

Walsh, J., M. Berdahl, Application of fluidics to instrumentation in hostile environments, AIAA, SAE, and ASME, Joint Propulsion Conference, 19th, Seattle, WA, June 27 – 29, 1983.

Weigl, B. H., A. Hatch, A. E. Kamholz, P. Yager, Novel immunoassay formats for integrated microfluidic circuits: diffusion immunoassays, Proc. SPIE Vol. 3912, p. 50 – 56, 03/2000.

Xu, H.H.G., J.R. Kelly, S. Jahnmir, V.P. Thompson, E.D. Rekows, Enamel subsurface damage due to tooth preparation with diamonds, Dent. Res. 76 (10), pp. 1698 (1997).

13
(Some) Future Trends

Saulius Juodkazis and Hiroaki Misawa

> *"The best way to predict (your) future is to create it"*
> R. Anthony

13.1
General Outlook

It is the most daunting and unforgiving task to speculate on the future trends in a fast changing world where a technology cycle is approximately 7 years. We can more reliably discuss the issue of how current achievements in 3D laser microfabrication projects onto the prediction roadmap in some of the technologies. One obvious current trend is the expanding functionality of a variety of optical far- and near-field microscopic techniques. The ordinary optical microscope has become an integral part of many 3D microfabrication setups. Due to nonlinear character of light–matter interaction, the volume of photomodification can be made smaller than is defined by the diffraction laws (Ch. 3). Moreover, in the applications where the optically-induced dielectric breakdown is a working principle of microfabrication, an even higher light localization can be achieved. It is defined by the absorption volume, which is given by the skin-depth in the ionized focal volume (Ch. 2). The lateral and axial cross-sections of the absorbing volume are typically 50–70 nm in ionized dielectrics and define the precision of structuring. This is already in the nano-realm (the 100 nm feature size is considered as the landmark value).

In the interest of increased efficiency multi-beam/pulse direct laser writing setups will evolve. Also, large area processing should be developed for use in practical applications. Here, multi-beam interferometric (holographic) techniques are the most promising (Ch. 10). With the ever increasing output power of ultra-short lasers these two methods for obtaining increased efficiency are the most practical. Direct laser writing and holographic recording are sometimes called the 3D maskless lithography. The image projection principle, on which standard lithography is based, can be incorporated into maskless lithography. It can be realized by utilizing novel image formation methods based on spatial light modulators (SLMs), digital micro-mirror devices (DMDs), or diffractive optical elements (DOE). The image formed by these devices can be projected into the fabricated material at different depths. In addition, the image can be dynamically changed during exposure, creating complex 3D patterns. The phase and amplitude of the image-forming beams/pulses allows one to create effective phase holograms, e.g., using the Gerchberg–Saxton algorithm [1]; hence, all the available laser pulse energy can be used to form the image.

Modern optical microscopy demonstrates an increasing resolution, 3D imaging by elaborated confocal methods, and new conceptual methods [2] for a basically unlimited resolution increase towards a < 10 nm molecular level [3]. We might expect that, after imaging, fabrication (in the general meaning of photomodification) will follow at molecular resolution. Such an imaging-to-fabrication transition has occured in current 3D laser microfabrication. One would expect a basic breakthrough when the two principles of nano-technology, namely the "top-down" and "bottom up" merge and become complementary. Indeed, there are two principles to manufacturing on the molecular scale: [4] self-assembly of structures from basic chemical building blocks (biomimetic route), the so-called, "bottom-up" approach, and assembly by manipulating small components with much larger tools the "top-down" approach. Whilst the former is considered to be the ideal method by which nano-technology will ultimately be implemented, the latter approach is more readily achievable using current technology. The concept of nano-technology (the term coined by Taniguchi in 1974 [5]) was introduced by R. Feynman in his lectures "There's plenty of room at the bottom" (1960) [6] and "Infinitesimal machinery" (1983) [7]. In this, he considered that, by developing a scaleable manufacturing system, a device could be made which could make a miniature replica of itself, and so on down to the molecular level.

3D laser microfabrication is already serving as a tool of the "top-down" approach. Once the fabrication at the molecular level is achieved by the "top-down" method, the "bottom-up" route can be implemented to provide the required efficiency and productivity for a future technology. As an example, we might imagine a controlled photomodification of the DNA sequence with subsequent reproduction; the ultimate merger of the "top-down" and "bottom-up" routes of nano-technology. 3D laser microfabrication will serve as an interface between the living and nonliving worlds, a trend already discernible in laser grafting, surgery, and other medical applications.

13.2
On the Way to the Future

3D laser microfabrication based on the usage of ultra-short (sub-1 ps) pulses is already making its way into efficient and practical applications in very different fields: micro-machining of complex composite structures, micro-processing of printer heads, machining of metals for the molds, fabrication of complex 3D intra-vascular stents, eye surgery, dentistry, etc., (Ch. 12). The understanding of physical and chemical principles of light–matter interaction in the case of in-bulk

3D structuring of materials should help to tailor materials for expected goals (Ch. 11; Ch. 7). The material can be prepared to acquire different functionality according to its treatment: exposure, annealing, and etching. This can be most easily realized in ceramics, glasses, and polymers. For example, the thresholds of the dielectric breakdown of polymers and void formation can be engineered by doping the host polymer by gas-releasing moieties [8]. In the case of exposure by ultra-short pulses, the white light continuum (a black-body type radiation according the Plank's law) generated by high temperature, $T_e \sim 10^4$ K, electrons can be used as a 3D freely-positioned exposure source, which makes photo-modification of material under processing from the inside.

3D photonic crystals and their templates formed in polymer materials (resists and resins) is another field where 3D laser microfabrication is expected to challenge solid state technology and already provides practical solutions for the IR-spectral region. The fundamental trend of the miniaturization of devices inherently provides faster (more efficient) operation, allows for a larger degree of integration and complexity (functionalization), increased sensitivity, and smaller consumption of material and energy resources. Hence, it is obvious that this trend will be followed by the fabrication of modern MEMS and micro total analysis system (μ-TAS) devices using laser microfabrication. The inherent 3D character of laser microfabrication is expected to prove indispensable for such applications.

13.3
Example: "Shocked" Materials

A nano-void can be created inside dielectrics at the focus of a tightly focused fs-pulse (Chs. 2, 10). The void results from shock-compression and rarefaction waves launched from the ionized focal spot, whose dimensions are comparable with the skin-depth in a metallic plasma state of an ionized dielectric. The shock-compressed ("shocked") micro-volume has unique physical and chemical properties as demonstrated below by an example of its etchability. New states of materials [9–11] with altered chemical properties [12, 13] are expected to be formed as a result of optically induced micro-explosions. Also, such micro-explosions are used in gene transfer through the cell's membrane, which is opened by passing a pressure wave.

As a tentative example, here we show how 3D laser microfabrication can contribute to solving the problems of processing of wide-bandgap dielectric materials. The breakthrough is expected in the field of solid-state lighting of public and private spaces [15]. The efficiency of modern white light solid-state sources based on light emitting diodes (LEDs) has reached that of luminescent lamps. Transition to solid-state lighting promises, in theory, close to a 100% electricity-to-light conversion and will revolutionize this field as well as contribute to the less energy consuming technology world wide. As the blue LEDs become less expensive and the fully solid state white lighting becomes an economically feasible option, the processing of substrate materials (Al$_2$O$_3$, SiC) on which GaN-based blue LEDs are grown becomes critical, because the most limiting factor of emissivity of GaN-based LEDs is due to poor light extraction.

In the field of UV-lithography, increasingly shorter wavelengths are used (following Moore's law) where wide band-gap fluoride materials for optical elements are required (e.g., a blue 400 nm wavelength is equivalent to a 3.1 eV bandgap). As a rule the wide-bandgap dielectrics are highly chemically inert, and usually brittle and hard. Hence, there are no technological ways to process these materials mechanically or chemically. The 3D-controlled optically-induced dielectric breakdown of dielectrics is expected to open new processing routes for natures hardest materials and other technologically important substances by, e.g., amorphization of crystalline Al$_2$O$_3$, TiO$_2$, SiO$_2$, decomposition of SiC, GaN, carbonization of diamond, etc. The demand to process these materials with sub-micrometer precision is prompted by a number of modern technologies: UV-lithography, light extraction from Al$_2$O$_3$ and SiC, 3D structuring of catalytic TiO$_2$ and silica for microfluidics and μ-TAS.

The amorphization of sapphire by dielectric-breakdown-generated shock waves makes it wet etchable in an aqueous solution of hydrofluoridic acid [16] as shown in Fig. 13.1. Sapphire, nature's hardest oxide, can be amorphized by a micro-explosion triggered shock wave. The amorphous regions were etched out, revealing strong structural modifications of sapphire along the line of consecutive pulses. Obviously, the pressure generated by the explosion should be much larger than the Young modulus (a cold pressure) of sapphire. This can only be achieved using tightly focused sub-picosecond laser pulses. It is noteworthy, that there were no cracks along the scanned line before the voids were cleave-opened. This phenomenon can be used to process the back-side sapphire surface of blue LEDs or to form Fresnel lenses into substrates in order to improve light extraction. The mechanism of a wet-etching enhancement by a factor of more than 10^3 is mostly

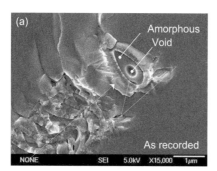
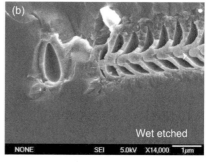

Fig. 13.1 SEM images of sapphire "shock-processed" by 180 fs pulses of 800 nm wavelength and 50 nJ energy (at focus) with a 500 nm intra-pulse separation (along the line). The single voxel and line of connected voxels were recorded at 20 μm depth. Images of two halves of the same sample: as recorded (a) and wet etched (b) for 5 min in a 5% aqueous solution of HF. Arrows mark direction of laser beam scan; triangle approximately depicts the focusing cone angle by objective lens of numerical aperture NA = 1.35. No cracks were observed before cleaving.

of a chemical nature. The shock-compressed amorphous regions of higher density should have smaller Al–O–Al angles, which make the oxygen more exposed and reactive (so-called Lewis base mechanism).

A similar enhancement of wet etching but to a smaller degree was observed in crystalline quartz [16] and in silica glass (Fig. 13.2) [17, 18], where the reduced Si–O–Si angle makes the material more basic inside the shock-compressed regions. One of the consequences of the mass density increase is a refractive index augmentation, which causes a polarizability increase via the Lorentz–Lorenz relation. The polarizability is directly related to the optical basicity [19], which is a phenomenological constant defining the chemical basicity (or acidity) of glasses. An increased polarizability signifies an increased basicity (i.e., an enhancement of ionicity) and vice versa. Hence, it is possible to link the change of refractive index-induced, e.g., by shock propagation, to chemical processing. The regions of silica with larger refractive index around the void become more basic (ionic) and hence show enhanced wet etching as compared with the un-irradiated regions. Silica of higher mass-density shows an increased chemical etchability (Fig. 13.2), a kind of counter-intuitive finding observed experimentally [14,18].

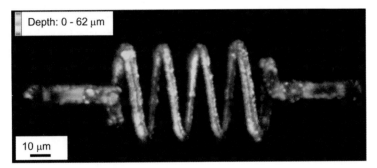

Depth: 0 - 62 μm

10 μm

Fig. 13.2 Confocal image of rhodamine photoluminescence from a high-aspect ratio structure wet-etched inside *Viosil* (Shin-Etsu Chemical Co.) silica (mass density of 2.2 g cm^{-3}; OH content 1200 ppm) [14]. Recording was carried out using 800 nm, 200 fs pulses of approximately 40 nJ (at focus); wet-etched in P-etch (HF(48%): HNO$_3$(70%): H$_2$O 15:10:300 by volume) for 840 min at the etching rate of 12 nm min^{-1}. The contrast of etching along the recorded pattern with that of un-irradiated silica was approximately 100.

13.4
The Future is Here

Studies of tightly focused laser pulses inside a transparent solid in well-controlled laboratory conditions encompass a very exciting field for both applied and fundamental science. The microscopic nature of phase transitions on the nanometer space scale and picosecond time scale, is poorly understood. The rise and fall rates of temperature in a confined micro-explosion is 10^{18} K s^{-1} which is impossible to achieve by any other means. Confined micro-explosion presents amazing evidence of the self-similarity of the processes occurring in laboratory table-top experiments and in nuclear explosions but at energies differing by 10^{21} times, and space and time scales by 10^7 (Ch. 2).

It is difficult to spot a particular moment in time when a new branch of science and technology makes its distinction from other closely related research, especially in a highly interdisciplinary field. This will become more obvious from the longer time perspective in the future. However, we can state that 3D laser microfabrication has become a highly active and dynamic field of research, is presented in all major optical and material science conferences and scientific journals and will develop into one of the major forces behind future nano-/micro-science and technology. Future reviews and books will certainly widen the scope of this first attempt to capture some of the main features and highlights of this dynamic field.

References

1 R. W. Gerchberg and W. O. Saxton, "A practical algorithm for the determination of phase from image and diffraction plane pictures," *Optik* **35** (2), pp. 237–46, 1972.

2 S. Hell and J. Wichmann, "Breaking the diffraction resolution limit by stimulated emission: stimulated-emission-depletion fluorescence microscopy," *Opt. Lett.* **19**, pp. 780–2, 1994.

3 V. Westphal and S. W. Hell, "Nanoscale resolution in the focal plane of an optical microscope," *Phys. Rev. Lett.* **94**, p. 143903, 2005.

4 K. E. Drexler, *Nanosystems: molecular machinery, manufacturing and computation*, Wiley, New York, 1992.

5 N. Taniguchi, "On the basic concept of nanotechnology," in *Proc. ICPE*, 1974.

6 R. Feynman *J. Microelectromechanical Systems (*Reprinted from 1960) **1**, pp. 60–66, 1992.

7 R. Feynman *J. Micromechanical Systems (*Reprinted from 1983) **2**, pp. 4–14, 1993.

8 K. Yamasaki, S. Juodkazis, T. Lippert, M. Watanabe, S. Matsuo, and H. Misawa, "Dielectric breakdown of rubber materials by femtosecond irradiation," *Appl. Phys. A.* **76**, pp. 325–9, 2003.

9 Q. Johnson and A. C. Mitchell, "First x-ray diffraction evidence for a phase transition during shock-wave compression," *Phys. Rev. Lett.* **29**, pp. 1369–71, 1972.

10 A. B. Belonoshko, "Atomistic simulation of shock wave-induced melting in argon," *Science* **275** (5302), pp. 955–7, 1997.

11 S.-D. Mo and W. Y. Ching, "Electronic and optical properties of θ-Al_2O_3 and comparison to a-Al_2O_3," *Phys. Rev. B* **57**, pp. 15219–28, 1998.

12 G. E. Duvall, K. M. Ogilvie, R. Wilson, P. M. Bellamy, and P. S. P. Wei, "Optical spectroscopy in a shocked liquid," *Nature* **296**, pp. 846 – 7 (DOI: 10.1038/296846a0), 1997.

13 A. N. Dremin, "Discoveries in detonation of molecular condensed explosives in the 20th century," *Combustion, Explosion, and Shock Waves* **36**, pp. 704–15, 2000.

14 A. Marcinkevicius, R. Waki, S. Juodkazis, S. Matsuo, and H. Misawa, "Processing of sapphire and glass materials after femtosecond exposure," in *OSA annual meeting & exhibit (Sept. 29–Oct. 3 2002, Orlando, USA)*, p. WF4.

15 A. Žukauskas, M. S. Shur, and R. Gaska, *Introduction to Solid-State Lighting*, John Wiley & Sons – IEEE Press Publication, 2002.

16 S. Juodkazis, Y. Tabuchi, T. Ebisui, S. Matsuo, and H. Misawa, "Anisotropic etching of dielectrics exposed by high intensity femtosecond pulses," in *Ad-*

vanced Laser Technologies (Sept. 10–15, 2004, Rome & Frascati, Italy), SPIE Proc. vol. 5850, I. A. Shcherbakov, A. Giardini, V. I. Konov, and V. I. Pustovoy, eds., pp. 59 – 66, 2005.

17 A. Marcinkevicius, S. Juodkazis, M. Watanabe, M. Miwa, S. Matsuo, H. Misawa, and J. Nishii, "Femtosecond laser-assisted three-dimensional micro-fabrication in silica," *Opt. Lett.* **26**(5), pp. 277–9, 2001.

18 S. Juodkazis, K. Yamasaki, V. Mizeikis, S. Matsuo, and H. Misawa, "Formation of embedded patterns in glasses using femtosecond irradiation," *Appl. Phys. A* , pp. 1549 – 53, 2004 (DOI: 10.1007/ s00339-004-2845-1).

19 J. A. Duffy, "The electronic polarisability of oxygen in glass and the effect of com-position," *J. Non-Crystal. Sol.* **297**, pp. 275–84, 2002.

Index

3D Laser Microfabrication. Principles and Applications.
Edited by H. Misawa and S. Juodkazis
Copyright © 2006 WILEY-VCH Verlag GmbH & Co. KGaA, Weinheim
ISBN: 3-527-31055-X